Introduction to Flight Test Engineering Volume II

Don Ward

Thomas Strganac

Rob Niewoehner

KENDALL/HUNT PUBLISHING COMPANY
4050 Westmark Drive   Dubuque, Iowa 52002

Cover Photos: Copyright 2007 Shutterstock

Copyright © 2007 by Donald Ward, Thomas Strganac, and Ron Niewoehner

ISBN 978-0-7575-4224-4

Kendall/Hunt Publishing Company has the exclusive rights to reproduce this work,
to prepare derivative works from this work, to publicly distribute this work,
to publicly perform this work and to publicly display this work.

All rights reserved. No part of this publication may be reproduced,
stored in a retrieval system, or transmitted, in any form or by any
means, electronic, mechanical, photocopying, recording, or otherwise,
without the prior written permission of the copyright owner.

Printed in the United States of America
10 9 8 7 6 5 4 3 2 1

CONTENTS

CHAPTER 10   Simulation and Modeling in Flight Test 1

10.1   Foundations 1
    10.1.1   Historical Perspective 1
        10.1.1.1   Evolution of Aircrew Training Simulators 1
        10.1.1.2   Rapid Increases in Computational Power and Software Capability 3
    10.1.2   Elements and Terminology of Simulation 5
        10.1.2.1   Elements 5
            10.1.2.1.1   Mathematical Model 5
            10.1.2.1.2   Implementation 6
            10.1.2.1.3   Regime of Application 7
        10.1.2.2   Definitions and Classifications 7
            10.1.2.2.1   Non-Real-Time (Analytical) M&S 7
                10.1.2.2.1.1 Subsystem Modeling 8
                10.1.2.2.1.2 Critical Envelope Expansion Models 8
                10.1.2.2.1.3 M&S Interaction with OT&E .. 9
            10.1.2.2.2   Real-Time M&S 9
                10.1.2.2.2.1 Engineering Man-in-the-Loop MITL Simulation 9
                10.1.2.2.2.2 Hardware-in-the-Loop (HIL) Simulation 12
                10.1.2.2.2.3 Iron Bird 14
                10.1.2.2.2.4 Inflight Simulation 15

10.2   Integrating M&S and Flight Test 17
    10.2.1   Test Planning 19
        10.2.1.1   Top Level Planning Support 19
        10.2.1.2   Detailed Test Planning Support 23
        10.2.1.3   Planning for End Item Verifications, Validations, and Demonstrations 23
    10.2.2   Systems and Subsystems Evaluations 24
    10.2.3   Verification Versus Validation 25
        10.2.3.1   Verification of a Simulation 25
        10.2.3.2   Validation of Simulation Elements 26
        10.2.3.3   Validation of the Integrated Simulation 27
        10.2.3.4   Sufficiency of a Validation (Accreditation) 30

10.3   Conducting a Simulation-Based Flight Test Project 30
    10.3.1   Setting and Understanding Test Objectives 31
        10.3.1.1   Integrated Teaming 31
        10.3.1.2   Concept of Operations 32

CONTENTS

	10.3.2	Practicalities of Providing an Appropriate Simulation 32	
		10.3.2.1 Type, Architecture, and Fidelity 32	
		10.3.2.2 Necessary Data 32	
		10.3.2.3 Updating Data Sets 33	
	10.3.3	Draft and Refine Test Matrix and Test Maneuvers 33	
		10.3.3.1 Run Simulation to Explore Test Envelope....................33	
		10.3.3.2 Refine Test Matrix 34	
		10.3.3.3 Develop Test Maneuvers..34	
	10.3.4	Carry Out a Failure Modes and Effects Analysis (FMEA)...........36	
	10.3.5	Train All Members of Test Team...36	
		10.3.5.1 Utilize Simulation to Prepare Crew, Test Conductors, and Specialists..36	
		10.3.5.2 Establish Training Standards and Desired Proficiency Levels 37	
		10.3.5.3 Consider Need for In-flight Simulation 37	
	10.3.6	Conduct Flight Tests and Compare Results to Simulation...........37	
		10.3.6.1 Integration of Simulation with Test Missions 37	
		10.3.6.1.1 Comparison with Previously Calculated Simulation Results 38	
		10.3.6.1.2 Run Simulations Parallel to Test Mission Events 38	
		10.3.6.1.3 "Shadowing" Test Missions with MITL Tools..39	
		10.3.6.2 Attain Desired Level of Matching with Simulation Results..39	
10.4	Summary 39		
References 39			

CHAPTER 11 Aeroelasticity: Theoretical and Analytical Foundation............43

11.1	Definitions..43	
	11.1.1	Aeroelastic Divergence...43
	11.1.2	Lift Effectiveness...43
	11.1.3	Control Reversal 44
	11.1.4	Flutter..44
	11.1.5	Nonlinear Aeroelasticity 45
11.2	Aeroelastic Analysis Foundations..45	
	11.2.1	Equations of Motion..45
	11.2.2	Solution Approaches..45
	11.2.3	Aeroelastic Similarity Parameters...47

CONTENTS

	11.2.4	Static Aeroelastic Phenomena	49
		11.2.4.1 Aeroelastic Divergence	49
		11.2.4.1.1 Aeroelastic Divergence - A Simple Example	50
		11.2.4.1.2 Aeroelastic Divergence - Altitude and Compressibility	51
		11.2.4.2 Lift Effectiveness and Sweep Effects	52
		11.2.4.3 Control Reversal	54
	11.2.5	Dynamic Aeroelastic Phenomena	55
		11.2.5.1 Bending-Torsion Flutter	56
		11.2.5.2 Example: Two Degree-of-Freedom Aeroelasticity	58
11.3	Aeroelastic Analysis Techniques		61
	11.3.1	Modal Methods	61
	11.3.2	Solution - "K" Method	62
	11.3.3	Matched Point Analysis	66
11.4	Nonlinear Aeroelasticity		67
	11.4.1	Divergence of the Nonlinear System	68
	11.4.2	Aeroelastic LCO	69
	11.4.3	Residual Pitch Oscillation (RPO)	72
	11.4.4	Internal Resonance (IR)	73
11.5	Summary		74
References			75

CHAPTER 12 Aeroelasticity: Ground and Pre-Flight Preparation ... 77

12.1	The Flight Envelope	77
12.2	FAR Advisory Circulars: Parts 23 and 25	78
	12.2.1 V-g Diagrams	78
	12.2.2 FAR Flight Envelopes	79
	12.2.3 Velocity-Frequency Diagrams	80
12.3	Ground Vibration Tests (GVT)	80
	12.3.1 Vibration Characteristics - Frequencies and Modes	80
	12.3.2 The Modal Survey	81
	12.3.3 The Frequency Spectrum	84
	12.3.4 Determination of Damping	85
12.4	Wind Tunnel Tests	86
12.5	Subcritical Response Techniques	88
12.6	Summary	90
References		90

CHAPTER 13 Aeroelasticity: Flutter Flight Tests ... 91

13.1	Flight Test Methods	92
13.2	The Flight Test Approach	92

13.3	Flutter Excitation in Flight	94
	13.3.1 Natural Atmospheric Turbulence	95
	13.3.2 Mechanical Devices	95
	13.3.3 Pilot Control Inputs and Stick Raps	95
13.4	Flutter Envelope Expansion	95
	13.4.1 Procedures Requiring Dives	96
	13.4.2 Case Study: the X-29	96
	13.4.3 Case Study: F/A-18 E/F Super Hornet	99
	13.4.4 Case Study: CF-18 Hornet with Pylon-Mounted GBU-24s	103
	13.4.5 Commercial and Cargo Aircraft	103
	13.4.6 General Aviation	103
	13.4.7 Flutter Mitigation: Procedural Rules of Thumb	104
	13.4.8 Flutter Mitigation: Design Rules of Thumb	104
13.5	Store-Induced Limit Cycle Oscillation	105
	13.5.1 Effect of Stores on LCO Potential	105
	13.5.2 Other Pathologies	106
13.6	Summary	106
References		107

CHAPTER 14 Stall, Post-Stall, and Spin Tests 109

14.1	Theoretical Foundations	110
	14.1.1 Post-Stall Definitions	110
	14.1.1.1 Stall	110
	14.1.1.2 Departure	111
	14.1.1.3 Post-Stall Gyration	112
	14.1.1.4 Spin	112
	14.1.1.5 Deep Stall	112
	14.1.2 Spin Modes	113
	14.1.3 Spin Evolution	114
	14.1.3.1 Incipient Phase	114
	14.1.3.2 Developing Phase	115
	14.1.3.3 Developed Phase	115
	14.1.4 Flight Path in a Spin	115
	14.1.4.1 Assumptions	116
	14.1.4.2 Balance of Forces	117
	14.1.5 Aerodynamic Factors in a Spin	117
	14.1.5.1 Autorotative Couple of the Wing	118
	14.1.5.2 Effect of Forebody Shape	119
	14.1.5.3 Effect of Damping Derivatives	120

	14.1.6	Effect of Mass Distribution 120
		14.1.6.1 Principal Axes 120
		14.1.6.2 Relative Density 121
		14.1.6.3 Center and Radius of Gyration 121
		14.1.6.4 Relative Magnitude of Airplane Moments of Inertia 122
	14.1.7	Simplification of the Post-Stall Equations of Motion 122
		14.1.7.1 Simplifying Assumptions 123
		14.1.7.2 Simplified Large Amplitude Moment Equations 123
	14.1.8	Aerodynamic Conditions for Dynamic Equilibrium 125
		14.1.8.1 Pitching Moment Equation 125
		14.1.8.2 Rolling and Yawing Moment Equations 127
		14.1.8.3 Estimation of Spin Equilibrium States 128
14.2	Post-Stall/Spin Test Preparation and Flight Safety 130
	14.2.1	Pre-Test Planning and Preparation 130
		14.2.1.1 Test Objectives 131
		14.2.1.1.1 Governing Principles 131
		14.2.1.1.2 Military Requirements 131
		14.2.1.1.3 Civil Requirements 134
		14.2.1.1.4 Spin Requirements 139
		14.2.1.2 Data Requirements 140
		14.2.1.3 Instrumentation for Post-Stall Flight Tests 140
		14.2.1.4 Test Team Training 141
	14.2.2	Safety Precautions and Special Equipment 143
		14.2.2.1 Emergency Recovery Devices 143
		14.2.2.2 Back-up Power 146
		14.2.2.3 Chase Aircraft 146
		14.2.2.4 Procedural Precautions 147
		14.2.2.5 Real-Time Monitoring of Post-Stall Tests 148
14.3	Summary 148
References 149

CHAPTER 15 Stores Certification Testing 151

15.1	Introduction 151
	15.1.1	The Key Questions 151
	15.1.2	Historical Background 152
		15.1.2.1 Early Days 152
		15.1.2.2 World War II 160
		15.1.2.3 Cold War Onward 166
	15.1.3	Acknowledgements 170

CONTENTS

- 15.2 Stores Certification ... 170
  - 15.2.1 Terminology, Organization, and Scope ... 171
  - 15.2.2 Importance of Analysis and Ground Tests ... 176
- 15.3 Captive Carry Compatibility ... 177
  - 15.3.1 Planning and Preparation ... 177
    - 15.3.1.1 Analysis ... 178
    - 15.3.1.2 Ground Testing ... 178
      - 15.3.1.2.1 Test Article Selection ... 179
      - 15.3.1.2.2 Fit and Function ... 179
      - 15.3.1.2.3 Vibration and Acoustic ... 179
      - 15.3.1.2.4 Electromagnetic Compatibility and Hazards of Electromagnetic Radiation to Ordinance ... 180
      - 15.3.1.2.5 Environmental ... 180
    - 15.3.1.3 Captive Compatibility Flight Profiles ... 180
      - 15.3.1.3.1 Weight and CG Review ... 181
      - 15.3.1.3.2 Structural Review ... 181
      - 15.3.1.3.3 Flutter Analyses ... 182
      - 15.3.1.3.4 Aerodynamic Analyses ... 182
    - 15.3.1.4 Captive Flight Testing ... 182
    - 15.3.1.5 Test Instrumentation ... 183
    - 15.3.1.6 Telemetry ... 183
  - 15.3.2 Risk Mitigation ... 184
    - 15.3.2.1 Test Limits ... 184
    - 15.3.2.2 High Speed Risks ... 184
    - 15.3.2.3 Safety Chase ... 184
    - 15.3.2.4 Configuration Limits ... 184
      - 15.3.2.4.1 Asymmetric Risks ... 184
      - 15.3.2.4.2 Heavyweight Risks ... 185
  - 15.3.3 Execution Lessons ... 185
    - 15.3.3.1 F-16 Conformal Tank Degraded Flying Qualities ... 186
    - 15.3.3.2 F-16 Joint Standoff Weapon (JSOW) Degraded Flying Qualities ... 186
    - 15.3.3.3 Store Structural and Fastener Failures ... 187
  - 15.3.4 Summary of Lessons Learned from Captive Carry Examples ... 188

CONTENTS

- 15.4 Separation Compatibility 188
  - 15.4.1 Planning and Preparation 188
    - 15.4.1.1 Analysis 189
      - 15.4.1.1.1 Flow Field Mapping and Carriage Load Predictions 189
      - 15.4.1.1.2 Theoretical Trajectory Predictions 189
    - 15.4.1.2 Ground Testing 190
      - 15.4.1.2.1 Articulated Dual Sting Wind Tunnel Tests 190
      - 15.4.1.2.2 Dynamic Drop Model Tests 191
      - 15.4.1.2.3 Grid Data Wind Tunnel Tests 191
      - 15.4.1.2.4 Flow Angularity Wind Tunnel Tests 191
      - 15.4.1.2.5 Influence Function Wind Tunnel Tests 192
      - 15.4.1.2.6 Static Ejection Tests 192
      - 15.4.1.2.7 Aeroelastic Ground Vibration Tests (GVTs) and Aeroelastic Effects Tests 193
      - 15.4.1.2.8 Gun/Rocket/Missile Firing Tests 193
    - 15.4.1.3 Instrumentation and Data Collection 193
      - 15.4.1.3.1 Special Instrumentation 194
      - 15.4.1.3.2 Test Data 194
    - 15.4.1.4 Test Conduct Considerations 194
      - 15.4.1.4.1 Test Articles 194
      - 15.4.1.4.2 Photo Chase 194
      - 15.4.1.4.3 Control Room 195
  - 15.4.2 Risk Mitigation 195
    - 15.4.2.1 Separation Collisions 195
      - 15.4.2.1.1 Store-to-Aircraft Collisions 196
      - 15.4.2.1.2 Store-to-Pylon/Rack Collisions 196
      - 15.4.2.1.3 Store-to-Store Collisions 196
    - 15.4.2.2 Build-Up Approach 197
    - 15.4.2.3 Probability of Impact 197
    - 15.4.2.4 Use of Inert Stores 197
    - 15.4.2.5 Store Subsystem Failure Modes 198
    - 15.4.2.6 Preflight and Postflight Inspection 199
    - 15.4.2.7 Forward-Firing Weapons 199
    - 15.4.2.8 Safety Chase 199
    - 15.4.2.9 Hung Stores 199
  - 15.4.3 Execution with Discipline (Professionalism) 199
    - 15.4.3.1 Clearing the Range 200
    - 15.4.3.2 Radio Communications 200
    - 15.4.3.3 Dynamic Separation Conditions 200
    - 15.4.3.4 Aircraft and/or Performance Limitations 200
    - 15.4.3.5 Aircraft Response to Separation 201
    - 15.4.3.6 Mission Impact 201

CONTENTS

    15.4.4 Stores Separation Lessons to be Learned 201
        15.4.4.1 A-10 Secondary Gun Gas Ingestion 201
        15.4.4.2 F-15E/GBU-12D/B Impact 202
        15.4.4.3 CBU-89 Tactical Munitions Dispenser (TMD)
                Releases from the B-1B 203
        15.4.4.4 F-16 Fuel Tank Jettison Store-to-Aircraft Collision ... 203
        15.4.4.5 F-15E/AGM-130 Near Miss 204
        15.4.4.6 F-15C/AGM-120A Jettison Near Miss 205
        15.4.4.7 TA-4J Safety Chase Impact 206
    15.4.5 Summary of Lessons Learned from Store Separation
        Examples 207
15.5 Conclusion 207
References 208

CHAPTER 16 Special Handling Flight Tests 211

16.1 Minimum Control Airspeed 211
    16.1.1 Theoretical Foundations 211
        16.1.1.1 Review of Steady Heading Sideslips (SHSS) 211
        16.1.1.2 Engine-Out Operations and Static Minimum
                Control Airspeed 215
        16.1.1.3 Determining the Critical Engine 221
        16.1.1.4 Other Factors 221
        16.1.1.5 Dynamic V_{mca} 222
    16.1.2 Flight Testing for V_{mca} 222
        16.1.2.1 Determining Static V_{mca} 222
        16.1.2.2 Determining Dynamic V_{mca} 223
        16.1.2.3 Extrapolating V_{mca} to Sea Level 224
    16.1.3 Minimum Control Airspeed Ground (V_{mcg}) 225
16.2 Cross-Wind Landing 226
16.3 Handling Qualities During Tracking (HQDT) 227
16.4 Aerial Refueling 229
16.5 Summary 232
References 233

CONTENTS

CHAPTER 17 Propulsion Systems Testing 235
17.1 Theoretical Foundations 236
17.1.1 Thrust as Momentum Transfer 236
17.1.2 The Propeller 237
17.1.3 The Internal Combustion Aircraft Engine 238
17.1.3.1 The Basic Engine 239
17.1.3.2 Superchargers and Turbochargers 241
17.1.3.3 Fuel-Air Mixture 243
17.1.3.4 Ignition Systems 244
17.1.3.5 Cooling 244
17.1.3.6 Water Injection 245
17.1.3.7 Atmospheric Water Vapor 245
17.1.3.8 Propeller Integration 245
17.1.4 The Gas Turbine 245
17.1.4.1 The Brayton Cycle 245
17.1.4.2 Gas Turbine Configurations 246
17.1.4.3 Inlets 247
17.1.4.4 Compressors 248
17.1.4.5 Compressor Bleed Air 249
17.1.4.6 Combustion Chambers 249
17.1.4.7 Fuel Controls 249
17.1.4.8 Turbines 250
17.1.4.9 Nozzle 250
17.1.4.10 Thrust Accounting 250
17.1.4.11 Engine Anti-Ice 251
17.1.4.12 Gas Turbine Operating Characteristic 251
17.1.4.12.1 Compressor Stall 251
17.1.4.12.2 Engine Start 252
17.1.5 Aviation Fuels 253
17.1.5.1 Jet Fuels 254
17.1.5.2 Aviation Gasoline 254

- 17.2 Flight Test Practices 255
  - 17.2.1 General 255
  - 17.2.2 Operability Testing and Propulsion Subsystems 255
    - 17.2.2.1 Negative-g Operability 255
    - 17.2.2.2 Reversing Systems 256
    - 17.2.2.3 Fuel Feed 256
  - 17.2.3 Flight Testing Propellers 257
    - 17.2.3.1 Feathering Systems 257
    - 17.2.3.2 Inflight Thrust Measurement 257
  - 17.2.4 Reciprocating Engines 257
    - 17.2.4.1 Instrumentation 257
    - 17.2.4.2 Cooling 258
    - 17.2.4.3 Induction System Heating 259
    - 17.2.4.4 Induction System Losses 259
    - 17.2.4.5 Airstart Testing 260
  - 17.2.5 Flight Testing Gas Turbine Engines 260
    - 17.2.5.1 Instrumentation 260
    - 17.2.5.2 Operating Characteristics 261
    - 17.2.5.3 Gas Turbine Cooling 262
    - 17.2.5.4 Airstart Testing 263
    - 17.2.5.5 Inflight Thrust Measurement of Turbofan and Turbojet Engines 264
    - 17.2.5.6 Fatigue Monitoring 265
- 17.3 Summary 265
- References 266

INDEX 269

LIST OF FIGURES

CHAPTER 10

Fig. 10.1	Antoinette Barrel Trainer	1
Fig. 10.2	Link Trainer	2
Fig. 10.3	Terrain Board for Night Vision Goggle Training	3
Fig. 10.4	Exponential Growth of Software Tools in Early 1990s	4
Fig. 10.5	Typical Simulation "Waterfall" Evolution	5
Fig. 10.6	Simulation Spectrum	6
Fig. 10.7	FSL Cockpit with Reconfigurable Displays and "Out the Window" Displays	11
Fig. 10.8	F-15 ACTIVE Aircraft and Its Integrated Subsystems	13
Fig. 10.9	F/A-18 HARV Configurable Simulation	13
Fig. 10.10	F/A-18 HARV in the RAIF	14
Fig. 10.12	RAIF Displays and Cockpit	14
Fig. 10.13	Variable Stability Inflight Simulator Test Aircraft (VISTA)	16
Fig. 10.14	North American JF-100C	16
Fig. 10.15	General Flow of Simulation-Based Flight Testing	18
Fig. 10.16	STEP Process	19
Fig. 10.17	Comparison of Development Cycles for Complex Systems	20
Fig. 10.18	Evolutionary Acquisition Impact on Testing	21
Fig. 10.19	Validation Comparison of a Turboprop Climb Performance Model	28
Fig. 10.20	Turboprop Climb Performance with Mathematical Elevator Input	28
Fig. 10.21	Net Propeller Thrust Comparison for Turboprop Thrust Model	29
Fig. 10.22	"Validated" Climb Profile after Correction for Propeller Thrust Anomaly	30
Fig. 10.23	Bottom-Up Planning Iteration	35
Fig. 10.24	Use of Modeled S&C Parameters in X-29 Envelope Expansion	38

CHAPTER 11

Fig. 11.1	Interdisciplinary Nature of Aeroelasticity	43
Fig. 11.2	Sweep Effects	44
Fig. 11.3	Oscillatory Motion	46
Fig. 11.4	Decaying Motion in the Time Domain	47
Fig. 11.5	Divergent Motion in the Time Domain	47
Fig. 11.6	Undamped Motion in the Time Domain	47
Fig. 11.7	Mass Ratio Geometry	48
Fig. 11.8	Free-Body Diagram for Aeroelastic Divergence Example	49
Fig. 11.9	Divergence Boundaries -- Mach Number and Altitude Effects	52
Fig. 11.10	Aerodynamic Eccentricity and Sweep	52
Fig. 11.11	The Wing with Sweep-Twist-Bending	53

LIST OF FIGURES

CHAPTER 11 (continued)

Fig. 11.12	Control Reversal	54
Fig. 11.13	Bending-Torsion Flutter	56
Fig. 11.14	Coupling of Flutter Modes	60
Fig. 11.15	Velocity-Damping (V-g) Diagram	65
Fig. 11.16	Frequency-Velocity Diagram	65
Fig. 11.17	Frequency-Velocity Diagram (Indicating Divergence)	66
Fig. 11.18	Solution for the Matched Point	67
Fig. 11.19	Nonlinear Stiffness	67
Fig. 11.20	Subcritical Bifurcation	70
Fig. 11.21	Supercritical Bifurcation	73

CHAPTER 12

Fig. 12.1	The VD/MD Operating Envelope	77
Fig. 12.2	The Altitude-Mach Operating Envelope	78
Fig. 12.3	V-g Diagram for a Five-Mode Analysis	78
Fig. 12.4	Envelopes	79
Fig. 12.5	Velocity-Frequency (V-ω) Diagram for the Five-Mode Analysis	80
Fig. 12.6	GVT Measurements - Input and Output	80
Fig. 12.7	An Illustration of the Fundamental Wing Modes	82
Fig. 12.8	Vibration Modes	83
Fig. 12.9	Typical Frequency Response	84
Fig. 12.10	Response in Time	85
Fig. 12.11	Mean Square Spectral Density (MSSD) of Accelerometr Output	86
Fig. 12.12	The Flutter Boundary	87
Fig. 12.13	Results of Peak-Hold Method	89

CHAPTER 13

Fig. 13.1	Flutter Damage Sustained During E-6A Envelope Expansion	91
Fig. 13.2	Flutter Expansion Flow Diagram	92
Fig. 13.3	Flutter Test Profile	93
Fig. 13.4	Flutter Excitation Schemes	94
Fig. 13.5	Location of Accelerometers for the X-29	97
Fig. 13.6	Predicted X-29 Flutter Boundaries	97
Fig. 13.7	X-29 Midflaperon Flutter Boundary Prediction	98
Fig. 13.8	X-29 Flutter Flight Test Matrix	99

LIST OF FIGURES

CHAPTER 14

Fig. 14.1   Classical Aerodynamic Stall 109
Fig. 14.2   Operational IAF Su-30MKI with Thrust Vector Control 110
Fig. 14.3   Spin Phases 114
Fig. 14.4   Helical Flight Path in a Fully Developed Spin 116
Fig. 14.5   Changes in Lift and Drag Coefficients at High Angle of Attack 116
Fig. 14.6   Aerodynamic Mechanisms for Autorotative Moments 118
Fig. 14.7   Lift and Drag on Advancing and Retreating Wing Panels 119
Fig. 14.8   Aerodynamic Contributions from the Forebody Shape 120
Fig. 14.9   Relationship Between Body and Principal Axes 121
Fig. 14.10 Relative Magnitude of Aircraft Moments of Inertia 122
Fig. 14.11 Pitching Moment Coefficient in a Steady Spin 120
Fig. 14.12 Angular Velocity Components in a Typical Upright Spin 121
Fig. 14.13 Stabilizing and Destabilizing slopes for C_l and C_n 122
Fig. 14.14 Comparison of Aerodynamic and Inertial Pitching Moments 124
Fig. 14.15 Emergency Parachute Recovery Subsystem Design Parameters 139
Fig. 14.16 Post-Stall Test Team Communications Net 142

CHAPTER 15

Fig. 15.1   Vickers FB.5 Gunbus with Gunner's Station in Forward Cockpit 152
Fig. 15.2   S.E.5a with Lewis Machine Gun on Top of Upper Wing (Foster Mount) 153
Fig. 15.3   Fokker E.III with Forward-Firing, Interruptor-Equipped Machine Gun 153
Fig. 15.4   Forward-Firing Machine Guns on Goering's Albatross D.III and Sopwith Camel 154
Fig. 15.5a "Ballon Buster" Rockets on French Nieuport 11 155
Fig. 15.5b Stick Grenades Being Loaded on German Halberstadt CL.II 155
Fig. 15.6   Handheld Bomb Released by Observers 155
Fig. 15.7   Carbonit Bomb Used by Germans in Early Years 156
Fig. 15.8a French Bomb 156
Fig. 15.8b British 1650-lb. 156
Fig. 15.8c German PuW Bomb 156
Fig. 15.9   French Caudron G.4 Bomber 157
Fig. 15.10 Preparation of a Handley Page O/100 for a Bombing Mission 158
Fig. 15.11 Size Comparison: Handley Page O/100 Bomber and Spad Fighter 158
Fig. 15.12a Gotha G.IV Ground Handling 158
Fig. 15.12b Loading Bombs on a aGotha G.IV 158
Fig. 15.13 Russian Ilya Mourometz 159

LIST OF FIGURES

CHAPTER 15 (continued)

Fig. 15.14	Messerschmitt Bf 109G	160
Fig. 15.15	Lockheed P-38J	161
Fig. 15.16	Kawanishi N1K2-J (Code Name "George")	161
Fig. 15.17	Messerschmitt Me 262A (*Schwalbe* or "Swallow")	161
Fig. 15.18	Messerschmitt Me 262A-1A "Happy Hunter II"	162
Fig. 15.19a	Republic P-47 Thunderbolt	162
Fig. 15.19b	Junkers Ju-87 Stuka	162
Fig. 15.20	Douglas SBD "Dauntless"	163
Fig. 15.21	Lancaster Dropping Incendiary Bombs and a 4000-lb "Cookie" plus J-Incendiaries	164
Fig. 15.22	Sketch of Heinkel 177 "Grief" Carrying Hs 293	165
Fig. 15.23	B-29 Enola Gay Landing after Dropping Atomic Bomb on Hiroshima	165
Fig. 15.24a	F-100C with Bombs	167
Fig. 15.24b	F-100D Firing a Bullpup	167
Fig. 15.25	F-4E Dropping bombs	167
Fig. 15.26	Sukhoi Su-25 "Frogfoot"	168
Fig. 15.27	Grumman A-6 Intruder	169
Fig. 15.28a	B-52 with Conventional Bombs	169
Fig. 15.28b	Loaded with AGM-129s	169
Fig. 15.28c	AGM-129	169
Fig. 15.29	Lockheed F-117A Nighthawk	170
Fig. 15.30	Aircraft/Stores Certification Process	173
Fig. 15.31	F-4E Separation Wind Tunnel Tests	177
Fig. 15.32	F-16 Conformal Tank Configuration	186
Fig. 15.33	AGM-154 JSOW	187
Fig. 15.34	GBU-10 Tail Section Failure	187
Fig. 15.35	JDAM Fin Trailing Edge Movement During Captive Carriage	188
Fig. 15.36	Probability of Occurrence Matrix	198
Fig. 15.37	GBU-12D/B Pitch-Up and Impact with Horizontal Tail	202
Fig. 15.38	CBU-89 Ripple Release from the B-1B	203
Fig. 15.39	PIDS Pylon Compared to Earlier F-16 Pylons	204
Fig. 15.40	Impact of the 370-gallon Fuel Tank with the F-16 Ventral Fin	204
Fig. 15.41	Comparison of the Fuel Tank Jettison from Opposing Wing Stations with JASSM on Adjacent Stations	205
Fig. 15.42	PIDS Pylon Compared to Earlier F-16 Pylons	205
Fig. 15.43	Near Miss of the F-15 Wing Tip by a Jettisoned AGM-120A	206

LIST OF FIGURES

CHAPTER 16

- Fig. 16.1 Steady Heading Sideslip (with Symmetric Power) 212
- Fig. 16.2 Evolution of the B-17 Vertical Tail 214
- Fig. 16.3 Modifications to the Canberra Vertical Tail and rudder 214
- Fig. 16.4 F-14 Tomcat 215
- Fig. 16.5 Steady Heading Sideslip (Asymmetric Power) 216
- Fig. 16.6 F-14 Lateral-Directional Trim with Asymmetric Power (Right Engine - Dry Maximum Power) 219
- Fig. 16.7 F-14 Lateral-Directional Trim with Asymmetric Power (Right Engine - Maximum Afterburner Power) 220
- Fig. 16.8 Mini Guppy Turbo (Guppy 101) 226
- Fig. 16.9 KC-10 Fueling F-117 with a Flying Boom 229
- Fig. 16.10 KC-130 Refueling F/A-18s Using Probe and Drogue 230

CHAPTER 17

- Fig. 17.1 Local Flow at Blade Section 237
- Fig. 17.2 Propeller Efficiency Map 238
- Fig. 17.3 Lycoming IO-540 Opposed Six-Cylinder Aircraft Engine 239
- Fig. 17.4 Piston Pressure Profiles 240
- Fig. 17.5 Engine Power Chart for Normally Aspirated Engines 241
- Fig. 17.6 Critical Altitudes for Supercharged Reciprocating Engines 242
- Fig. 17.7 Example Engine Power Chart for a Supercharged Engine 243
- Fig. 17.8 Effect of Mixture Setting 243
- Fig. 17.9 T-s Diagram fo Brayton Cycle Engines 246
- Fig. 17.10 Split-Spool Turbojet 246
- Fig. 17.11 Low-Bypass Ratio Afterburning Turbofan 247
- Fig. 17.12 High-Bypass Ratio Turbofan 247
- Fig. 17.13 Turboprop 247
- Fig. 17.14 Compressor Flow Geometries 248
- Fig. 17.15 Example Compressor Map 251
- Fig. 17.16 Gas Turbine Acceleration/Deceleration Fuel Schedules 252
- Fig. 17.17 Carburetor Heat Data Reduction 259

LIST OF TABLES

CHAPTER 10
Table 10.1 Growth of Modeling and Simulation Computing Capability – 1984-2004 4
Table 10.2 Integration of M&S into T&E Tasks 10

CHAPTER 11
None

CHAPTER 12
None

CHAPTER 13
Table 13.1 Mode Type, Frequency, and Nature of Instability 96

CHAPTER 14
Table 14.1 Spin Mode Modifiers 113
Table 14.2 F-4E Spin Modes 113
Table 14.3 F3H Demon Computed Spin Modes Versus Estimated Spin Modes 125
Table 14.4 Progressive Test Phases for Post-Stall Demonstration Maneuvers ..127
Table 14.5 Definitions of Departure and Spin Susceptibility and Resistance 128
Table 14.6 Typical Sensors Needed for Stability and Control Data 136

CHAPTER 15
Table 15.1 Types of Captive Compatibility Flight Profiles (CFP) 181

CHAPTER 16
None

CHAPTER 17
Table 17.1 Basic Fuel Properties 254

LIST OF SYMBOLS AND ABBREVIATIONS

CHAPTER 10

ACETEF	Air Combat Environment Tactical Evaluation Facility (at PAX River)
ACTIVE	Advanced Control Technology for Integrated Vehicles (at NASA Dryden)
AGARD	Advisory Group for Aerospace Research & Development (now RTA)
ASTRA	Advanced Stability Training and Research Aircraft (at ETPS)
ATTAS	Advanced Technologies Testing Aircraft (at DLR)
ATTHeS	Advanced Technologies Testing Helicopter System (at DLR)
BOAC	British Overseas Airways Corporation
CAD	Computer-Aided Design
CCTV	Closed Circuit Television
CD	Concept Definition
CDD	Capabilities Development Document
CDR	Critical Design Review
CFD	Computational Fluid Dynamics
COI	Critical Operational Issues
CPD	Capabilities Production Document
CRT	Cathode Ray Tube
CTP	Critical Technical Parameters
DAB	Defense Acquisition Board
DAG	Defense Acquisition Guidebook
DIA	Defense Intelligence Agency
DLR	German Institute fur Flugsystemtechnik
DoD	Department of Defense
DOE	Design of Experiments
DT&E	Developmental Test and Evaluation
ETPS	Empire Test Pilots' School
FC	Flight Control
FHS	Flying Helicopter Simulator (at DLR)
FMEA	Failure Modes and Effects Analysis
FOT&E	Follow-on Operational Test and Evaluation
FSL	Flight Simulation Laboratory (at Texas A&M University)
HARV	High Angle of Attack Research Vehicle
HIL	Hardware-in-the-Loop
HPC	High Performance Computing
HUD	Head Up Display
ICD	Initial Capabilities Document
IOT&E	Initial Operational Test and Evaluation
IPPD	Integrated Product and Process Development
IPT	Integrated Product Team
IT	Information Technology
ITEA	International Test and Evaluation Association
JCIDS	Joint Capability Integration and Development System
JROCS	Joint Requirements Oversight Council

LIST OF SYMBOLS AND ABBREVIATIONS

CHAPTER 10 (continued)

LCO	Limit Cycle Oscillation
LFOT&E	Live Fire Operational Test and Evaluation
LRIP	Low Rate Initial Production
MDA	Milestone Decision Authority
MITL	Man-in-the-Loop
MS X	Milestone X (A, B, or C)
MTD	Maneuver Technology Demonstrator
MTFM	Model-Test-Fix-Model
NASA	National Aeronautics and Space Administration
NCW	Network Centric Warfare
NRL	National Aerospace Laboratory (the Netherlands)
OTA	Operational Test Agency
OT&E	Operational Test and Evaluation
OTW	Out-the-Window
PM	Program Manager
RAIF	Research Aircraft Integration Facility (at NASA Dryden)
RAC	Risk Assessment Code
SA	Sensitivity Analysis
SAE	Society of Automotive Engineers
SBA	Simulation-Based Acquisition
SEP	Systems Engineering Plan
SEMP	Systems Engineering Management Plan
SCME	Singapore Center for Military Experimentation
STEP	Simulation, Test, and Evaluation Process
STOL	Short Takeoff and Landing
T&E	Test and Evaluation
TEMP	Test and Evaluation Master Plan
TIFS	Total In-Flight Simulator (at CALSPAN)
UAV	Unmanned Aerial Vehicle
USD	Undersecretary of Defense
USDA&T	Undersecretary of Defense for Acquisition and Testing
USAFTPS	United States Air Force Test Pilot School
VBS	Virtual Battle Station
VSS	Variable Stability System
VSTOL	Vertical/Short Takeoff and Landing
VV&A	Verification, Validation and Accreditation
WIPT	Working-Level Integrated Product Team

CHAPTER 11 (only those symbols not previously introduced are listed)

a	distance (fraction of semichord) from midchord to elastic axis	none
a	factor delineating spring hardening or softening	none
a_∞	freestream speed of sound	fps

LIST OF SYMBOLS AND ABBREVIATIONS

CHAPTER 11 (continued)

Symbol	Description	Units
c	chord, reference length	ft
$C(k)$	Theodorsen's Function	none
$\mathbf{C}$	damping matrix	lb-sec/ft, ft-lb-sec
$\mathbf{C}_A$	aerodynamic damping matrix	lb-sec/ft, ft-lb-sec
C_L	lift coefficient	none
C_{L_δ}	change in lift per control surface deflection	per ° or radian
C_{M_δ}	change in moment per control surface deflection	per ° or radian
C_{L_α}	slope of lift curve for incompressible flow	per ° or radian
$C_{L_\alpha,0}$	slope of lift curve for incompressible flow	per ° or radian
e	distance from elastic axis to aerodynamic center	ft
$F(k)$	real part of the Theodorsen function	none
$\mathbf{F}$	external force vector	lbs, ft-lb
$\mathbf{F}_A$	motion-dependent aerodynamic force vector	lbs, ft-lb
$\mathbf{F}_t$	external loads (like turbulence, gusts, and buffeting)	lbs, ft-lb
g, G	damping	none
$G(k)$	imaginary part of the Theordorsen function	none
$H_0^{(2)}$	Hankel function of the second kind	none
$H_1^{(2)}$	Hankel function of the second kind	none
I	mass moment of inertia	slugs-ft^2
$\mathbf{I}$	identity matrix	none
k	reduced frequency	none
k	an integer	none
K	spring constant	lbs/ft, ft-lbs/rad
K_N	nonlinear spring constant	lbs/ft, ft-lbs/rad
$\mathbf{K}$	stiffness matrix	lb/ft, ft-lb/rad
$\mathbf{K}_A$	aerodynamic stiffness matrix	lb/ft, ft-lb/rad
$\underline{\mathbf{K}}_A$	aerodynamic stiffness matrix with q_∞ removed	ft, ft^3/rad
K_y	stiffness in plunging motion	lb/ft
K_α	stiffness torsional motion	ft-lb/rad
L	lift	lbs
$L_{flexible}$	lift for flexible structure	lbs
L_{rigid}	lift for rigid structure	lbs
M	aerodynamic moment	ft-lbs
M_{ac}	aerodynamic moment about the aerodynamic center	ft-lbs

CHAPTER 11 (continued)

Symbol	Description	Units
M_{EA}	aerodynamic moment about the elastic axis	ft-lbs
M_∞	Mach number	none
$M_{\infty,DIV}$	Mach number at divergence	none
m	mass	slugs
$\mathbf{M}$	mass matrix	slugs, slugs-ft^2
q	dynamic pressure	psf
q_a	reference dynamic pressure for a_∞ at altitude	psf
q_{div}	dynamic pressure at divergence	psf
q_{rev}	dynamic pressure at control reversal	psf
$\bar{q}$	nondimensional dynamic pressure	psf
$\bar{q}_D$	dynamic pressure at divergence normalized to no sweep	psf
q_∞	freestream dynamic pressure	psf
$q_{\infty,DIV}$	freestream dynamic pressure at divergence	psf
$\mathbf{q}$	generalized coordinate vector	various
r	distance from elastic axis to center of mass	ft
r_α	radius of gyration	ft
s	wing span	ft^2
U	component of velocity in x direction	knots, mph, or fps
$\bar{V}$	flutter speed index	none
V_D	design dive speed	knots, mph, or fps
V_L	limit speed	fps
V_∞	freestream velocity	knots, mph, or fps
$\mathbf{x}$	vector of coordinates	various
$\dot{\mathbf{x}}$	first time derivative of vector of coordinates	various/sec
$\ddot{\mathbf{x}}$	second time derivative of vector of coordinates	various/sec^2
y	plunge deflection	ft
$\dot{y}$	first time derivative of plunge deflection	ft/sec
$\ddot{y}$	second time derivative of plunge deflection	ft/sec^2
α	pitch deflection, or angle of attack	°, radians
$\dot{\alpha}$	first time derivative of pitch deflection, or angle of attack	°/sec
$\ddot{\alpha}$	scanned time derivative of pitch deflection, or angle of attack	°/sec^2
$\alpha_{effective}$	effective angle of attack	°, radians
β	eigenvector	various
δ	damping ratio, or control surface deflection	none, radians
Φ	matrix of eigenvectors	various

LIST OF SYMBOLS AND ABBREVIATIONS

CHAPTER 11 (continued)

Γ	dihedral angle	°, radians
Λ	sweep angle	°, radians
Λ	matrix of eigenvalues	psf, (rad/sec)^2
μ	mass ratio	none
ρ_∞	freestream density	slugs/ft^3
θ	elastic twist	radians
ω	frequency	rad/sec
$\omega_{bending}$	natural frequency in bending	rad/sec
ω_h	natural frequency in plunge - uncoupled	rad/sec
$\omega_{torsion}$	natural frequency in torsion	rad/sec
ω_α	natural frequency in pitch - uncoupled	rad/sec
Ω_f	frequency of external forcing function	rad/sec

Abbreviations and acronyms
- IR   Internal Resonance
- RPO   Residual Pitch Oscillation
- V-g Velocity-damping

CHAPTER 12

M_D	design dive Mach number	none
V_E	equivalent airspeed	knots, mph, or fps

Abbreviations and acronyms
- AC   Advisory Circular
- FAR   Federal Aviation Regulations
- GVT   Ground Vibration Test
- MSSD   Mean Square Spectral Density
- PSD   Power Spectral Density

CHAPTER 13 (only those symbols not previously introduced are listed)

M_{DF}	Design Flutter Mach number	none
V_{DF}	design flutter speed	knots, mph, or fps
V_L	limit velocity	knots, mph, or fps
V_{max}	maximum level flight speed	knots, mph, or fps

Abbreviations and acronyms
- ASE   Aeroservoelastic
- D   Divergence Instability
- F   Flutter Instability
- FCS   Flight Control System
- FECU   Flutter Exciter Control Unit

LIST OF SYMBOLS AND ABBREVIATIONS

CHAPTER 13 (continued)

GBU	guided bomb unit
HARV	high angle of attack research vehicle
KEAS	knots equivalent airspeed
MFD	multi-function display

CHAPTER 14 (only those symbols not previously introduced are listed)

Symbol	Description	Units
C_D	drag coefficient	none
$C_{\mathcal{L}_p}$	roll damping derivative	none
C_{mb}	pitching moment coefficient based on wing span	none
$C_{m,rb}$	pitching moment coefficient from rotary balance data	none
$C_{m,rb_{spin}}$	potential equilibrium pitching moment coefficient from rotary balance data	none
C_{n_r}	yaw damping derivative	none
D_A	drag force due to advancing wing panel	lbs
D_R	drag force due to retreating wing panel	lbs
F	force vector	lbs
I_x	mass product of inertia about x stability axis	slug-ft^2
I_{xy}	mass product of inertia about x and y stability axes	slug-ft^2
I_y	mass product of inertia about y stability axes	slug-ft^2
I_{yz}	mass product of inertia about y and z stability axes	slug-ft^2
I_z	mass product of inertia about z stability axes	slug-ft^2
k_x	radius of gyration about x stability axis	ft
k_y	radius of gyration about y stability axis	ft
k_z	radius of gyration about z stability axis	ft
L_A	lift force due to advancing wing panel	lbs
L_R	lift force due to retreating wing panel	lbs
$\mathcal{L}$	total rolling moment	lbs-ft
M	total pitching moment	lbs-ft
N	total yawing moment	lbs-ft
P	total rolling angular velocity	rad/sec
Q	total pitching angular velocity	rad/sec
R	resultant aerodynamic force	lbs
R	total yawing angular velocity	rad/sec
R_A	advancing wing panel contribution to resultant aerodynamic force	lbs
R_R	retreating wing panel contribution to resultant aerodynamic force	lbs
R_y	y component of resultant aerodynamic force	lbs

LIST OF SYMBOLS AND ABBREVIATIONS

CHAPTER 14 (continued)

Symbol	Description	Units
R_z	z component of resultant aerodynamic force	lbs
r	helix radius in a fully developed spin	ft
S	wing reference area	ft²
V	magnitude of the 3velocity vector for the aircraft cg	fps
$\mathbf{V}$	velocity vector for the aircraft cg	fps
W	weight of aircraft	lbs
X_A	aerodynamic force in x direction on advancing wing panel	lbs
X_R	aerodynamic force in x direction on retreating wing panel	lbs
Z_A	aerodynamic force in z direction on advancing wing panel	lbs
Z_R	aerodynamic force in z direction on retreating wing panel	lbs
ΔC_{D_1}	increment in drag coefficient between advancing and retreating wing at low angle of attack	none
ΔC_{D_2}	increment in drag coefficient between advancing and retreating wing at high angle of attack	none
ΔC_{L_1}	increment in lift coefficient between advancing and retreating wing at low angle of attack	none
ΔC_{L_2}	increment in lift coefficient between advancing and retreating wing at high angle of attack	none
α_A	local angle of attack of an advancing wing section	°
α_R	local angle of attack of a retreating wing section	°
α_{A_1}	angle of attack of the advancing wing at low angle of attack	°
α_{A_2}	angle of attack of the advancing wing at high angle of attack	°
α_{R_1}	angle of attack of the retreating wing at low angle of attack	°
α_{R_2}	angle of attack of the retreating wing at high angle of attack	°
α_s	stall angle of attack	°
β	sideslip angle	°
λ	helix angle in a spin	°
μ	relative density parameter	none
ρ	density of air	slugs/ft³

Abbreviations and acronyms

AOA	Angle of Attack
FAA	Federal Aviation Administration
FRL	Fuselage Reference Line
IAF	Indian Air Force
PSG	Post-Stall Gyration
US	United States
USN	United States Navy
USAF	United States Air Force

CHAPTER 15 (only those symbols not previously introduced are listed)

Abbreviations and acronyms

AC	Aircraft Aerodynamic Center or Neutral Point
AEDC	Arnold Engineering Development Center
ALCM	Air Launched Cruise Missile
AFB	Air Force Base
AGM	Air-to-Ground Missile
AIAA	American Institute of Aeronautics and Astronautics
AIM	Airborne Intercept Missile
ARA	Aircraft Research Association
AUR	All-up-round
BLU	Bomb Launcher Unit
BRU	Bomb Release Unit
CCMT	Captive Carry Miniature Telemetry (unit)
CDEP	Compatibility Engineering Data Package
CEP	Circular Error Probable
CFD	Computational Fluid Dynamics
CFP	Compatibility Flight Profile
CG	Center of Gravity
CTS	Captive Trajectory System
DOE	Design of Experiments
DOF	Degree of Freedom
E^3	Electromagnetic Environmental Effects
EMC	Electromagnetic Compatibility
EMI	Electromagnetic Interference
ERU	Ejector Release Unit
GAM	Guided Aircraft Missile
HERO	Hazards of Electromagnetic Radiation to Ordnance
HVAR	High Velocity Aerial Rocket
JDAM	Joint Direct Attack Munition
JSOW	Joint Standoff Weapon
JSSAM	Joint Air-to-Surface Stand-off missile
MER	Multiple Ejector Rack
MPI	Mean Point of Impact
NATO	North Atlantic Treaty Organization
NLR	National Research Laboratory
OFP	Operational Flight Program
PIDS	Pylon Internal Dispenser System
SRAM	Short Range Attack Missile
STV	Separation Test Vehicle
TER	Triple Ejector Rack
TMD	Tactical Munitions Dispenser
WCTMD	Wind-Corrected Tactical Munitions Dispenser

LIST OF SYMBOLS AND ABBREVIATIONS

CHAPTER 16 (only those symbols not previously introduced are listed)

Symbol	Description	Units
b_w	wingspan	ft
$C_{\mathcal{L}_\beta}$	dihedral effect coefficient	none
$C_{\mathcal{L}_{\delta_a}}$	roll moment coefficient due to aileron (aileron control power)	none
$C_{\mathcal{L}_{\delta_r}}$	roll moment coefficient due to rudder	none
C_{n_β}	yaw stiffness coefficient	none
$\Delta C_{n\,eng}$	yaw moment coefficient due to engine asymmetry	none
$C_{n_{\delta_a}}$	yaw moment coefficient due to aileron (adverse yaw)	none
$C_{n_{\delta_r}}$	yaw moment coefficient due to rudder (rudder control power)	none
C_{Y_β}	side force coefficient due to sideslip	none
$C_{Y_{\delta_a}}$	sideforce coefficient due to aileron	none
$C_{Y_{\delta_r}}$	sideforce coefficient due to rudder	none
C_T	thrust coefficient	none
S_w	wing Area	ft²
T	thrust	lbs.
V_{mc}	minimum control airspeed	KCAS
V_{mc_a}	minimum control airspeed, air	KCAS
V_{mc_g}	minimum control airspeed, ground	KCAS
V_r	rotation speed	KCAS
$V_{refusal}$	refusal speed	KCAS
y_{eng}	engine distance from centerline	ft
β	sideslip angle	radians
ϕ	bank angle	radians
η, η_p	propeller efficiency	none

Abbreviations and acronyms

BFL	Balanced Field Length	
HQDT	Handling Qualities During Tracking	
PIO	Pilot Induced Oscillations	
SHSS	Steady Heading Sideslip	
SHP	Shaft Horsepower	HP
THP	Thrust Horsepower	HP

CHAPTER 17

A_9	nozzle area (station 9 = exhaust plane)	ft²
c	specific fuel consumption	(lbm/hr)/HP
D	Propeller diameter	ft
J	advance ratio	
$\dot{m}$	mass flow rate	slugs/sec or lbm/sec
N	compressor speed	rev/sec
N_c	corrected compressor speed	rev/sec
n	propeller rotation rate	rev/sec
p_0	pressure (station 0 = free stream)	psf
p_2	pressure (station 2 = compressor face)	psf
p_3	pressure (station 3 = compressor exit)	psf
p_9	pressure (station 9 = exhaust plane)	psf
r	radius of interest (from axle of turbine or compressor)	ft
V_e	exhaust velocity speed	knots, mph, or fps
β	blade pitch angle	degrees
δ	pressure ratio	
θ	temperature ratio	
ω	rotation rate	radians/sec

Abbreviations and acronyms

BHP	Brake Horsepower	HP
FADEC	Full Authority Digital Engine Control	
FTH	Full Throttle Height	ft
GAP	General Aviation Propulsion	
LVDT	Linear Variable Displacement Transducer	
MAP	Manifold Pressure	inches Hg
MCP	Maximum Continuous Power	HP
OAT	Outside Air Temperature	°F or °C
TSFC	Thrust Specific Fuel Consumption	lbf/(lbm/hr)
VLJ	Very Light Jet	

PREFACE

The purpose of this book is to consolidate the fundamental principles used in classical performance and flying qualities flight testing of manned aircraft. It is intended for use as an introductory text for undergraduate students or as a self-teaching reference for an engineer newly engaged in flight tests. Worked examples are included in each chapter.

The first half of the book covers performance measurements. Chapter 1 is devoted to an overview of why flight tests are conducted and what the most important constraints are. Chapter 2 reviews the standard atmosphere and applies basic aerodynamic equations to the basic measurement system found in virtually every airplane, the pitot-static system. Techniques used to calibrate this system are described and illustrated. Chapter 3 reviews basic point performance equations and applies them to explain how to measure climb, descent, and turn performance. Both steady state and energy approximation expressions are used to lead into discussion of the sawtooth climb, the check climb, the level unaccelerated turn, and the level acceleration flight test methods. Chapter 4 briefly introduces propulsion systems, both for propeller-driven and jet-powered airplanes and summarizes useful relationships for determining range and endurance for such airplanes. Speed power flight tests commonly used to collect cruise performance data are introduced to close out this chapter. Chapter 5 rounds out the performance section of the book by outlining the equations used to estimate takeoff and landing performance and then describing measurements that must be taken to document appropriate measures of merit during these critical phases of flight.

The second half of the book deals with aircraft stability and control, concentrating on measurements that must be made to ascertain flying qualities. Chapter 6 covers the foundations of longitudinal static stability, concluding with a discussion of the flight test techniques often used to obtain such data. Chapter 7 does the same thing for maneuvering stability. Chapter 8 summarizes both the theoretical differences in the equations of motion and the flight test methods used to measure static lateral-directional stability. Chapter 9 is a concise summary of dynamic stability and control for both symmetric (longitudinal) motions and asymmetric (lateral-directional) ones. The importance of qualitative pilot ratings in describing aircraft flying qualities and their usefulness as a communications tool between the test pilot and the test engineer are stressed. The chapter concludes with a discussion of typical flight test techniques, including several practical ways of interpreting dynamic response data. Chapter 10 introduces post-stall flight tests, intending to interest the beginning flight test student in more advanced topics.

Gratitude is due to many people. First and foremost, for their unflagging support, I thank my family, especially my wife, Joyce. It is due to her patience with my early and late hours that the manuscript is finally finished. She also encouraged me throughout my twenty-seven years in the flight test profession, even though she undoubtedly often wondered why I was so obsessed. I thank her and dedicate this effort to her. All the students who have been exposed to my attempts to teach this subject at the United States Air Force Test Pilot School, at Texas A&M, at the University of Kansas, and at the United States Air Force Academy have contributed in no small way to this effort. Finally, thanks are due to all of the colleagues who reviewed these pages. Dr. Richard Howard of the Naval Postgraduate School deserves special mention for his contributions both as a PhD student who understood the practicality of the subject and as a peer who made insightful suggestions. The faculty of the Aerospace Engineering Department at Texas A&M University were wholly supportive, especially Dr. Walter Haisler, Dr. John Junkins, and Dr. Tom Pollock. I hope this subject will provide as much pleasure and challenge to the reader as the writing of this book has brought to me.

Donald T. Ward
March 1993

Preface to the Second Edition

The second edition of **Introduction to Flight Test Engineering** has been expanded to include a new chapter (Chapter 10) outlining the background and techniques used to prepare for aeroelastic flight tests, a subject usually ignored in such an introductory volume. This addition is the major change in this edition. Of course, the other material has been slightly rearranged to accommodate this insertion and known errors in these other chapters have been corrected. Dr. Thomas W. Strganac, joins me in providing this new material. This new addition to the text is based largely on the teaching experience of both authors, notably on their short course, taught since 1991, titled "Hazardous Flight Tests". This course, sponsored by the Continuing Education Division of the University of Kansas, has provided much of the new material. We are very appreciative of the assistance provided by Mrs. Jan Roskam and her staff in this endeavor.

Donald T. Ward
July 1998

Of course, both of us are indebted to even more individuals than were mentioned in the original preface for their help in producing this text. I am indebted to those individuals who provided the many invaluable experiences afforded me during my 15 years as an engineer with NASA and 9 years on the faculty at Texas A&M University. The development and presentation of the "Hazardous Flight Test" short course with my co-author has been an extremely rewarding experience and collaboration. I dedicate my efforts as an engineer, researcher, and professor to my family. Sincere appreciation is given to my wife, Kathy, who has selflessly supported my pursuit of my professional interests. To my son, Christopher, who has begun to build his educational foundation in his desired profession (Paleontology) at the University of Texas, and to my daughter, Kasey, who continues to follow her love of life and horses, I hope the two of you will pursue your dreams with devotion.

Thomas W. Strganac
July 1998

Preface to the Third Edition, Volume I

The third edition of **Introduction to Flight Test Engineering** is being expanded to two volumes, with this first volume to be followed shortly with a second one addressing more advanced topics. This first volume is comprised of the same subject matter as the first nine chapters of the original book. The chapters on aeroelastic flight tests and post-stall flight tests have been moved to Volume 2. This reorganization of the material allowed us to insert additional information into Volume 1, notably a risk management section in Chapter 1

PREFACE

and section on instrumentation for each of the other chapters. Dr. (Capt.) Robert J. Niewoehner of the United States Naval Academy has joined Dr. Strganac and me in refreshing this book (and correcting more errors). Rob brings a wealth of practical experience from the naval aviation community and has taught these topics both at the Naval Academy and has joined us on several occasions in teaching short courses for the Continuing Education Division of the University of Kansas. Welcome, Rob.

Donald T. Ward
February 2006

Like Don, I am pleased to welcome Rob Niewoehner to our team; may you enjoy the fruits of your labor on this book as much as we have for over a decade. As we continue to present this material, I am impressed by the knowledge and curiosity of students who attend our short courses. We owe a big thank you to all of them for asking questions and forcing us to learn enough to answer them. Sincere appreciation is due all the ladies at University of Kansas who help us, prod us, and keep us on track to respond to our students. As always, my largest vote of thanks goes to my wife, Kathy. She continues to be my support in every thing I do. Kathy and our children, Christopher and Kasey, continue to be the joy of my life.

Thomas W. Strganac
February 2006

It has been a considerable pleasure for me to use Don Ward and Tom Strganac's book for the past six years while teaching undergraduate flight test engineering. To be included as a partner both in the classroom and in print has been both an honor and delight. Flight test engineering constitutes the intersection of my professional life as a pilot, and my scholastic life as an engineer and educator. Educating future engineers and aviators is my daily concern, and there is no subject to which I would like them drawn as flight test engineering. I hope that my contribution to my colleagues' work faithfully represents their voice and interests. My professional tribute must be directed to Professor David F. Rogers, who patiently re-cast this fighter pilot into an engineering educator (and taught a Naval Aviator how to flare). While Dave endured only my poor landings, my wife, Natalie, has endured both my hours and my vocation. She and our sons are my delight.

Robert J. Niewoehner
February 2006

Preface to the Third Edition, Volume II

As explained in the Preface to the Third Edition, Volume I, this second volume is a collection of advanced topics to supplement and add to the material in the Second Editions last two chapters. We have expanded the material on aeroelasticity and added four new chapters touching on the integration of modeling and simulation with flight test, stores certification flight tests, minimum control speed tests, aerial refueling certification, and propulsion system testing. Producing this second volume in less than 18 months has taxed all three of us. We sincerely thank those who have helped us meet this challenge, many of whom (like

PREFACE

our wives and families) have now indulged us twice over. I am particularly grateful to my co-authors for their patience and good humor when all of us had little patience to spare. A special note of thanks also goes to Mr. Eddie Roberts and his colleagues for their assistance with the stores certification chapter. We solicit your comments and suggestions for improving this book and look forward to meeting as many of you in person as is feasible.

I salute a remarkably unique individual who flew on to a higher realm this year: Thomas U. McElmurry (Mac to everyone who knew him). I have never met nor flown with a finer man; he embodied everything one could want in a friend and a colleague.

Donald T. Ward
July 2007

Our new edition represents substantial growth from our first manuscript; our experience has grown significantly by the addition of our newest author, Rob; our knowledge has grown from many new experiences and challenging questions presented by our students; and, our text has grown with the addition of several new chapters, necessitating a split into two volumes.

Chapters 11, 12 and 13 represent an expansion in the arena of aeroelasticity. In particular, new material is presented in nonlinear aeroelasticity, and has been motivated by this instructor's interests in store-induced limit cycle oscillations. Rob has provided excellent and relevant discussion related to his experience as a test pilot associated with the F-18 E/F flutter clearance.

This author is grateful to many individuals, including the staff of the University of Kansas Continuing Education Program who have afforded me the opportunity to instruct (and learn from them) many students at professional short courses. My students have represented a very diverse background – including a substantial international experience, as well as participants from government facilities, universities, and industry. I am also grateful for the experiences and lessons learned from my two co-authors. It is rewarding to listen to the dialogue (especially filtered by translation through e-mail) of two military officers, especially test pilots from two different branches (and aircraft eras) of our military. As the son of a military officer who successfully flew 35 missions in Eastern Europe from 1944-45, my attention is piqued whenever members of our military are present. As I write this third edition, I find myself ever mindful of the service and sacrifices our military personnel provide for the United States. I am grateful.

This past year saw the loss of a friend who represented the best of men. Thomas McElmurry had three notable careers in which he touched the lives of many. He served as a military officer, as a public servant with NASA, and as an instructor for our department at Texas A&M University. He was truly a gentleman who had a positive influence on all who knew him. He will be missed.

Thomas William Strganac
July 2007

PREFACE

The principle is well known among educators that you really don't start to know a subject until you've taught it. Well, I've now further learned how much there is to learn from the process of writing. Keeping up with my colleagues forced this fast-jet pilot to dig into both the civil regulations, and a variety of technologies (such as reciprocating engines) with which I'd only been superficially familiar. It was tremendously rewarding, and I'm grateful to the two of them for the chance to learn and further contribute to their already successful team.

I've now been teaching long enough that my first crops of students have completed combat tours flying from ships at sea, and are toiling through Test Pilot School. I pray that this work would inform and further fuel their fascination and appreciation for the flight test trade. While it's the airplanes that bring flight test teams to work, it's the quality and devotion of the people that make the profession really rewarding.

Rob Niewoehner
July 2007

Chapter 10
SIMULATION AND MODELING IN FLIGHT TEST

Simulation and modeling play a significant role in many facets of flight test. Test preparation and planning depend heavily on the modeling effort, largely because that effort suggests how to more efficiently conduct flight tests. Risk management for specific envelope expansions like flutter, stores separation, and post-stall tests, depends heavily on credible models specifically developed for these tasks. At the same time, flight test data are one of the most reliable means of validating and updating the models and the simulations that form the basis of operating manuals and training devices for aircraft.

In this chapter we emphasize one of the central themes in current government policy and commercial practice, a comprehensive Modeling and Simulation (M&S) effort, well-integrated with the Test and Evaluation (T&E) strategy. For U.S. military projects, DoDI 5000.2 states bluntly: "Projects that undergo a Milestone A decision shall have a T&E strategy that shall *primarily address* M&S, including identifying and managing the associated risk, and that shall evaluate system concepts against mission requirements."[1] (italics added for emphasis). Furthermore, program managers receive the following guidance: "...Modeling and Simulation (M&S) is integral to and inseparable from T&E in support of acquisition."[2] In commercial developments, the marketplace and regulatory agencies increasingly emphasize M&S, as we will note in our examples.

10.1 FOUNDATIONS

10.1.1 Historical Perspective

The first use of M&S was aircrew training, but the new discipline soon expanded to other areas. It has become an important analysis and predictive tool for designers and engineers as well as business and military strategic planners. We briefly outline the early history of M&S, but the main emphasis in this chapter is on the importance of using M&S to support flight test (and vice versa).

Fig. 10.1 Antoinette Barrel Trainer (Adapted from 1910 catalog)[3,4,5,6]

10.1.1.1 Evolution of Aircrew Training Simulators. Several historians refer to the Sanders Teacher (circa 1910), essentially a complete airplane mounted on a universal joint and facing into the wind, as one of the first training devices. It was able to rotate and tilt freely. A similar design by Eardley Billings was used at Brooklands Aerodrome in England at about this time[3,4]. Another early "flight simulator", dubbed an Antoinette barrel trainer (Fig. 10.1) was photographed for the Antoinette catalog in about 1910[3,4,5,6]. The trainee sat in the top part of two half sections of a barrel free to rotate in pitch and roll. These motions were generated manually by "assistants" as shown in the photograph. The prospective pilot was tasked to align a reference bar with the horizon. These early examples hardly suggest the enormous potential of M&S both in engineering design development and in flight tests but they are representative of the first crude efforts to develop this flight training technology.

Less than a decade later, fueled by the urgency of World War I aviation and its explosive growth several inventors were producing "coupled" designs that fed back the trainee's control movements to the drive the "cockpit" to an orientation that followed the pilot's inputs. Lender and Heidelberger[3,5] patented several such devices in 1917 that were driven by compressed air motors. The propeller gave some semblance of airflow and engine noise. An instructor could also introduce "turbulence" to simulate rough air for the student. Another version of these this group of devices used the propeller, along with vanes inserted into the propeller slipstream and controlled by the trainee to tilt the fuselage of the device[6].

Fig. 10.2 Link Trainer (USAF photo)[7]

By 1929 Roeder set down the first detailed outline of requirements for computer control of aircraft and submarines moving in three dimensions. His German patent application could describe a modern flight simulator with little modification. At about the same time several inventors, notably Link and Buckley, patented devices built around the principles of the Antoinette "barrel" trainer – except the motions were generated mechanically and the device moved proportional to pilot inputs. Roeder's German patent was the forerunner to several pioneers in simulations. Edwin Link filed his patent in the United States in 1930[8] for such a machine, driven by electrical signals and powered by pneumatically driven bellows. Link's contract with the U.S. Army Air Corps in the early thirties was to provide such devices to help train pilots carrying mail by air. This work led to widespread use by the Air Corps of the famous Link Trainer about a decade and more later (Fig. 10.2 shows a later version of Link's trainer similar to the C-10 in which one of the authors trained in the late 50s). These ideas were expanded upon by teams led by Dr. R.C. Dehmel (Bell Laboratories) and A.E. Travis, making broader use of electronic subsystems to add radio navigational signals and electronic signals to emulate aircraft navigation. Travis' device added a visual subsystem based on a loop of film showing the pitch, roll, and yawing motions of the fuselage. These advances were very important steps in the evolution of simulation. Mueller advocated an electronic analog computer for real-time simulation of airplane longitudinal dynamics for design purposes; Mueller suggested extending the time scale and adding a man in the the loop. By the early 1940s, several electronic simulators extended this capability[3,5,6,8].

Curtiss-Wright developed the first full aircraft simulator used by an airline for Pan American Airways, installed in 1948. Though it included no motion and no visual system, it otherwise replicated the appearance and behavior of the Boeing 377 Stratocruiser. Redifon in England built similar Stratocruiser synthetic crew trainers for BOAC, and later produced a Comet I trainer for BOAC utilizing fixed-spring control loaders[5].

The 1950s saw a research effort at the University of Pennsylvania that led to one of the first simulators, manufactured by Sylvania Corporation, based on early digital computers. Link's company developed a competing design in the early 1960s and from that time forward, simulation was tied to growth in computing power and capability[6].

Fig. 10.3 Terrain Board for Night Vision Goggle Training (US Navy photo)[9]

10.1.1.2 <u>**Rapid Increases in Computational Power and Software Capability.**</u> Computer-generated imagery represented the next major innovation, enabled by an explosion in computer capabilities. Figure 10.3 illustrates such a display for a modern simulator training military crews in the use of night vision goggles. This picture suggests the intense role that simulation now plays in training aircrew members; the current FAA certification of commercial airline pilots for the airline transport rating depends heavily on simulator training, especially to learn aircraft systems and to practice emergency procedures. It is possible to get a type rating for specific airplanes with most of the training done in high fidelity simulators rather than in the aircraft itself. This training yields a considerable cost saving, a superior result enabled by the simulation of severe environments, and the avoidance of risks otherwise posed by training to those environments. These themes motivate the continued expansion of M&S into the T&E enterprise.

The explosive growth of computing power since the 1940s gave the M&S world computing capacity that pushed other areas necessary to make these tools more useful to engineers and designers. In the late 1950s, Englebart introduced the idea that computers were not just for crunching out numbers; he advocated graphical display and manipulation of data. In 1962, Sutherland introduced Sketchpad, a way to use a light pen to sketch on a computer screen, which led to the first Computer-Aided Design (CAD) software. By 1968 Engelbart had produced a word processor, a primitive hypertext system, and a collaborative tool. In 1970 Sutherland introduced the first version of a head-mounted display and

Englebart a first-generation "mouse"[10], used to move text around on the screen. Up to this point in time, flight simulators used "terrain boards" and moving cameras to generate screen displays, but in the 1970s computer-generated imagery began to replace this approach and provided much more flexibility and versatility for displaying scenery and emulating motion. Figure 10.8 illustrates such a terrain board with a CCTV camera used to capture the necessary scenery images for the simulator. This sort of ancillary equipment is not often used now; computer-generated imagery now provides a much less complicated way to put up realistic, textured images – yet another of the innovations needed to make M&S a better tool for both training and for analysis.

Since the advent of the personal computer in the 1970s and 1980s, typical personal computer capabilities have grown exponentially which has likewise fostered significant expansion of engineering software analysis tools. Table 10.1 illustrates how various personal computer hardware characteristics have grown during the two decades following 1984. These capabilities have been widely exploited with software analysis tools as suggested by Fig. 10.4 throughout the aerospace industry. Figure 10.4 illustrates this point by noting the growth in number of software patents[11] (from 4 different sources) over the first two decades of that period. More directly, modeling and design software usage followed a similar exponential growth pattern, as suggested in Table 10.1[12].

Table 10.1 Growth of Modeling and Simulation Computing Capability – 1984-2004

Computational Capability	1984	2004	Ratio - 2004/1984
Clock Speed	5 MHz	3 GHz	600
Memory	256 Kbytes	512 Mbytes	2000
Disk Storage	10 Mbytes	80 Gbytes	8000
Communication speeds	1 Kbytes	3 Mbytes	3000
Flops	17 K	1 G	60000

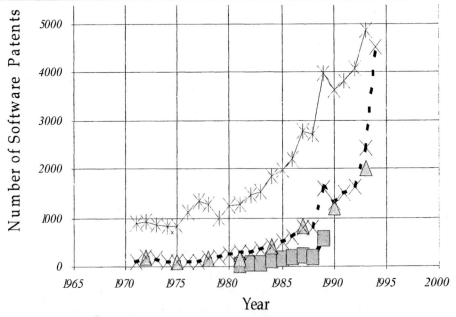

Fig. 10.4 Exponential Growth of Software Tools in Early 1990s[11]

Chapter 10 — Simulation and Modeling in Flight Test

The increased capability of M&S software to represent results graphically is not captured in this chart. The software tools available today to the engineer at his desktop computer are remarkable in their analysis and display capability.

Having considered the historical context of M&S, the increased capabilities, and the trend toward wider use of these tools, we now turn to the task of classifying models used and relating them to flight test needs and objectives.

10.1.2 Elements and Terminology of Simulation

This chapter specifically treats the use of simulation to facilitate aircraft flight test. While kin to several other uses of flight simulation (training, for example), flight test applications present unique characteristics and employ specialized terms.

10.1.2.1 Elements.
Nonetheless, one or more of the three basic elements[4] of flight simulation generally are still needed to produce the results sought. A mathematical, dynamic model of the aircraft or the aircraft subsystem to be evaluated is the first of these elements. Second, there must be a specified application evaluated and/or regime of flight appropriate to the purpose of the simulation activity planned. Finally, for most simulation tasks, some kind of device, a cockpit with appropriate controls and displays or some other lesser fidelity means of implementing the model, transforms the mathematically produced dynamic effects into cues for the analyst or for the subject. The first two elements are common to almost every form of simulation; the hardware element frequently require tailoring to specific flight test applications.

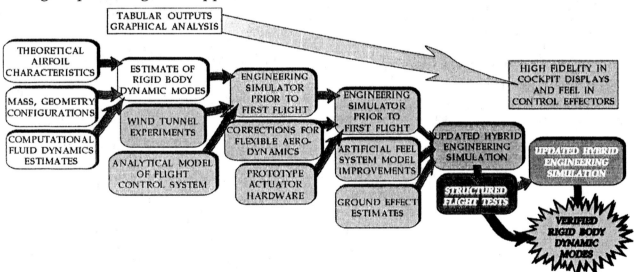

Fig. 10.5 Typical Simulation "Waterfall" Evolution

10.1.2.1.1 Mathematical Model.
Mathematical modeling undergirds almost every facet of designing a complex system like an airplane. Such models characterize the propulsion subsystem, the flight control subsystem, the hydraulic subsystem, the flight management subsystem – essentially every subsystem that makes up the airplane. But from the flight test perspective, we care most about how a mathematical model can be used to suggest the kinds of ground and flight tests needed to most efficiently verify and validate the vehicle design. The modeling of a complex system like an airplane inevitably requires an iterative process, as Fig. 10.5 suggests. (Notice that such a description could be developed

for many different aspects of the simulation effort; this specific waterfall chart is illustrates how a flight dynamics model might develop, though the diagram is not comprehensive. It does not illustrate how man-in-the-loop war games might be developed to evaluate the operational effectiveness of a new system's features.) In its early stages the model may deliver only batch results with no direct human interaction, while later stages may include hardware-in-the-loop (HIL), and high fidelity environmental effects intended to prepare the test pilot for what to expect in the behavior of the airplane. As depicted in Fig. 10.5, simulation and aircraft dynamics are expected to eventually converge, especially after adjustments in the simulation model are made from carefully designed flight test experiments. Thus, model refinement becomes an integral deliberate goal of the T&E process.

10.1.2.1.2 *Implementation*. The implementation of a simulation flows from the intended use. Our primary concern in most flight test activities is to assist the design and certification process by providing data that can be analyzed at every phase of the development. A primary concern of the flight test team is to provide verification and validation of the design as early and as efficiently as possible. Fig. 10.5 further suggests that the demands on the model implementation are likely to change from essentially pure mathematical output, either graphical or tabular, to real-time, pilot-in-the-loop qualitative results as the test program evolves to support the overall developmental effort.

Similarly, a model or simulation that estimates operational effectiveness parameters for OT&E can take one of three forms: (1) strictly mathematical representations (constructive simulations), (2) systems test beds (virtual or hybrid simulations) which contain some, but not necessarily all, of the actual hardware used in the system, and (3) live exercises (live simulations) where troops use equipment under actual environmental conditions approaching combat situations but under peacetime operations[11]. Obviously, modeling and simulation support can grow from (1) through (3) as the system evolves. This almost continuous spectrum of simulation growth is illustrated in Fig. 10.6; it culminates in validation of the operational effectiveness of the system.

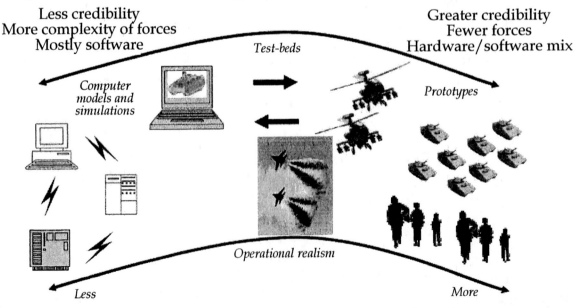

Fig. 10.6 Simulation Spectrum (Adapted from[13])

10.1.2.1.3 *Regime of Application*. The intelligent use of modeling and simulation demands that the mathematical model and the type of implementation be carefully tailored at each step in the design process. Considerable thought and, inevitably, some compromises, are necessary. Model implementation updates must occur whenever better data become available. For example, it would be premature to invest in a first class, high fidelity feel system in the early stages of the development when the aerodynamic model remains immature. By the same token, though, it may be foolish not to make such an investment in time to train for the first flight. Above all, mathematical models necessarily neglect some details to make the simulation tractable, the designer electing to omit or simplify those features or phenomena deemed to be insignificant. Significant behaviors may not appear in the modeled system; they are revealed only in flight test. So, M&S results are not a wholesale substitute for actual test data. The human decision-making process and the proficiency of personnel in performing their tasks are two such "immeasurables".

Using models to develop test plans requires considerable care in choosing exactly what level of model and simulation evolution is appropriate to the planning activity at hand. The first model is likely developed at a very early stage (and thus be of low fidelity) while the Test and Evaluation Strategy (TES) and its derived T&E Master Plan (TEMP) are still evolving. Some flexibility must be built into the TEMP so that detailed test plans can be revised when higher fidelity M&S allow more detailed planning. Moreover, as design iterations take place, performance predictions receive increased emphasis. This kind of flexibility is considerably facilitated in higher fidelity models, provided the architecture of the original (low fidelity) model deliberately accommodates growth.

10.1.2.2 Definitions and Classifications. So far, we have not distinguished 'models' from 'simulations', as commonly used. "A model is a physical, mathematical, or otherwise logical representation of a system, entity, phenomenon, or process."[13] Mathematical models are written in diverse languages and may be grouped together to run either serially or in parallel in a specified order by an executive module. This executive package often also controls the data transfer between models and spells out how results from a simulation sequence are presented to the user. They may or may not include direct interaction with real components or human operators. "A simulation is a method of implementing a model *(or models)* over time."[13] The executive module may control the time-stepping (frequency and speed of execution) for a simulation, creating two general categories of M&S: "non real-time" and "real-time".

10.1.2.2.1 *Non-Real-Time (Analytical) M&S*. Analytic models and simulations are common in aerospace system development today. These simulations do not run in clock-time, but run either faster or slower than clock time. Some studies are computationally accelerated, permitting hundreds of scenarios to be evaluated in a brief period of time (e.g.- flight dynamics or campaign models). Others, due to the complexity of the modeled physics, may run much slower than clock time (for example, unsteady computational aerodynamics). Trade studies, subsystem and system performance predictions, operational limitations, system safety analysis, test planning, anomaly analysis, and risk management are some of the elemental activities that benefit from analytic M&S.

Usually these M&S packages do not operate in real-time use a standalone or batch mode without direct operator interaction. The departure/spin testing of the F/A-18 Super

Hornet was preceded by a batch simulation evaluation involving several hundred test points, varying flight condition and combinations of control deflections. The executive test module could process the entire test matrix in several days. Results identified the highest risk points for flight investigations, and points most sensitive to inaccuracies in the aerodynamic data base. The entire batch simulation was re-run following significant updates to the aerodynamic database to both validate the database (Did the simulation reproduce the flight behavior?) and more accurately predict tests yet to be flown (What are the points we really need to fly in order to convince ourselves that we know this airplane?).

For some development and experimental flights it may be advantageous to run the simulation in parallel in the control room as flight test data are telemetered to the control room displays. For such M&S displays to be synchronized with telemetered data it is often necessary to optimize the computer code. Said another way, engineers usually utilize standalone models to create high fidelity predictions of system and subsystem behavior. Computational Fluid Dynamics (CFD) codes with very fine grid structures or very detailed Computer Aided Design (CAD) descriptions of the vehicle shape and structure are two examples of such high fidelity codes that may not readily allow synchronization with flight test data in real-time. These kinds of tools support specific aspects of the design process and may not be useful during the conduct of flight tests. However, the test engineer must at least appreciate what these codes can do and know what they predict for the behavior of the subsystem or system undergoing test.

10.1.2.2.1.1 *Subsystem Modeling.* One of the more important uses of non realtime analytic M&S is in the early stages of testing – before an end item (the aircraft) is ready for flight tests. Gaining an understanding of how the subsystems should perform is a first step for the flight test team in evaluating the integrated design. But, this important step is often neglected or given short shrift in the planning. The challenge is to be sure the integrated team structure commonly used in the aerospace industry encourages rather than discourages dialog between designers and test engineers so that appropriate fidelity is used. Hines says: "It makes little sense to have a high fidelity aerodynamic model matched with only an approximation of the flight control system characteristics."[14] Results from subsystem modeling that includes poorly matched components are of little value to the flight test planner. If the intended use of data from a subsystem model is to predict the flight dynamics behavior in the "heart of the envelope" (where linear models may be quite adequate), a simple propulsion module may suitably serve this purpose. However, simplified propulsion models that neglect gyroscopic effects, inlet distortion effects, and other detailed powerplant characteristics inaccurately models the behaviors in the stall and post-stall flight regime. The flight test planner interested in such an envelope expansion task often must consider the limitations of the simulation providing his information.

10.1.2.2.1.2 *Critical Envelope Expansion Models.* Planning for flight envelope expansion tests is often heavily dependent on carefully evolved, highly credible models specific to a discipline (like structures and their finite element models or FEMs) or optimized for a portion of the flight regime (stall, post-stall models). The waterfall chart of Fig. 10.5 is one illustration of how an evolving model should be used in laying out and refining the test matrix for high AOA, flutter, stores separation, and other such hazardous test operations. Chapters 11-15 discuss such envelope expansion activities and models that are usually built up for each of these critical envelope expansions. In almost every case, models like these demand analytic detail beyond that feasible for real-time execution. Likewise, each of these specialized analysis tools can be enhanced when flight test data is fed back to the M&S specialists. Once again, M&S and T&E are mutually dependent.

10.1.2.2.1.3 *M&S Interaction with OT&E.* Operational test and evaluation (OT&E) is also heavily dependent on models that evolve with time to guide planning and to make both OT&E and live fire test and evaluation (LFT&E) as efficient and complete as possible. The Defense Acquisition Guidebook (DAG) underscores the integrated process necessary to achieve this goal[13]: "...the integrated approach should not compromise DT&E, OT&E, or LFT&E objectives. The program manager, in concert with the user and test communities, without compromising rigor, is required to integrate modeling and simulation (M&S) activities with government and contractor DT&E, LFT&E, system-of-system interoperability and performance testing into an efficient continuum." This form of testing relies heavily on computer-generated scenarios (including complementary friendly forces and enemy threats) to lay out and conduct realistic operational and live tests. The responsible operational test agency (OTA) is responsible for "accrediting" all M&S used in OT&E, as is the program manager for M&S supporting DT&E effort. A robust verification, validation, and accreditation (VV&A) process is mandated by DoDI 5000.61[15]. Special emphasis is placed on validation of threat representation in terms of credibility and currency (as approved by the Defense Intelligence Agency). The entire integrated product and process development (IPPD) team approach is designed to produce, among other things, a process for continuously updating and validating the threat representation for the T&E working-level integrated product team (T&E WIPT) that includes both DT&E and OT&E expertise. Accreditation of the simulated threat is the responsibility of the OTA, though the verification and validation that typically precede and support accreditation are joint responsibilities. This integrated conduct M&S and T&E often is made possible and becomes visible in the planning documents, the TES and the TEMP.

10.1.2.2.2 **_Real-Time M&S_**. Several forms of real-time M&S contribute to T&E productivity and safety. The most useful M&S for engineering developments, certainly as the vehicle approaches first flight, is real-time simulation with man-in-the-loop (MITL). Real-time hardware-in-the-loop (HIL) simulation further extends realism and validity. Table 10.2 (next page, expanded from Hines[14]) underscores the value of real-time M&S.

10.1.2.2.2.1 *Engineering Man-in-the-Loop (MITL) Simulation.* MITL simulations play an important role in simulation-based testing, as the second column of Table 10.2 indicates. They contribute to every one of the listed T&E tasks, and are important to every phase of the development of an air vehicle (military or commercial) today. Commercial aircraft certification regulations allow (even encourage) the use of MITL results in meeting some documentation requirements. Pilot training and procedural standardization depend largely on MITL simulations. While the focus in this chapter is largely on the interactions between M&S and T&E activities, M&S pervades the entire development process. This fact justifies the resources and time spent in developing appropriate simulation tools capable of contributing to all facets of new system development.

Table 10.2 Integration of M&S into T&E Tasks

T&E Tasks	Analytic	MITL	HIL	Iron Bird	In-Flight	Operational Scenarios
Planning	XX	XX				XX
Maneuver Definition	X	XX			X	
Anomaly Investigation	X	XX	X	X	X	
Test Scenario Development		XX			X	X
Software VV&A		X	XX		X	X
Failure Modes	X	X	XX			
T&E Training		XX	X			
Flutter/Limit Cycle	XX	XX	X	XX		
Post-Stall Sensitivities	X	X				
Threat Representation	X	X				XX
Human Factors		XX	XX	X	XX	XX

Note: X suggests some usefulness and XX implies highly relevant to the task

MITL simulations supporting design, crew training, and T&E consist of the three basic elements previously discussed, typically including specialized implementation to provide a tactile and visual interface with the human operator. Tactile interfaces provide a way for the crew member to control the aircraft or its subsystems; the control stick/wheel/rudder pedals, the throttles, and the autopilot knobs or keypad are examples of such interfaces. Modern MITL simulations include visualizations of cockpit displays, motion bases, and often out-the-window (OTW) scenery to cue and stimulate crew member reactions. The required fidelity of these features depends on other basic elements of simulation – intended use (regime of application) and mathematical approach chosen to provide dynamic cues to simulation subjects. There are many examples of very high fidelity MITL simulations used for flight test planning, maneuver definition, crew training, and anomaly prediction/correction, but there are also a number of examples of lower fidelity facilities that are quite useful for specific purposes (an avionic workstation for software verification, or an iron bird for hydraulic subsystem testing). There is a clear correlation between cost, complexity, fidelity, and utility of a simulation facility.

The Air Combat Environment Tactical Evaluation Facility (ACETEF) at Patuxent River Naval Air Station in Maryland is one example of a high fidelity simulation used for multiple purposes. It primarily supports the Naval Air Warfare Center's mission of acquiring aircraft and systems for the US Navy and the US Marine Corps by testing "…installed aircraft systems in an integrated multi-spectral warfare environment using state-of-the-art simulation and stimulation technology. Aircraft platforms, typically placed in an anechoic chamber, are made to behave as if they are in a real operational environment through a combination of digital simulations and stimulation by computer-controlled environment generators. The ACETEF has several laboratories providing signal generation, man-in-the-loop cockpits, high-performance computing (HPC), and warfare environment. These laboratories can work autonomously or collectively to provide varying levels of test and analysis capabilities."[16] ACETEF "…is the world's only fully operational, multi-role Installed System Test Facility (ISTF) utilizing Modeling and Simulation (M&S) that can integrate a simulated warfare environment with state of the art simulators, stimulators, actual weapon systems/aircraft, High Performance Computing (HPC), high fidelity cockpit trainers (Manned Flight Simulator), and external facilities/capabilities to provide a cost effective RDT&E capability to complement the entire acquisition process."[17] Though ACETEF is not always used in MITL mode, many projects do use its MITL capability. The Joint Strike Fighter (now the F-35) conducted MITL campaign-level war games in the ACETEF facility well before the prototypes' first flights, an example of applying M&S early in the acquisition cycle to help nail down requirements and reduce program risk[17].

The National Aerospace Laboratory (NLR) in Amsterdam operates another very sophisticated, high fidelity MITL facility. This research simulation has supported both civil aviation (glass cockpits for transport aircraft) projects and military upgrades (F-16 cockpit Mid-Life Update). The reputation of NLR's MITL simulation has revolved around handling qualities studies and human-machine interface studies that depend on high-fidelity cockpits which can be readily reconfigured, along with motion systems that are state of the art and a variety of detailed mathematical representations of subject airplanes and helicopters, including the Boeing 747-200/400, a Fokker 100, a Fokker 50, a Cessna Citation, a McDonnell-Douglas DC-10, an MBB BO-105 helicopter, and a Eurocopter Puma[18]. NLR has both 4-DOF and 6-DOF motion bases with appropriate cabs[18].

Most companies and many universities have somewhat less generic and less sophisticated MITL facilities performing applied and basic research. The Singapore Center for Military Experimentation (SCME), for example, utilized a workstation with an augmenting display panel to study early air combat experiments with a subject pilot[19] and to explore Network Centric Warfare (NCW) concepts. Cost was a strong driver in choosing this approach: "The underlying experimentation philosophy in SCME is to conduct the most disruptive experiments with the least amount of resources." The resulting Augmented Virtual Battle Station (VBS) cockpit is described by Chia[19].

Fig. 10.7 FSL Cockpit with Reconfigurable Displays and "Out the Window" Displays[20]

Another example of low cost MITL simulations is the fixed-base simulator installed in 1993 at Texas A&M University's Flight Simulation Laboratory (FSL)[21] with a rudimentary virtual cockpit. Figure 10.7 illustrates the configuration of this device in 2001. This facility has been used for a number of conceptual studies of human-machine-software interface studies[21,22] – exactly the kind of evaluations for which such a simple, low cost facility is well-suited.

These examples illustrate how pervasive engineering/MITL simulation now is within the industry. All aspects of the design process are affected, and none more noticeably than the T&E phase (the verification and validation processes) of system development. Whether the system is a general aviation aircraft, a system meant to serve the commercial airline industry, or an element in the most complex military system-of-systems, MITL simulation is integral to the whole process and must be closely tied to the T&E effort.

10.1.2.2.2.2 *Hardware-in-the-Loop (HIL) Simulation.* In one sense, HIL simulations extend engineering/MITL models. By using actual hardware when available for specific subsystems, a higher fidelity response can be obtained. Instead of mathematically modeling subsystem behavior (like that of an actuator or a pneumatic component), using the actual hardware can make the simulation more like the end item. Flight control systems and their software are often are developed in this manner. Avionics issues such as computational latency errors in computation, processor bus delays or throughput problems are frequently revealed while conducting tests of other systems.

There are, of course, pitfalls to be avoided: (1) there are likely to be some sensors and/or inputs that must still be modeled; (2) synchronization between the hardware used and the remaining digital or analog models must be carefully orchestrated; and (3) latency between the data streams must be minimized. The issue here is latencies introduced by the simulation, as opposed to latencies in the actual integrated hardware. The former degrades the fidelity of the simulator and can unrealistically degrade the apparent system performance. The latter, actual latencies are phenomena that the test team wants to expose if they lead to degradations in the actual integrated system.

These challenges strongly argue for a tightly integrated, cross-disciplinary team charged with designing ground tests that precede flight tests and with evaluating flight test results to improve any HIL simulation. Hines[14] cites the F-18 HARV flight control and avionics systems integration and the F-15 Short Takeoff and Landing (STOL) Maneuver Technology Demonstration (MTD) programs as two examples of efficient use of HIL M&S. Indeed, Smolka[23], speaking for the NASA/Air Force/contractor team wrote: "First, for an open-loop test program of this nature, accurate ground based simulation was essential to provide the test team with a tool with which to develop flight test techniques. Second, the simulator proved its worth in allowing the test team to identify potential catastrophic test conditions that were not predictable analytically, were unanticipated during the development of the test maneuvers, and for which FC [flight control] protection logic could not be designed without excessive development time and cost. This underscored the need to collectively weigh software limits, envelope limits, and maneuver restrictions to ensure flight safety." The actual hardware used in the HIL simulation at Dryden Flight Research Center included most of the critical avionics hardware/software, up to the actual Heads-Up Display (HUD) unit. Smolka's conclusions rather succinctly summarize how important HIL simulation is to such a flight test research program.

Quadruple redundant digital flight control surfaces

Thrust vectoring nozzles

Electronic air inlet controllers

Dual channel nozzle controllers

Fig. 10.8 F-15 ACTIVE Aircraft and Its Integrated Subsystems (Adapted from [23, 24])

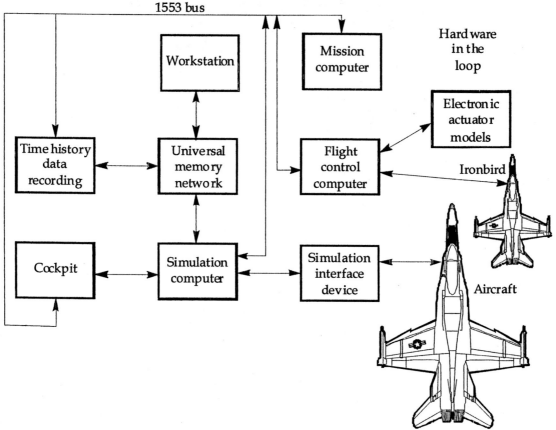

Fig. 10.10 F/A-18 HARV Configurable Simulation (Adapted from [25])

The F/A-18 High Angle of Attack Research Vehicle (HARV) also benefited from NASA Dryden's use of HIL simulation. Similar to the ACETEF at Patuxent River, this research center's HIL facility allows the test team to simulate research missions with the actual aircraft to be tested connected to the Dryden fixed-base simulation cab. The test airplane's sensors and systems can be stimulated by the simulation to represent diverse flight conditions. Figure 10.9, adapted from Chacone, et al[25], illustrates this configurable simulation facility. This collection of devices is more than just another HIL simulation; it is made up of several components that are part of a larger M&S facility at NASA Dryden, the Walter C. Williams Research Aircraft Integration Facility (RAIF). Built at a cost of over $22 million dollars, the RAIF opened in 1992 and was renamed after the first director of the NASA Dryden Flight Research Center in 1995. The RAIF is designed to accept a mix of small commercial aircraft and military fighter aircraft (see Fig. 10.10). Systems test and evaluations, along with ground vibration tests for flutter evaluations are carried out in this facility. Except for engine runs, all in-flight functions that simulate a flight can be carried out in the RAIF; moreover, a simulated flight can be monitored and controlled by a pilot and/or the flight test team (Fig. 10.10). This latter capability provides a rather complete training opportunity for the test team before actual flights are made.

Fig. 10.10 F/A-18 HARV in the RAIF[25]

10.1.2.2.2.3 *Iron Bird.* Figure 10.9 refers to the use of an "iron bird" as part of the F/A-18 HARV configurable simulator in the NASA Dryden RAIF. In this facility, the actual aircraft is used to provide hardware (actuators and other mechanical/electrical components), it is more common in the industry for these components to be mounted in a rigid steel frame laid out in the identical geometry as test aircraft (hence the name). Hines[14] notes, the"... intent of an iron bird simulator is to verify and validate that all of the mechanical/electrical components will function together as an integrated system." As Figs. 10.9 and 10.10 suggest, an iron bird simulation is often connected to a high fidelity MITL simulator. The Boeing 777 iron bird mockup even used the aircraft electrical power generators to allow examination of electrical power quality and loading of this subsystem. Hydraulic lines and control cables for the 777 were also part of this iron bird facility. The only models used in such a simulation are those developed for the aerodynamics and for the propulsion subsystem.

Fig. 10.12 RAIF Displays and Cockpit (NASA Dryden photo[26])

An iron bird emulator of this type has several uses during the integration of the systems. Most of the maintenance models (procedures, training manuals, and the like) can be examined with such a tool. Failure states can be induced and the consequences of taking various corrective actions can be evaluated. Flight control system susceptibility to limit cycle oscillations (LCOs – see Chapter 11) can be examined and the flight control parameters set to minimize the probability of undesirable pilot-machine interactions.

An iron bird mockup/simulator requires careful planning and is relatively expensive. If dynamic flight loads on actuators are modeled then the iron bird can be very expensive. Care must be taken to insure that the components of the iron bird are identical to those used on the airplane, including the layout of hydraulic lines, and the brackets used for their support (key to discovery of vibrational resonances). Space for such a facility is usually at a premium. Consequently, such simulations tend to remain active throughout the life of a given design and are often updated as the design evolves. Since such simulations are most useful to the integrating contractor, it is customary for all members of the test team having a need for the iron bird to go to the manufacturer's facility.

Despite the expense of such a facility, the investment often shows considerable return. Many program managers who agonized over the expense later found themselves relieved when problems were identified and corrected on the iron-bird early in development. The 777 development completed 4047 hours of ground testing compared to 667 hours of ground testing on the 767. But the 777 also sustained a much higher flight rate than did the 767, presumably because more problems were found on the ground before flight tests began. During initial testing, the 777 flight rate was approximately 5 times that of the 767: 75 hours/month versus 15 hours/month[14].

10.1.2.2.2.4 *In-Flight Simulation.* Starting in the mid-1940s variable stability airplanes appeared with the capability to emulate other airplanes in flight[27]. This development followed Heinkel and Fischel who introduced stability augmentation in 1940 using one section of a split rudder to control the poorly damped Dutch roll in the Henschel Hs-129[27], producing the first effective yaw damper. By 1947 both the NACA at Moffett Field and Cornell Aeronautical Laboratory (now CALSPAN) recognized the need for variable stability aircraft to help in solving design problems like the low dihedral effect evident in some of the U. S. Navy's carrier fighters during the approach to a carrier landing. The NACA installed its first variable stability system (VSS or VariStab) on a Grumman F6F-3 Hellcat, first flown in 1948. CALSPAN, led by William Milliken, built a yaw damper for a Vought F4U-5 Corsair, modified with a split rudder similar to the Hs-129 in Weingarten's Fig. 1. The CALSPAN-modified Corsair first flew in March 1949. The original modifications were soon expanded to full lateral-directional control on these two aircraft.

Second-generation variable stability aircraft at CALSPAN included a C-45, an F-94, an NT-33, and two B-26 aircraft, the latter two of which flew training and research missions until the mid-1980s. In 1981 one of the B-26 VS aircraft was lost during a training sortie for the USAF Test Pilot School (USAFTPS), killing the crew of three. A wing structural failure unrelated to VSS operations was the cause of this tragedy, but the accident and subsequent investigation led to the retirement of the second B-26 to the Air Force Museum in 1986.

CALSPAN presently operates three variable stability aircraft under USAF sponsorship for customers worldwide. The oldest of these special purpose vehicles is the NC-131 Total In-Flight Simulator (TIFS)[27, 28]. This aircraft has been used since 1970 to support over 30 different programs, including the Space Shuttle, at least one version of a proposed supersonic transport, and the F-16 fighter.

In the late 1960s CALSPAN, originally with the sponsorship of both the USAF TPS and the USN TPS, developed a new variable stability system for a Lear Jet 24 which first flew in 1981. A second variable stability Learjet 25[28] went into service in 1991 and a third one was scheduled to enter service in 2006.

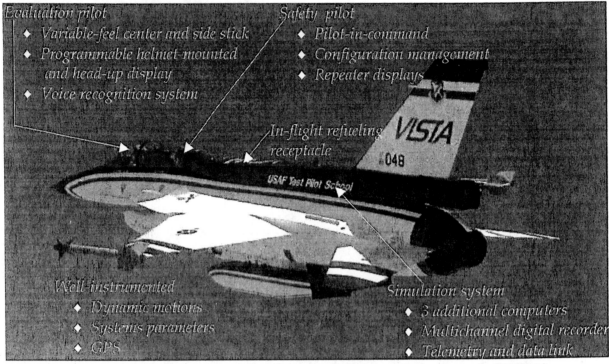

Fig. 10.13 Variable-Stability In-Flight Simulator Test Aircraft (VISTA)
(Adapted from USAF Photo)[28]

The most recent high performance variable stability aircraft developed by CALSPAN is the Variable-Stability In-flight Simulator Test Aircraft (VISTA)/NF-16D (Fig. 10.13) that was delivered to USAF Test Pilot School in 1995.

Fig. 10.14 North American JF-100C (NASA Photo[31])

Other variable stability aircraft flown in the United States and Europe included an NF-106B operated by the Aerospace Research Pilots' School (now USAF Test Pilots' School) at Edwards in the mid-to-late 1960s. NASA Dryden also flew a JF-100C (Fig. 10.14) in support of the X-15 and the Supersonic Transport developments in the early 1960s. A Beagle

Bassett was modified as a variable stability trainer (VST) at Cranfield in 1972[29] under contract to the Empire Test Pilot's School (ETPS). The Advanced Stability Training and Research Aircraft (ASTRA) Hawk was also modified for ETPS by Cranfield and has been in service at the ETPS since 1988[29]. In the United Kingdom, a variable stability Harrier has been used since the early 1990s as an experimental test bed for VSTOL development, and was recently instrumental in the design of control strategies for the F-35B, including embarked operations at sea demonstrating F-35B control laws.

The German Institute für Flugsystemtechnik (DLR) has also operated in-flight simulation vehicles for some time, with the VFW 614 Advanced Technologies Testing Aircraft (ATTAS) developed there from 1981 to 1986. This system uses a model-following approach[32], has served several different projects, and is continually updated. DLR also operated a rotary wing in-flight simulator, the Bo 105 Advanced Technologies Testing Helicopter System (ATTHeS), for several years before embarking on an upgraded system, the EC 135 Flying Helicopter Simulator (FHS) in 1995. The EC 135 (FHS) allows "…in-flight evaluation of new control technologies, cockpit designs, and man-machine interfaces in a real environment and with the pilot in the loop…". It includes a core fly-by-light computing subsystem, along with up-to-date sensors and programmable displays[33].

10.2 INTEGRATING M&S AND FLIGHT TEST

The most common use of M&S is to provide training for flight crews in the use of complex systems like an aircraft, as was underscored in the historical perspective provided of Section 10.1. Aircrew training simulators may remain the single most important output of simulator development in the aerospace industry. But, if we focus on support of flight test activities through M&S, then these simulations have a slightly different emphasis, requiring features not found in operational flight training simulators. We address four ways that M&S supports flight test: (1) by facilitating test planning and definition of test maneuvers, (2) by allowing test teams to rehearse test missions, (3) by allowing both quantitative and qualitative assessments of aircraft systems (or subsystems), and (4) by improving the prediction and analysis of critical performance measures for specific types of tests (predicting dynamic behavior during envelope clearance, anticipating interactions that may cause abnormal aircraft or subsystem operation, and exploring sensitivities of the aircraft and subsystems to off-nominal operating conditions).

The previous paragraph must not be misunderstood to suggest that M&S only "supports" testing; rather "…Modeling and Simulation (M&S) is integral to and inseparable from T&E in support of acquisition."[12] It is a prime concern in this chapter to underscore the integrated role of M&S and T&E to the flight test team in any developmental effort. This integration requires careful planning and anticipation of priorities that fit both the overall and specific flight test objectives. Figure 10.15 illustrates a general flow of steps in planning for dynamic flight tests; this flow chart may not apply to every project that seeks to conduct a simulation-based flight test, but it is a reasonable starting point for such tests.

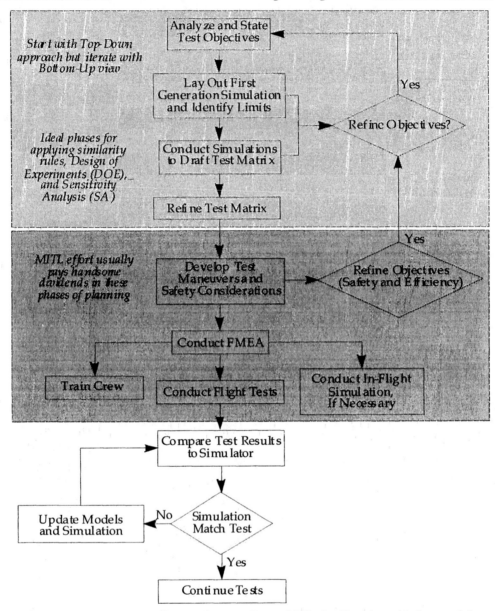

Fig. 10.15 General Flow of Simulation-Based Flight Testing (Adapted from [14])

In the mid-1990s Dr. Paul Kaminski, then Undersecretary of Defense for Acquisition and Testing (USD A&T), strongly supported the notion of simulation-based acquisition (SBA) and specifically called for implementation of an iterative process similar to the one suggested in Fig. 10.5. In a speech to the ITEA Convention in Huntsville, Alabama, he said: *"I am requiring that the simulation, test, and evaluation process — let's call it STEP — shall be an integral part of our test and evaluation master plans [TEMPs]. This means our underlying approach will be to model first, then test, and then iterate the test results back into the model."*[34] Naturally, this process (Fig. 10.16) was not implemented without reservations by the acquisition community; O'Bryon has articulated some of the concerns[35]. Nonetheless, the concept is still included in DoD guidance and has validity for the practice of simulation based acquisition (SBA) and is germane to our discussion of integrated M&S/T&E efforts.

Chapter 10 Simulation and Modeling in Flight Test

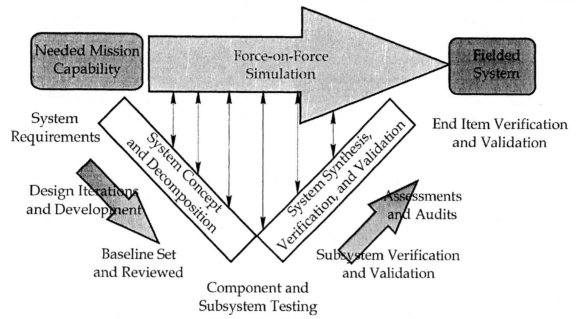

Fig. 10.16 STEP Process[13]

The F-22 test effort is a prime example of how integration of M&S and flight test can be improved. Teaming between the contractor, the ground test agency, and the flight test agency is illustrated and lessons learned have been well-documented. Quoting from the summary conclusions of Webb, Kidman, and Malloy[36]:

- *The United States Air Force has a long history of using modeling and simulation (M&S) in the test and evaluation (T&E) process. While most M&S usage to date has been in the aircraft performance and flying quality areas, advancing technology and complex integration requirements are resulting in increased M&S use across a broader spectrum of technical disciplines, including all aspects of aircraft propulsion systems.*
- *The M&S-based approach for simultaneous validation of propulsion ground- and flight-test data and calibration of the engine model is capable of detecting and identifying sensor anomalies as they occur, and of distinguishing these anomalies from variations in component and overall engine aerothermodynamic performance.*

10.2.1 Test Planning

Planning flight test activities is one of the most important tasks facing the test team and appropriate models and simulations can provide significant help if used with common sense and tempered with good engineering judgment.

10.2.1.1 Top Level Planning Support

Planning for flight tests should be initiated at a very early stage in the life cycle of an aircraft design. Most top level management plans, often called a Project Plan or an Integrated Project Plan, are underpinned by subordinate plans. The systems engineering function of the organization is usually tasked to produce a Systems Engineering Management Plan (SEMP) or a Systems Engineering Plan (SEP)[13] that tailors all developmental activities to the risk inherent in the system development. A SEP "...defines how the program will be organized, structured, and conducted and how the total engineering process will be controlled to provide a product that satisfies customer requirements..."[13]. Inevitably, the SEP requires verification and validation efforts to drive the major risk elements in a development down to an acceptable level. Verification measures conformance to specified requirements and is an integral part of the development process. Verification is controlled and is typically funded by the developing agency, not the customer. It answers the question: Did we build the system "right", that is, according to the design specifications? Validation is tied closely to the customers' needs; this set of evaluations is based on operational considerations, not simply contractual ones alone. Validation addresses the questions: Does the system provide the capabilities sought by the customer? Is it the right system for the job?

All testing activities, whether done for verification or validation, are aimed at providing data to reduce uncertainty. Reducing uncertainty drives down risk simply because risk is made up of uncertainty. Larger test organizations nowadays typically operate under a Test and Evaluation Master Plan (TEMP) that supports the SEP. This TEMP is usually in place prior to starting the major developmental effort for a program (second major phase of the pre-acquisition phase of the development cycle as shown in Fig. 10.17). It should be noted that the size of the blocks does not suggest a time line; the Sustainment Phase is typically longer than all of the other phases combined. The emphasis on the Acquisition Phase indicates where the interest from the test community is most important. The chart also suggests the similarity between military and commercial developments for a complex system like an aircraft, though the commercial cycle shows more clearly when the test activity starts and when it becomes more intense. In military developments prototype models (or perhaps even preproduction models) of the final design often begin developmental testing (verification) during the Pre-Acquisition Phase (SDD as it is now called). Flight tests of an end-item article may begin as early as the product development phase in a commercial effort. Low rate initial production (LRIP) and full rate production (FRP) decisions are not usually made until major demonstrations of capability (operational testing or validation) are completed for military systems. These decisions, though perhaps named differently, are also held up until at least some amount of testing (validation) that specifically targets customer needs is completed satisfactorily for commercial systems. The software elements often term this kind of work "beta" testing.

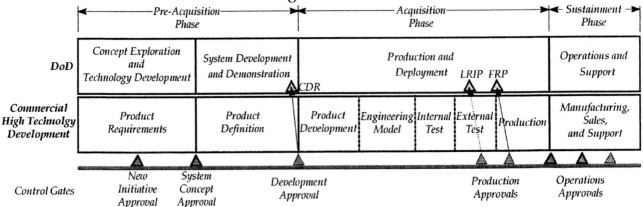

Fig. 10.17 Comparison of Development Cycles for Complex Systems

Considering each of these decision (or control) gates shown in Fig. 10.17, an early start for overall test planning is absolutely essential in any major development effort. Both the SEP (or SEMP) and the TEMP should be completed well in advance of the Critical Design Review (CDR), Development Approval, or similar milestone. Since hardware and software designs are far from frozen at this stage of the development, models of the system are about all that is available for the test planners as they produce the TEMP.

To facilitate TEMP development, early models must be flexible enough to include alternative options for systems and subsystems architecture. Flexibility and reusability are key features for the M&S effort at this point, usually at the expense of fidelity.

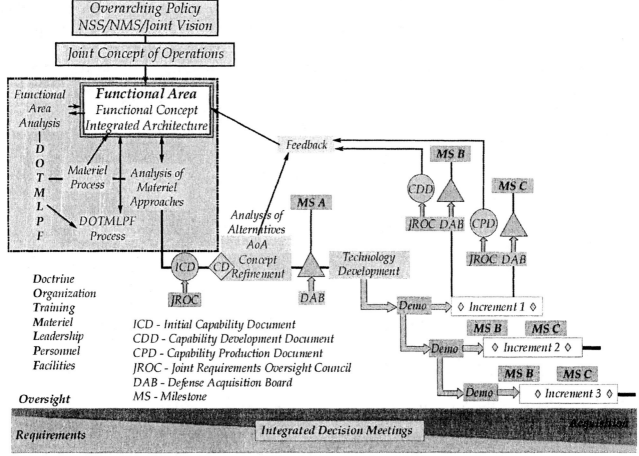

Fig. 10.18 Impact of Evolutionary Acquisition on Testing
(Adapted from Figure 2, DoDI 5000.2[1])

The emphasis on evolutionary acquisition by the DoD 5000-series instructions revised in 2003 impacts test planning and the modeling effort. Since this incremental capability approach to system acquisition came to the DoD regulations from commercial practices within the aerospace industry, commercial developments will likely take a similar approach. However, certification of commercial aircraft under Part 23 or Part 25 of the FAA regulations or the European Certification Standards adds another dimension to the T&E process. In both commercial and military developments a team of engineers (DoD guidance[13] calls this team a T&E Working-level Integrated Product Team or T&E WIPT) is tasked with producing both the T&E Strategy (TES) and the TEMP for program management. The WIPT starts with a broad set of desired capabilities during the Pre-Acquisition Phase of the development. The TES, then, is a rather general discussion of the program's approach to the verification, validation, and accreditation process and an "approved" TES is due at Milestone A in the DoD environment[13]. This time line means the TES should be submitted 45 days before Milestone A to allow time for this upper level approval. The TES depends heavily on preliminary M&S products since the system architecture and many of its technology risks are ill-defined at this point. The companion Initial Capabilities Document (ICD in Fig. 10.18) dictates TES content, but there should be considerable interaction between the M&S group, the T&E WIPT, and the customer to produce a Capability Development Document (CDD) that allows the TES to morph into a TEMP that documents: (1) specific desired operational capabilities, (2) test progress with quantified results that evaluate success or failure of the proposed design, (3) the organizational structure and manning recommended for the T&E WIPT and the rest of the test team, and (4) the necessary resources, facilities, and equipment to carry out the planned testing. The TEMP is due by Milestone B and "... is considered a contract among the program manager, OSD, and the T&E activities..."[2]. Though we have used terms (Milestones A and B, TES, and TEMP) and offered due dates based on DoD practice, commercial developments follow the same general pattern, with the added complexity of having to insure that the FAA certification proceeds in consonance with the program schedule.

Evolutionary acquisition is an interesting challenge to the test community. It is desirable to avoid test duplication for a capability that was previously confirmed in appropriate tests. However, when hardware and/or software configuration changes significantly, the test team owes program management reconfirmation that previously verified capabilities and previously validated operational suitability have not been degraded. These two competing principles lead to updates of the TEMP at Milestone C or at the Full Rate Production (FRP) decision even for systems developed outside the evolutionary acquisition framework. As Fig. 10.25 suggests, the TEMP is likely to be updated at every increment (Milestone B, Milestone C, and FRP decision points).

Each of these top level test planning documents (and the effort required to produce them) is tightly integrated with the evolving mathematical models and the simulations used to implement them. The Defense Acquisition Guidebook (DAG) bluntly states that M&S and T&E are inseparable. Commercial aerospace development practices suggest the same strong tie. To design a test effort that reduces risk in near-optimum fashion depends on a carefully crafted modeling and simulation activity. Any cost-effective support from the M&S organization must be structured around test data collected by an active test team. The M&S WIPT and the T&E WIPT must have close ties for either of them to function well, especially in the early planning efforts for both groups. The following paragraph from the DAG (and, again, the principle applies to commercial developments) indicates the importance of tight integration between M&S and T&E planning[2]:

"An initial goal for the T&E manager is to assist in developing the program M&S strategy. One focus should be to plan for architectures providing M&S interoperability and reusability across the program's life cycle. For example: integrate program M&S with the overall T&E Strategy; plan to employ M&S tools in virtual evaluations of early designs; use M&S to demonstrate system integration risks; supplement live testing with M&S stressing the system; and use M&S to assist in planning the scope of live tests and in data analysis."

Early planning efforts facilitate integration; they must begin in the formative stages of the program and must continue throughout the development cycle. The "model-test-fix-model" (MTFM) philosophy that drives these integrated actions was introduced in Fig. 10.10. There are five primary ways that M&S is used in direct support of T&E:

- Planning to identify and recommend resource allocation
- Rehearsing to make T&E more efficient
- Optimizing to focus on reducing high-risk exposures
- Executing complex procedures
- Analyzing results to explain anomalies and predict performance

We have so far addressed only the underlying philosophy and a part of the planning effort. Next, we look at more detailed planning interactions and the other ways M&S and T&E complement one another.

10.2.1.2 Detailed Test Planning Support. The TEMP has five major parts: (1) system introduction, (2) integrated test program summary, (3) developmental T&E outline, (4) operational T&E outline, and (5) T&E resource summary. To decompose the activities outlined in the TEMP and produce an integrated detailed test plan with clear, unambiguous test objectives, the program manager's staff must establish thresholds for each technical parameter (with Critical Technical Parameters identified as "…measurable system characteristics that, when achieved, allow the attainment of operational performance capabilities[2]…") at each stage in the development cycle, and set priorities within these detailed activities is a formidable challenge. Detailed test planning opens up the matrices in the TEMP and spells out individual tasks necessary to produce the required data.

Subsystem models give designers an opportunity to exercise those elements. These kinds of analysis tools allow sensitivity studies that point to technology risks to be addressed as early as possible if they lie on a critical path in the development. This sort of testing (subsystem and assembly levels) are often overlooked by T&E planners, but this kind of activity is basic to solid engineering. Detailed modeling efforts at these levels can identify those areas where iterations on the design are most useful and can indicate where later testing emphasis should be placed. Often, this phase of the testing is the most productive developmental testing done. Technical challenges revealed and corrected in this phase also return the greatest value to the development team; the cost of a corrective actions increases throughout the development.

10.2.1.3 Planning for End Item Verifications, Validations, and Demonstrations. End Item Verifications, Validations, and Demonstrations are ultimately focused on evaluating the operational suitability, operational effectiveness, and survivability of a system. The terms are defined generally (for both commercial and military systems) in the standards[37] that define systems engineering for complex systems like an airplane. An "end product" is defined as: *The portion of a system that performs the operational functions and is delivered to an acquirer.* An "End Product Verification" is: *Confirmation by examination and provision of objective evidence that the specified requirements to which an end product is built, coded, or assembled have been fulfilled.* An "End Product Validation" is: *Confirmation by examination and provision of objective evidence that the specific intended use of an end product (developed or purchased), or an aggregation of end products, is accomplished in an intended usage environment.* Demonstration carries the standard connation of the word, though military terminology generally reserves demonstrations for the operational phase of testing. For example, Live Fire Demonstrations for a military system are specifically detailed in DoDI 5000.2[1]. By and large, flight tests of new or modified aircraft (or even major aircraft subsystems) fall within the meaning of an end product or an aggregation of end products.

All of the primary uses of M&S are exercised in the end-item phase of testing. Arguably the planning task is most important from a cost-effective test perspective; however, distinctions between functions become a bit vague. In the modern product development cycle, neither a TES nor a TEMP is conceived without first exercising early system models to bound the anticipated risk areas. A bit later, perhaps after initial aerodynamic analyses and/or wind tunnel testing, a more detailed model or simulation is run to help lay out the test matrix for end item verification flight tests. The planned capabilities are compared to threat models (military systems) and business models (commercial systems) to better define efforts to reduce uncertainty by collecting data on early development models, prototypes, and preproduction systems and subsystems. As first flight approaches, the test crew spends considerable time rehearsing the procedures and practicing the flight profiles for the test missions. By this stage the simulations are well-documented but simulation characteristics must be continually and aggressively updated. Early flight test data from the full-up subsystems must be a factor in accrediting the simulation(s) for these initial training tasks. Often it is useful to use hybrid (HIL) simulations with flight qualified hardware incorporated in the simulation[38,39]. Simulations are one of the best ways a test pilot can prepare for first flight or later flights in the envelope expansion. It is important that both flight crew and ground test team conducting the envelope expansion be fully trained in all possible emergencies. Simulation of complete sorties allows refinement of procedures with the attention to detail essential to such tasks as flutter testing, stores testing, high angle of attack testing, and other high workload regimes where the margin for error is small.

M&S is used to predict flight test results and to help design operating instructions even before the vehicle takes to the air. Collecting takeoff and landing data is often specifically tied to the modeling effort for this aspect of the airplane's performance. Since the test team cannot control ambient conditions such as winds and temperatures, flight test data are used to validate performance at conditions which do occur, including sensitivity to a range of factors that are observed. Then, if the simulation accurately predicts observed behaviors and trends, the modeled can be trusted to represent those conditions for which the weather did not cooperate. The test points are chosen to verify M&S predictions; for example, certification authorities accept the takeoff and landing model only when correlation with an adequate set of experimental data is demonstrated.

10.2.2 Systems and Subsystems Evaluations

As stated previously, the evaluation of subsystems and assemblies at an early stage and comparison with model predictions for these elements is imperative for efficient development of complex systems. There is no more important interaction between the M&S effort and the T&E effort; testing these building blocks and verifying that they meet their design goals is an essential step in the process. It cannot be ignored or underfunded by program management without incurring unacceptable risk to the program.

In avionics testing, particularly electronic warfare, HIL simulation be *more* accurate than that feasible during in-flight testing. This result is because HIL ground testing can include emitters, signals, and power levels that might interfere with civil communications. HIL ground tests can employ representative signal densities, environments and waveforms that simply cannot be replicated in a flight test. Furthermore, test productivity is boosted by a data generation rate per test day or test week far greater than that achievable in flight. As with other tests, the in-flight test points are devoted to points meant to verify ground test results, or those points not feasible on the deck or on the ground.

The use of M&S at this stage can be either a boon or a curse. Experience with the NASA Dryden HIL during the Automatic Formation Flight program shows that simulation results must be interpreted wisely to avoid complicated and costly fixes unnecessary to successful flight tests. But, use of simulation and the analysis of feedback from flight test results led to better understanding of the issues and better interpretation of simulation results. The lesson to be learned is: "Aircraft hardware-in-the-loop simulation is an invaluable tool to flight test engineers;..." even though engineers "...must always be skeptical of the results and analyze them within their proper context."[40]

10.2.3 Verification Versus Validation.

There are significant differences in how M&S and T&E interact depending on whether or not the systems and subsystems are in the verification or the validation phase of development. Generally speaking, the verification effort, based on its purpose, is more closely tied to M&S. The credibility of many of the models or simulations is typically not fully established until verification is well along. If unproven models are utilized during the validation part of the process, the results can be misleading – even dead wrong. For End Product Validation it is foolish to proceed without having completed End Product Verification and using that data in the established accreditation process for M&S. This need to be sure a realistic operating envelope has been established before conducting tests to validate capability, operation suitability, and operational effectiveness must always be balanced with the need to have early OT&E results so fixes can be incorporated in the production models that are delivered to the customer. Commercial systems (notably software intensive ones) often use "beta" versions to meet these competing objectives. The wise program manager and his test team carefully balance verification and validation efforts with careful attention to appropriate use of verification data in the M&S tools used to lay out validation tasks and to analyze operational data. There is a danger that the M&S tools may lag market environment or threat assumptions may be out of date for military systems. Then, validation test planning may not address appropriate questions.

10.2.3.1 Verification of a Simulation.
Verification determines "that a model implementation and its associated data accurately represents the developer's conceptual description and specifications"[15]. Even though this description of the verification process comes from a military document, it fits activities for both military and commercial systems. The important question for our purposes is: How is the process best carried out in the context of a simulation-driven test effort? We introduced a general flow of the work to be done in Fig. 10.15. In section 10.3, details are added to this outline, but before spelling out these details, we must examine the rest of the verification, validation, and accreditation process for M&S elements in a simulation-based test effort.

The first step in the verification process is agreement between the M&S WIPT and the T&E WIPT on requirements for the model to be constructed and on how that model will be used. The give and take between simulation and test personnel is essential before the M&S WIPT begins translating requirements into a model appropriate to the task. This initial step in the verification cannot ignore the downstream effects of the concept of operations that led the developer to this requirement. Thus, the ultimate user (or his representative – often from the operational test community for a military system or from marketing for a commercial system) also has a stake in guaranteeing that the concept of operations coincides with the intended use of the end product. Simulation specialists are responsible for correctly translating the specifications into mathematical expressions, block diagrams, and even electromechanical components that accurately emulate the physics of the system. All participants have the responsibility to evaluate outputs at each node of the simulation model to be sure the anticipated physical laws are adhered to, though the test engineers and the users are not always qualified or expected to carry out these calculations. Their job is to be sure the inputs to the model are complete and accurate and then that the output falls within reasonable bounds.

Once the requirements for a model or simulation are agreed upon, the verification effort must continually assess progress on each element and ensure feedback to the M&S specialists as the work progresses. This work must be a "closed-loop" process. The M&S developer is heavily dependent on timely and specific updates to the data bases used and changes to the concept of operations. Closing this loop demands a strong, yet flexible, configuration management scheme so the M&S elements can remain credible and useful. The flight test engineer usually does not have responsibility for any of the M&S elements, but the "closed-loop" part of this process cannot succeed with his or her active participation in and complete cognizance of the results and any questions that need to be answered. The answers to relevant questions often need hard data before they are adequate for the purposes of both T&E and M&S personnel.

Verifications should include sensitivity analysis, which is usually defined as quantitative evaluation of how the output of a model varies with differing parameters, inputs, or excitation[41]. This tool can be used in at least six different ways, but we are most interested in assessing the fidelity of the model to the physical process, the quality of the proposed model, the strongest factors contributing to variation in output, and any interactions between inputs and externally applied (sometimes stochastic) excitations. It is unacceptable to assume that all excitation of complicated systems will fall within anticipated bounds. Failing to examine the robustness of a simulation element or an integrated simulation is one of the more common flaws in a V&V undertaking. Sensitivity analyses are applicable to both verification and validation, but the most useful time to conduct these analyses is early on when design changes are less costly, making them especially important during the verification phase of the effort.

10.2.3.2 **Validation of Simulation Elements**. Validation is somewhat more complicated process than verification, but verification results often provide considerable indication of how best to approach validation. Validation evaluates "the degree to which a model or simulation is an accurate representation of the real-world from the perspective of the intended uses of the model or simulation."[42] However, such a general definition is only part of what is needed. It must be emphasized that the concern here is validation of the parts of the simulation, not validation of the integrated M&S package alone; we consider that latter task in the next paragraph. Hines uses the example of the flight control system model, pointing out emphatically that this model is one small part of the entire integrated simulation (although a very important one)[14]. The aerodynamic models, the propulsion

Chapter 10   Simulation and Modeling in Flight Test   27

model, the cockpit displays, the navigational system models all play a key role in the integrated simulation. Indeed, the interface (passing of data and synchronization of data streams) between these modules is extremely important, but this interface is not part of the validation of each element or module. Conversely, it is usually imprudent to attempt validation (or even verification) of the integrated simulation until each of the modules has been satisfactorily validated.

A fair (and important) question is: how does one validate an element or module within a simulation? The most common validation approach is to compare results from the element model with measured results from the actual hardware/software module. These kinds of results often are available only from component or subassembly manufacturer's ground tests and even when available the actual test conditions and limitations on the ground tests must be evaluated critically. When there are differences between model outputs and available data the validation process must discriminate between differences that matter and those that do not affect the integrated system. Sometimes the most useful information derived from attempts to validate a simulation element is the range and timing of possible outputs from the element. This set of possibilities feeds directly into the sensitivity analyses so important to the validation of the overall integrated simulation. Though the difficulty of obtaining such validation data for the elements of a simulation is considerable, this part of the validation effort may be the most useful in guaranteeing the credibility of the M&S effort and its usefulness to the flight test team.

10.2.3.3 **Validation of the Integrated Simulation**. Validation depends heavily on flight test results for comparison data. Often data latency between simulation elements may produce undesirable consequences and perhaps even make simulation results misleading. Tracking down reason for such discrepancies is almost impossible without having completed validation of the individual elements along with a sensitivity analysis identifying the important assumptions inherent in each elemental model. The flight test team is usually heavily involved in validation of the integrated simulation because of both test planning needs and training needs. In many cases flight test data are not available when the validation begins; the order of collecting flight test data is often affected by a need to validate this overall simulation so it can be used to predict the next stages in flight envelope expansion. Validation procedures always involve comparing outputs (such as time histories) from the simulation with the best available information.

Next, we turn to the basics to be examined in a set of validation comparison data between the integrated simulation results and the "best available truth" data. Figure 10.19 depicts data extracted from a mathematical performance simulation compared to flight test data. The simulation modules are based on a linearized aerodynamic model and a nonlinear propulsion module, each derived from flight tests flown specifically to identify the model parameters for the simulation developers. The simulation is driven by the actual flight test elevator time history. The ordinate is the altitude variation with time resulting from that elevator history. The example illustrates some of the points made above about how a validation process should be carried out. Naturally, if the purpose of and requirements for the simulation change, then the details of what has to be examined changes. But the elemental process is illustrated with the following turboprop performance example.

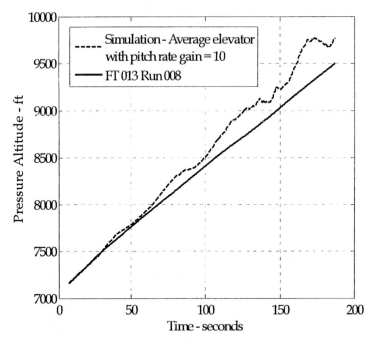

Fig. 10.19 Validation Comparison of a Turboprop Climb Performance Model

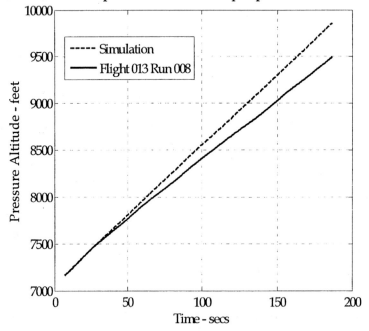

Fig. 10.20 Turboprop Climb Performance with Mathematical Elevator Input

Figure 10.19 suggests two concerns regarding this integrated simulation: (1) the simulation overpredicted rate of climb slightly, especially above 8500 feet and (2) elevator inputs produced an erratic rate of climb when simulated, rather than the smooth trace observed in flight. The primary purpose for this integrated thrust and aerodynamics model was to evaluate fairly small changes in lift and drag characteristics for the airplane configuration at altitudes even higher than the data band for this test. Therefore, the overprediction of rate of climb at this nominal climb band was not acceptable. The second concern, while not as directly associated with the primary purpose of the simulation, suggested a second source of error, one that often occurs in using measurements in a simulation. The elevator deflection measurements were quite noisy and, even after averaging, caused considerable fluctuation in climb rate, especially at the higher altitudes, when used as excitation for the simulation. Sensitivity of the elevator gearing was measured on the airplane; it was so sensitive that the noise spikes in the elevator measurements caused the large variation in rate of climb. When a mathematical approximation for the elevator input was used, rather than the elevator measurements, the climb profile smoothed considerably (Fig. 10.20).

Of course, the most significant problem – overprediction of climb rate – was not improved using a mathematical elevator input. To tackle this more serious anomaly, an examination of the validation of the simulation model elements was fruitful. The propulsion module of the simulation was the most complicated of the modules and the original flight test measurements were made anticipating that this nonlinear module was more difficult to model than the airframe aerodynamics. For example, separate measurements were made of the estimates for the jet thrust from the turbine and for the propeller thrust from the shaft of the turboprop. The M&S developers also found both coding errors in the model and modeling errors. After correcting the coding errors, the difference between propeller thrust from the simulation and measured propeller thrust still showed a considerable discrepancy (Fig. 10.21), accounting for most of the rate of climb errors at altitude.

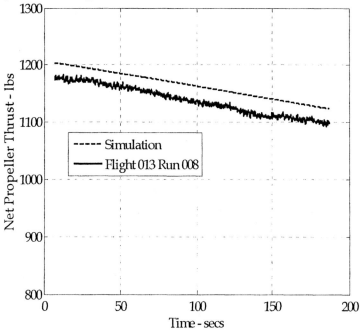

Fig. 10.21 Net Propeller Thrust Comparison from the Turboprop Thrust Module

Figure 10.22 depicts the estimated climb profile after correcting for the bias in propeller thrust as suggested by Fig. 10.21. This simplified example illustrates the kind of analysis and iteration that takes place during the validation of an integrated simulation and how element or subsystem V&V supports validation of the integrated simulation.

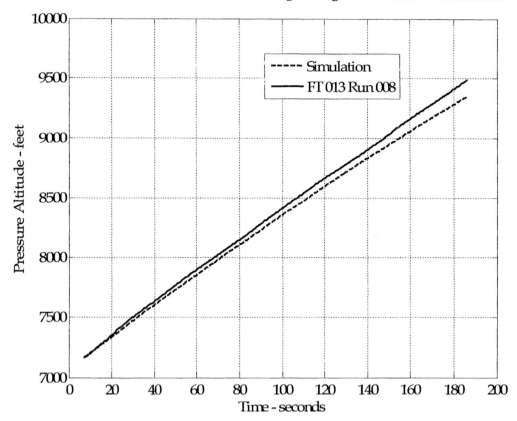

Fig. 10.22 "Validated" Climb Profile after Correction for Propeller Thrust Anomaly

10.2.3.4 **Sufficiency of a Validation (Accreditation)**. How long should this sort of analysis and iteration continue? Accreditation is defined as "...official certification that a model or simulation is acceptable for use for a specific purpose.."[42]. Pragmatically, accreditation attempts to define how much variances between simulation results and the best available "truth" data is acceptable for a given M&S effort and its objectives. As Hines puts it, "There is no magic answer on the sufficiency of the validation effort. It is up to the judgment of the test team [more correctly, the accreditation authority[15]] to decide when a simulation has been sufficiently validated."[13] The very definition of accreditation implies that as external influences (threat, market environment, and the like) change and as the system evolves and test results become available, accreditation of useful simulation tools will likely have to be repeated to capture and continue credibility of the these tools in support of the T&E effort. To expect accreditation to be a one-shot, never to be repeated process is simply unrealistic in the STEP environment. As the specific purpose for a simulation tool changes, accreditation must be expanded to include the capabilities added to the M&S tool and to recognize expanded objectives.

10.3 CONDUCTING A SIMULATION-BASED FLIGHT TEST PROJECT

Figure 10.15 depicts the general flow of a flight test effort built around and tightly integrated with modeling and simulation. We now turn to discussing specific elements important for a flight test specialist intending to maximize the contribution of simulation.

10.3.1 Setting and Understanding Test Objectives

The first step in laying out any series of tests is to set down the test objectives. Quite often the customer's test objectives explicitly dictate the simulation requirements, the complexity of the M&S tools, or perhaps waiving the need for such support. The key questions are: (1) How will M&S increase the likelihood of a successful test? (2) Can M&S improve tests so that they're safer, more productive, more informative, or less expensive? (3) What assumptions are appropriate at this stage of development? (4) What simulation results are essential to guide the test planning? (5) What combination of test maneuvers/data are useful to both the development effort (verification) and determining operational suitability and effectiveness (validation) of the final system. A new airplane development will usually necessitate a simulation capability designed to expand during the development process. A modification to an older aircraft or system, representing modest changes in performance or capability may not justify the cost of developing a brand new M&S tool. In this latter case, M&S integration with the T&E effort may not be economic nor efficient. For those cases where questions provide a compelling need for M&S effort justifying the cost, there are at least two areas to be emphasized – integrated teaming and an evolving concept of operations. Typically, the minimum M&S effort that should be pursued for a new aircraft or a new system is an analytic, real-time simulation which helps the test team spell out appropriate test conditions, select maneuvers, and logically sequence the order of the test maneuvers.

10.3.1.1 Integrated Teaming.

Successful integration of the M&S and flight test efforts hinges upon the integration of the M&S WIPT and the T&E WIPT (or their equivalents in the commercial framework). Each of these teams has separate responsibilities but they must have close ties and must follow the basic principles of integrated teaming of function efficiently. The first principle is that each team must be made up of cross-disciplinary membership. The T&E WIPT must not be made up solely of flight test specialists; there must be an understanding of how to model aerodynamics for the specific objectives of the test. At least one T&E WIPT member should appreciate how all the types of simulation can be tied together in support of the T&E effort (and the difficulties of doing so). By the same token, the M&S WIPT benefits from having at least one member who is familiar with (perhaps has conducted) some form of test planning and flight test operations. It is also quite useful for each of these teams to have at least part-time specialists in each of the most critical modeling areas – aerodynamics, flight controls, propulsion, electrical power, hydraulic systems, or pneumatic systems. If the integrated teams lack one of more set of inputs like those described above, there will inevitably be omissions in their planning and a higher risk to achieving the objectives of the test program.

The organizational structure for integrated teaming is also important to success. Teams must be lead (not just managed or manipulated). One of the primary benefits of integrated teaming is the opportunity to develop a sense of project ownership for individual specialists. Having a healthy mix of experienced team members and new engineers and technicians is one way to promote this sense of purpose and to grow synergy. Finally, and perhaps most important of all, each of the WIPTs must have continuing inputs from the using community, which leads us to the Concept of Operations.

10.3.1.2 **Concept of Operations**. Integrated teams that are not continuously aware of the user's needs and how the ultimate customer will operate a system have little hope of testing in a way that will lead to equipment that meets the user's needs. One of the most important points in Fig. 10.18 where we introduced the notions of evolutionary acquisition is that the acquisition process (including the test effort) is more likely to thrive and produce useful systems when joint decisions are made between the acquisition agents and the requirements developers throughout the development cycle. Obviously, the M&S WIPT and the T&E WIPT need the benefit of these joint decisions and the Concept of Operations and its supporting documents provide a conduit for this flow of changing threats, evolving tactics, and technology innovation. The organizational structure and the communications paths have to be continually checked and rechecked to guarantee that this information flow is not impeded or restricted.

10.3.2 **Practicalities of Providing an Appropriate Simulation**

Laying out the M&S effort needed for a given set of test objectives involves a number of practical considerations like the type and fidelity of simulation to be generated, where to collect input information and site the simulation, and a plan for the V&V of both the elements and the integrated simulation.

10.3.2.1 **Type, Architecture, and Fidelity**. The early planning for the test effort (the TES) should be laid out to force decisions on the types of M&S to be used at an early stage. The necessary resources must be set aside with appropriate lead time to complete the coding and the V&V for the M&S tools. For complex aircraft systems, the M&S development cycle is itself a systems engineering challenge. Intelligent choices must be made about the kind of architecture for the M&S tools, with flexibility and growth as two of the primary considerations. Most M&S tools ought to be planned and resources programmed for continual update and use by more than one facet of the project; in that sense, integrated M&S support implies not just integration of the M&S and T&E efforts but the design, manufacturing, training, and logistical support efforts as well. The simulation's geographic location depends strongly on its intended uses. Quite often the investment in a sophisticated simulation starts very early in the program and evolves as the design evolves. Consequently, it often makes sense to locate the simulation at the manufacturer's facility (especially for commercial aircraft with a heterogeneous set of users). On the other hand, the influence of changing threats on a military design sometime make it more rational to site the simulation at a central location for the primary service customer. Such locations can have the advantage of encouraging user input to the M&S effort and a closer interaction with the end user to keep the requirements evolution up to date.

10.3.2.2 **Necessary Data**. A simulation designed for specific purposes must have data both to generate a mathematical model and to verify/validate its usefulness. Naturally, the original source of data is unlikely to be trustworthy for the entire life of the simulation; as the system evolves with test information, so must the simulation if it is to produce useful information. As Fig. 10.5 strongly implies, the data sets typically start as analytical estimates and with every step in the "waterfall" the uncertainty in the data goes down as more measurements and corrections are made. Wind tunnel experiments usually improve upon CFD estimates, prototype hardware subsystem testing, and computer elements within the subsystems all lead to better data sets and allow the simulation to be in reasonably good shape by first flight. From the M&S developer's point of view, the simulation should be structured to grow only as fast as the data available to improve it become available. It makes no sense to install a sophisticated set of displays when the aerodynamic model is a built around theoretical estimates. Understanding that element test and ultimately, the integrated flight tests, must be designed to reduce uncertainty, it follows that the best sets of data for updating the simulation are the latest test results.

Notice that the M&S community seeks "sets" (plural) of data, not a single set. One of the axioms of simulation development is that a model cannot be adequately verified from the same set of data that produced the model. Said another way, it is strongly preferred to extract the parameters for a given model from one set of data and then, using the excitation from another set of results (within the limits of the model validity, of course), to show that the output responses for the second set of data can be predicted by the model. Naturally, care must be taken that external influences (atmospheric variables, instrumentation noise and drift, and the like) are as nearly identical as possible in the two comparable sets of data. Moreover, there must be enough data sets in a region to give confidence in a given model. These considerations are strong reasons for carefully considering techniques like Design of Experiments (DOE) and Sensitivity Analysis (SA) when laying out either a ground or flight test matrix[43,44,45].

10.3.2.3 **Updating Data Sets**. The feedback loops (Fig. 10.15) are important, both at the beginning of the process (before refining the test matrix) and in the last stage of comparison between integrated simulation output and flight tests of the end product. The diamonds in this figure are critical to success. The mechanisms for updating the simulation as long as new data are flowing from the tests and analysis must be continually exercised and reevaluated for timeliness by both the M&S WIPT and the T&E WIPT.

10.3.3 Draft and Refine Test Matrix and Test Maneuvers

This section on the Simulation-Based Flight Testing flow chart (Fig. 10.15) and the two following blocks constitute the meat of the planning exercise for the T&E team. It is also one of the tasks where a validated, properly exercised simulation can most assist the test effort. In this section we elaborate on some of the techniques used to integrate the M&S tools and the T&E planning.

10.3.3.1 **Run Simulation to Explore Test Envelope**. Assuming that an appropriate simulation exists, one that has been verified and validated, the test engineer can use it to explore the envelope and decide how to best satisfy the data needs of each discipline. It is usually in this phase of the test planning that the concept of a cross-disciplinary test team pays its biggest dividends. Team leaders with previous experience often are able to heuristically predict the magnitude of a test project; but "educated guesses" should not be used without examining the choices made with the simulation. Test time is simply too expensive and too valuable to the overall effort not to focus on such important issues. The simulation also offers a strong tool for unearthing ways to combine test conditions for more than one purpose. Hines[14] gives an example of test conditions for three different groups (structures, stability and control, and propulsion that were proposed for Mach numbers of 0.72, 0.75, and 0.77, respectively. Unless there is some compelling reason to choose one of the extreme points, it makes sense to choose the average value of M = 0.75 and to collect data for all three sets of specialists with this one test event.

To reinforce the point made earlier, DOE (design of experiments) and SA (sensitivity analysis) are very useful tools in deciding what simulation runs are appropriate. Gould's description[45] of the DOE process helps explain why using these tools is important:

DOE - a way of obtaining the maximum amount of information for the least amount of data (saves money, time, and resources)

Though the study of these techniques is not within the scope of this chapter, it is worth setting down the eight keys to success with DOE that Anderson and Kraber[44] advocate:

1. *Set good objectives.*
2. *Measure responses quantitatively.*
3. *Replicate to dampen uncontrollable variation.*
4. *Randomize the run order.*
5. *Block out known sources of variation.*
6. *Know which effects (if any) will be aliased.*
7. *Do a sequential series of experiments.*
8. *Always confirm critical findings.*

These suggestions are guidelines in this phase of the test planning effort; they provide useful guidance in deciding what set of simulations are most helpful to the test planner.

10.3.3.2 **Refine Test Matrix.** Formal techniques like DOE and SA are no substitute for experience and common sense in laying out the flight test matrix. By and large these tools (indeed, Fig. 10.15) follow the pattern of top-down test planning, that is, starting with the objectives and working down to detailed set of test plans. However, a completely different view of the test plan, one that is much more efficient in the use of test resources, is "bottom-up" iteration (Fig. 10.23) in addition to the top-down effort. After an initial plan is derived from a top-down point of view, the team carefully outlines (down to the detail of what graphs and charts are anticipated) the results needed. Then the process is reversed to see what software tools are necessary to manipulate the data stream to obtain these results. Backing up one level further, the inputs to this data reduction software dictate a minimum set of measurements to use the software and to diagnose anticipated anomalies. This information dictates the minimum instrumentation suite needed and allows planning for how the data stream can be managed and fed to all interested parties (including the M&S specialists). Constraints on telemetry bandwidth may necessitate careful choices between telemetered measurands and those stored onboard for post-flight processing. One flight test company reports that this "bottom-up" approach has reduced the size and cost of handling the data on several flight test projects by at least an order of magnitude[46] and has contributed significantly to more efficient test operations.

10.3.3.3 **Develop Test Maneuvers.** This stage of test planning typically demands a pilot-in-the-loop simulation with a representative cockpit and a fairly complete dynamic model of the aircraft and its flight control behavior. The previous step, in concert with DOE and SA efforts should provide a draft test matrix, for the test team. This step is typically where the pilot's perspective should add safety and realism, along with improving the test productivity. These two facets – safety and efficient ordering of the tests – deserve the test pilot's attention throughout the planning phase. Both open-loop and closed-loop maneuvers must be evaluated.

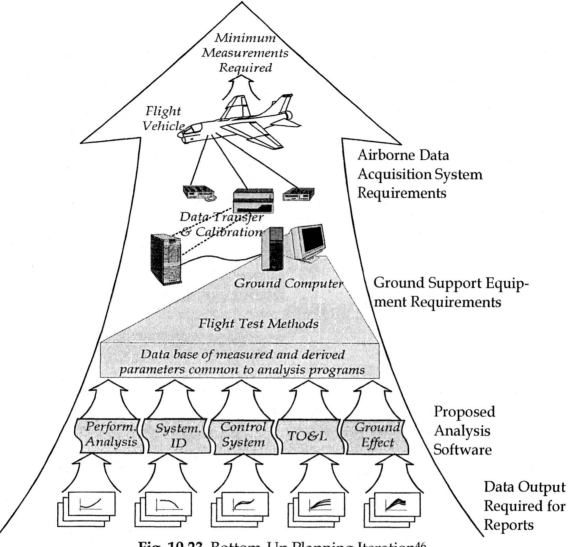

Fig. 10.23 Bottom-Up Planning Iteration[46]

Much of the open-loop (free response of the airplane) simulations can be carried out by flight test engineers but, if the team expects the pilot to provide excitation for these measurements, the pilot practices the maneuvers in a high fidelity simulator. Such preliminary work often identifies areas where additional design work is needed to safely accomplish the dynamic flight test tasks. This approach is applicable to any part of the test effort that requires the specific skills of a trained test pilot. Tasks like windup turns, maximum speed dives, dynamic inputs where aerodynamic damping is low, stall or post-stall maneuvering, closed-loop tracking tasks, autorotations (for rotary wing aircraft), and takeoff and landing maneuvers are some of the maneuvers that fall within this category. Often it is helpful to drive up the pilot's gain by ratcheting up the difficulty of the task. Offset approaches, off-nominal initial conditions for a tracking task or for aerial refueling simulations, and carrier landing simulations requiring pinpoint touch-downs are all ways to increase the intensity of the task in the simulator. Naturally, most of these high gain exercises are only effective if the visual simulation is tightly and credibly integrated with both the simulation feel system and the dynamics generated by the computed and displayed dynamic motions.

Practically, MITL simulations are also an ideal way to optimize the order of the test events and to fulfill more than one data collection objective with each event. Both the test engineer responsible for laying out the individual test cards and the event sequence and the test pilot ought to evaluate in the simulator whether or not one sequence or another of test events is more productive for collecting data. This effort can be tremendously fruitful in improving the efficiency of the test and can reveal opportunities to piggy-back tests, gathering data for multiple disciplines through a single maneuver or flight condition.

10.3.4 Carry Out a Failure Modes and Effects Analysis (FMEA)

The next layer in the flow of work in a simulation-based test program deals with discovering anomalies during the M&S evaluations. The most effective way to find anomalies is to use as much of the actual hardware and software as possible in the simulation; HIL and MITL for manned aircraft are strongly recommended in so far as is practicable. Hines[14] puts it rather succinctly: "...these tests are usually the last simulations that are required to be performed to verify that the aircraft is safe to fly. The object is to put the aircraft into various situations and then induce a failure of some sort." The pilot needs to be in the loop to add realism to the man-vehicle dynamics and to train for such failure states. It is much safer to develop an emergency procedure in a simulator than to do so during flight. Once again, the choices of failure states to simulate and order to consider them benefits from DOE and SA investigations done before a simulation matrix is laid out. This set of simulations and the resultant analysis begs for the use of cross-disciplinary teams; engineering specialists that see the problems uncovered have a much better chance of suggesting effective fixes. If they are not part of this phase of the effort and fixes are not addressed in a timely fashion, the cost of corrective action escalates.

10.3.5 Train All Members of Test Team

Training of the entire test team – flight crew, test conductors, and other control room personnel – is a major value of integrating M&S and flight test. The stages of simulation-based flight test just discussed, especially the last two layers in the planning effort, have significant value in training flight crew and, if so utilized, also provide the flight test control team insight into what to expect during the conduct of test missions. However, planning exercises, as essential as they are, do not provide enough "repetitions" (as professional athletes and coaches put it). The next section deals with these training issues.

10.3.5.1 **Utilize Simulation to Prepare Crew, Test Conductors, and Specialists**. We strongly emphasize the importance of leadership and teaming. The training of all participants in each set of test events is an outgrowth of that disciplined approach. At this stage, especially after the FMEA simulations are carried out to flesh out the test plan, there is usually a high regard for fidelity of the M&S. It is common (and strongly advocated) that the simulation effort be captured and used in "playback" mode to train all individuals involved in test sorties as completely as possible. Failing to insist upon such a course of action nowadays reflects weak leadership. The team leadership must insist that rehearsals to include all procedures are completed and recorded in personnel training records. Leadership must insist upon the kind of discipline that allows timely and correct reaction to anomalous events. Emergency procedures are not just for the pilot or flight crew; everyone in the control room must understand their fundamental responses when things do not go exactly as planned. These disciplined reactions are not necessarily natural; they are learned behaviors. Flight test is a very exciting, but terribly unforgiving, activity. While this advice applies to all kinds of test missions, some test activities have less margin for error than others. High angle of attack work, flutter envelope clearances, stores separation testing, takeoff and landing tests (especially vertical ones), and aerial refueling are some of these specialized kinds of tests that tax the structure and discipline of a test team. Training on realistic M&S equipment using realistic scenarios is the most effective way to embed that kind of discipline in the team.

10.3.5.2 **Establish Training Standards and Desired Proficiency Levels**. Test team leadership is also responsible for setting and maintaining realistic proficiency in the use of the tools provided for test conduct and test monitoring. Though these activities are examining new and unproven systems, the procedures and the discipline for handling unknowns in a new system are rarely unique. Holding to these standards and maintaining them for the duration of a multi-year test effort can be a daunting challenge to leadership. Establishing and upholding realistic standards is the starting point for discipline; continuity of personnel (or lack thereof) often stresses such a system. Having a replacement for every key individual in training at every step of the way is another hallmark of leadership versus "manager manipulation".

10.3.5.3 **Consider Need for In-Flight Simulation**. Though Fig. 10.15 shows in-flight simulation and training of the test team on the same level, in-flight simulation can be of a great deal more benefit that just a training device. It can provide information about the response of a new aircraft that is not attainable with any other M&S facility; ground-based simulations simply lack some of the motion cues that are available in the air. A neutrally stable ground simulation may provide correct engineering data to the test engineer and to the designer, but it cannot completely prepare the test pilot to react appropriately to such seldom experienced dynamics as the in-flight simulator can. Pilot gains can be much more easily excited in an in-flight simulator, giving a better indication of the pilot-vehicle interactions than can be obtained in a ground simulation.

10.3.6 Conduct Flight Tests and Compare Results to Simulation Predictions

It may almost seem anticlimactic to actually conduct the flights and collect the data, given the extensive preparation and planning described in the preceding pages for a simulation-based test program. It is now quite common for the M&S facilities to remain active and productively supportive during the flying portion of a test program.

10.3.6.1 **Integration of Simulation with Test Missions**. There are at least three relatively new techniques used today to bring the power of simulation to bear during the conduct of testing: (1) calculate simulated results *a priori* and have them available during tests for immediate comparison with each event flown, (2) run simulations in parallel with test events, utilizing actual inputs for a given maneuver to drive the simulation, or (3) "shadow" MITL simulations in the control room with test mission events.

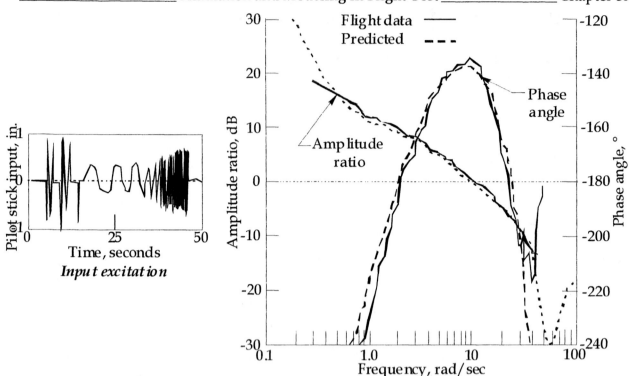

Fig. 10.24 Use of Modeled S&C Parameters in X-29 Envelope Expansion (Adapted from [47])

10.3.6.1.1 *__Comparison with Previously Calculated Simulation Results__*. This approach has been used to varying degrees for a long time for specific types of flight tests. Loads and flutter test teams have done this kind of test integration for quite some time. To some degree a number of programs in the mid-1970s followed this tack for other types of testing. The B-1 development included a very large number of parameters telemetered to the control room and analysis was done in near real-time to clear the test crew to the next planned test event in a series – not just for loads and flutter but for propulsion tests, stability and control tests, and for early stall preparations. The F-111, the B-1, and the B-2 followed similar approaches for much of their systems testing on their low altitude terrain following systems. Another very good example of this kind of work is the use of pre-computed values of stability and control parameters for the X-29 aircraft which flew from 1984-1989 at NASA Dryden. Figure 10.24 illustrates the use of predicted changes in the stability and control parameters as the envelope was expanded. These predictions were plotted on the CRT screens to allow quick interpretation of test results and/or any safety implications from the test event results. This approach facilitated decisions to move on to the next buildup test point in near real-time.

10.3.6.1.2 *__Run Simulations Parallel to Test Mission Events__*. British Aerospace used Flight Mechanics Reprediction, where the aircraft state and flight conditions are input to the appropriate simulation in the control room for the EF-2000 Typhoon. This excitation drove the simulation just prior to initiating a test maneuver and, if the conditions fit the planned ones, the maneuver was executed. In parallel, the simulation was run with the maneuver inputs from the aircraft data stream are processed through the simulator and again compared to predictions. If the measured results of the maneuver were within a prescribed tolerance band, the pilot was cleared to the next point. According to Hines[14], these computations and comparisons were completed within two minutes after the maneuver was finished. An early version of this technique was called Parallel Simulation[48]. While our examples so far are largely based on dynamics of the airframe, the approach is rather easily stretched to cover tests other than flight dynamic ones. Two key points in mechanizing such an approach are: (1) models must be able complete the computations in a short span of time and (2) noise and data dropout in flight test measurements input to the simulation must be taken into account.

10.3.6.1.3 *"Shadowing" Test Missions with MITL Tools*. The third approach to using simulation in the control room has been used rather less often than the other two, largely because it is manpower-intensive and rather expensive. A second flight crew flies a MITL simulation along the same flight profile as the test aircraft and the control room team continuously observes the differences in the "shadow" trajectory and the one produced by the test article. Such an approach has so far proven impractical for use by all except high-risk or high-visibility flight test projects.

10.3.6.2 **Attain Desired Level of Matching with Simulation Results**. The last loop in Fig. 10.15 completes the flow of flight test activities in a simulation-based effort. The choice of acceptable matching and the absolute need for updating all models with each set of data collected is reinforced in the blocks depicting this last stage of the flow diagram.

10.4 SUMMARY

The use of modeling and simulation to make a flight test program more efficient is but one part of a simulation-based acquisition. The growth and the utility of M&S in the recent past indicates that a flight test team that fails to take advantage of the benefits of tightly integrating the M&S effort and the T&E effort is likely to fall short of expectations. Indeed, the whole management thrust for system developments in the aerospace industry is expected to rely more and more heavily on M&S. But M&S that is not verified and validated with the kinds of data that only flight test can produce is also likely to disappoint both its proponents and delight its detractors.

REFERENCES

[1] DoD Instruction 5000.2, "Operation of the Defense Acquisition System," USD(ATL), May 13, 2003.

[2] "Defense Acquisition Guidebook," Version 1.6, Chapter 9, Defense Acquisition University, July 24, 2006. (https://akss.dau.mil/dag/)

[3] Breen, K. E., "Flight Simulators," 2006: http://www.oldbeacon.com/beacon/flight_simulators_history.htm

[4] Rolfe, J. M. and Staples, K. J. (Editors), **Flight Simulation**, Cambridge University Press, Cambridge, United Kingdom, 1986.

[5] Moore, K., "A Brief History of Aircraft Flight Simulation," 2006: http://homepage.ntlworld.com/bleep/SimHist1.html

6. Page, R. L., "Brief History of Flight Simulation," http://www.siaa.asn.au/get/2395364797.pdf

7. http://www.nationalmuseum.af.mil/shared/media/photodb/photos/051019-F-1234P-037.jpg

8. Hayes, M. J. D and Langlois, R. G., "ATLAS: A Novel Kinematic Architecture for Six DOF Motion Platforms," *Canadian Society for Mechanical Engineering (CSME) Transactions*, Vol. 29 (2005), no. 4, pp. 701-709. http://www.mae.carleton.ca/~jhayes/Papers/HayesLanglois_rev2.pdf

9. Ruffner, J. W., Antonio, J. C., Joralmon, D. Q., and Martin, E., "Night Vision Goggle Training Technologies and Situational Awareness ," School of Aviation Safety, Naval Air Systems Command, Pensacola, FL, 2001. https://www.cnet.navy.mil/nascweb/sas/nvt.htm

10. "Virtual Reality: History," 2006. http://archive.ncsa.uiuc.edu/Cyberia/VETopLevels/VR.History.html

11. "Exponential Growth of Software Patents," http://lpf.ai.mit.edu/Patents/ counts.html

12. Moler, C. "The Growth of MATLAB and the MathWorks over Two Decades," The MathWorks News & Notes, Jan. 2006. (http://www.mathworks.com/company/newsletters/news_notes/clevescorner/jan06.pdf)

13. "Modeling and Simulation Support to T&E," Test and Evaluation Management Guide, Fifth Edition, The Defense Acquisition Press, Fort Belvoir, VA 22060-5565, Jan. 2005. http://www.dau.mil/pubs/gdbks/T&E_MgmtGuide.pdf

14. Hines, D. O. (Editor), "Simulation in Support of Flight Testing," RTO AGARDograph 300, Flight Test Technique Series – Volume 19, North Atlantic Treaty Organization, Research and Technology Organization, Neuilly-sur-Seine Cedex, France, Sept. 2000.

15. DoD Instruction 5000.61, "DoD Modeling and Simulation (M&S) Verification, Validation, and Accreditation (VV&A)," USD(ATL), May 13, 2003.

16. "Evolution of the ACETEF to Support SBA and Testing," nd. http://acquisition..navy.mil/navyaos/acquisition_topics/program_management/ippd/integrated_product_process_development/evolution_of_the_acetef_to_support_sba_and_testing

17. Waite, W. F., Jolly, A. C., Swenson, S. J., Shepherd, S., and Gravitz, R. M., "Validation of Hardware-in-the-Loop (HIL) and Distributed Simulation Systems," invited paper for Foundations for V&V in the 21st Century Workshop (Foundations '02), Johns Hopkins University/Applied Physics Laboratory, Laurel, MD, Oct. 2002.

18. "Flight Simulation Research," National Aerospace Laboratory (NLR) – the Netherlands, Mar. 1998. (http://www.nlr.nl/eCache/DEF/244.html).

19. Chia, C. W., "Low Cost Virtual Cockpits for Air Combat Experimentation," Paper Number 1596, Interservice/Industry Training, Simulation, and Education Conference (I/ITSEC), Orlando, FL 2004.

[20] http://aero.tamu.edu/gallery/?id=21

[21] Painter, J. H., Kelly, W. E. III, Trang, J. A., Lee, K. A., Branhan, P. A., Crump, J. W., Ward, D. T., Krishnamurthy, K., Woo, D. L. Y., Alcorn, W. P., Robbins, A. C., and Yu, R. J., "Decision Support for the General Aviation Pilot," 1997 IEEE International Conference on Systems, Man, and Cybernetics, Orlando, FL, Oct. 1997.

[22] Painter, J. H., Rong, J., Valasek, J., and Ioerger, T., "Multi-Agent Flight Management Systems," 21st Digital Avionics Systems Conference (DASC) on Air Traffic Management for Commercial and Military Systems, Irvine, CA, Oct. 2002.

[23] Smolka, J. W., Walker, L. A., Johnson, G. H., Schkolnik, G. S., Berger, C. W., Conners, T. R., Orme, J. S., Shy, K. S., and Wood, C. B., "F-15 ACTIVE Research Program," Proceedings of the Society of Experimental Test Pilots 40th Symposium, Beverly Hills, CA, 1996.

[24] http://www.dfrc.nasa.gov/gallery/Photo/F-15ACTIVE/Small/EC96-43485-3.jpg.

[25] Chacone, V., Pahle, J. W., and Regenie, V. A., "Validation of the F-18 High Angle of Attack Research Vehicle Flight and Avionics Systems Modifications," NASA TM 101723, NASA Dryden Flight Research Center, Edwards, CA, Oct. 1990.

[26] "Fact Sheet: Walter C. Williams Research Aircraft Integration Facility," NASA Dryden Flight Research Center, Edwards, CA, Dec 2006. (http://www.nasa.gov/centers/dryden/news/FactSheets/FS-007-DFRC.html).

[27] Weingarten, N. C., "History of In-Flight Simulation & Flying Qualities Research at CALSPAN," *AIAA Journal of Aircraft*, Vol. 42, No. 2, March/April 2005.

[28] http://www.calspan.com/variable.htm

[29] http://www.af.mil/photos/index.asp?galleryID=3&page=24

[30] http://www.qinetiq.com/home_etps/school/aircraft.html

[31] http://www.dfrc.nasa.gov/gallery/photo/F-100/HTML/EC62-144.html

[32] http://www.dlr.de/ft/en/desktopdefault.aspx/tabid-1387/1915_read-3375/.

[33] http://www.dlr.de/ft/en/desktopdefault.aspx/tabid-1387/1915_read-3374/.

[34] Kaminski, P. G., Speech before the ITEA Convention, Huntsville, AL, Oct. 1995.

[35] O'Bryon, J. F., "Meet MASTER – Modeling and Simulation Test & Evaluation Reform: Energizing the M&S Support Structure," *Program Manager Magazine*, Defense Acquisition University, April, May 1999. http://www.dau.mil/pubs/pm/pmpdf99/ma-ap99.pdf

[36] Webb, A.T.; Kidman, D.S.; Malloy, D.J., "Knowledge Gained from F/A-22/F119 Propulsion System Ground and Flight Test Analysis", in *Flight Test–Sharing Knowledge and Experience*, meeting proceedings RTO-MP-SCI-162, Paper 23, pp. 23-1–23-14, Neuilly-sur-Seine, France, 2005. http://www.rto.nato.int/abstracts.asp.

[37] ANSI/EIA-632-1998, "Processes for Engineering a System," Electronic Industries Alliance, Approved: Jan. 1999.

[38] Norlin, K. A., "Flight Simulation Software at NASA Dryden Flight Research Center," NASA TM-104315, NASA Dryden Flight Research Center, 1995. (Also published as AIAA-95-3419.)

[39] Carter, J. F., "Production Support Flight Control Computers: Research Capability for F/A-18 Aircraft at Dryden Flight Research Center," NASA TM-97-206233, NASA Dryden Flight Research Center, 1997.

[40] Ryan, J., Hanson, C. E., Norlin, K. A., and Allen, M. J., "Data Synchronization Discrepancies in a Formation Flight Control System," NASA TM-2001-21070, NASA Dryden Flight Research Center, 2001.

[41] Breierove, L. and Choudhari, M., "An Introduction to Sensitivity Analysis," paper prepared for the MIT System Dynamics in Education Project, Sept. 1996 (Versim example added Oct. 2001: http://sysdyn.clexchange.org/sdep/Roadmaps/RM8/D-4526-2.pdf.

[42] DoD Manual 5000.59-M, "DoD Modeling and Simulation (M&S) Glossary," USD(ATL), Jan. 1998.

[43] Wakeland, W. W., Martin, R. H., and Raffo, D., "Using Design of Experiments, Sensitivity Analysis, and Hybrid Simulation to Evaluate Changes to a Software Development Process: A Case Study," 4th International Workshop on Software Process Simulation and Modeling, Portland, OR, May 2003. http://www.sysc.pdx.edu/faculty/Wakeland/papers/SPIP_Manuscript_WakelandMartinRaffo.pdf

[44] Anderson, M. J. and Kraber, S. L., "Keys to Successful Design of Experiments," http://www.qualitydigest.com/july99/html/doe.html

[45] Gould, N. M., " Design of Experiments," PowerPoint presentation at NDIA Conference, Dallas, TX, Oct. 2001. http://www.qualitydigest.com/july99/html/doe.html

[46] Schweikhard, W. S., "Bottom-Up Planning Process," PowerPoint presentation in Flight Test Principles and Practices, KUCE Short Course, , Orlando, FL, Oct. 2006.

[47] Clark, R., Burken, J. J., Bosworth, J. T., and Bauer, J. E., "X-29 Flight Control System: Lessons Learned," NASA TM 4598, NASA Dryden Flight Research Center, CA, June 1994.

[48] Schürer, S., Soijer, M. W., and Oelker, H. C., "Assessment of Parallel Simulation on the Basis of Eurofighter 2000 Flight Dynamics," Society of Flight Test Engineers – European Chapter Annual Symposium, Toulouse, France, 2003.

Chapter 11
AEROELASTICITY: THEORETICAL AND ANALYTICAL FOUNDATION

"Some fear flutter because they do not understand it, others fear flutter because they do"

.... Theodore von Karman

Aeroelasticity[1,2,3,4] is the study of the interaction between aerodynamic forces, structural (elastic) forces, and inertial forces. The three-ring diagram shown in Figure 11.1 helps to **underscore the interdisciplinary nature of aeroelasticity.**

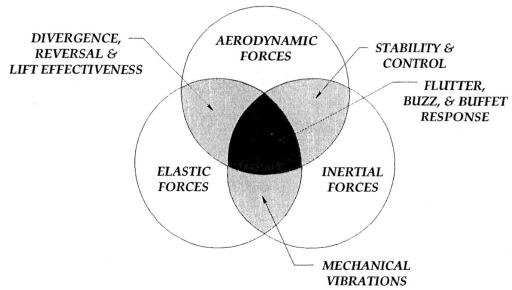

Fig. 11.1 Interdisciplinary Nature of Aeroelasticity

The combination of any two rings represents a discipline of aerospace engineering. The combination of elastic and inertial forces forms the study of mechanical vibrations. The combination of inertial and aerodynamic forces forms the study of rigid-body vehicle stability and control. The combination of aerodynamic and elastic forces forms the study of static aeroelasticity. The combination of all three rings forms the study of dynamic aeroelasticity. Other disciplines such as control (aero**servo**elasticity), thermal (aero**thermo**elasticity, and damage (for example, aging aircraft) issues may be considered as additional overlapping rings.

11.1 DEFINITIONS

As in most specialties within aerospace engineering, aeroelasticity has its own jargon and special use language. To begin, we need to establish a common set of terminology for the phenomena unique to aeroelastic analysis.

11.1.1 Aeroelastic Divergence

Aeroelastic divergence is a static aeroelastic instability leading to catastrophic failure of the structure, and occurs when the deformation-dependent aerodynamic forces exceed the elastic restoring forces (i.e., the stiffness) of the structure.

11.1.2 Lift Effectiveness

Lift Effectiveness is the change in aerodynamic loads due to wing flexibility, which typically leads to a change in the load distribution and a change in lift.

11.1.3 Control Reversal

> *Control reversal is the loss of the effectiveness of a control surface to produce maneuver loads due to deformation of the primary structure to which the control surface is attached.*

Two characteristics of aeroelasticity are evident in the preceding definitions: (1) aeroelasticity is dependent upon the flexibility of the vehicle structure, and (2) aero-elasticity involves a feedback between aerodynamic loads and structural deformations. That is, in all cases above, aerodynamic loads depend upon structural deformation and, in turn, this deformation depends upon the loads.

11.1.4 Flutter

Flutter is a dangerous structural dynamics phenomenon that involves an interaction of aerodynamic, elastic and inertial loads. From a flight test perspective it is definitely a hazardous type of flight envelope expansion.

> *Flutter is a dynamic aeroelastic instability, characterized by growing oscillatory motion involving the interaction of structural modes affected by motion-dependent aerodynamic loads, that leads to a catastrophic structural failure.*

Aerodynamic loads, especially those that are unsteady in character, are the source of excitation energy for this destructive type of structural dynamics. Typically, two or more interacting modes of motion, dependent upon flight conditions, coalesce to form a single aeroelastic frequency and mode of motion that resembles a combination of the participating modes. From an energy perspective, energy is extracted from the aerodynamic flow field and is absorbed by the structure. At conditions below flutter, structural and aerodynamic damping dissipates this energy; at flutter, system damping is lost thereby leading to a growing oscillatory motion.

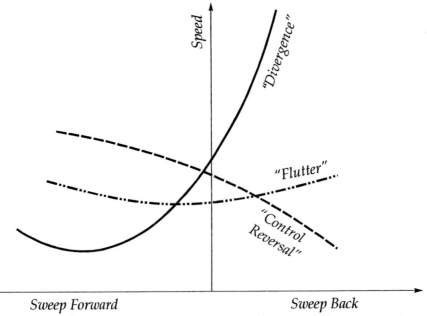

Fig. 11.2 Sweep Effects (Adapted from Bisplinghoff[1])

Wing sweep has a significant effect on certain aeroelastic instabilities (Fig. 11.2). Forward sweep adversely affects the divergence boundary and aft sweep adversely affects control reversal. Relative to these other instabilities, flutter is affected little by sweep.

Chapter 11 Aeroelasticity: Theory and Analytical Foundations

11.1.5 Nonlinear Aeroelasticity

Typically, in most engineering analysis, one prefers to assume the system behaves in a linear manner to efficiently and expeditiously reach an acceptable conclusion. And, fortunately, such assumptions are validated through experiments, including flight tests. Nonetheless, flight vehicles possess nonlinearities[5] which may lead to responses not predicted by standard analysis. Limit cycle oscillations (LCOs) are common in the aeroelastic environment. An LCO is difficult to define as they are characterized by many different features. In the literature, one finds them referred to as Limited Amplitude Flutter, suggesting a hazardous unstable situation; yet, others may refer to LCOs as stabilizing in that the motion is bounded. We offer the following definition:

> *Limit Cycle Oscillation is an oscillatory response of a component of the air vehicle like a wing or a control surface, with limited amplitude, but dependent on the nature of the nonlinearity as well as flight conditions, such as airspeed, altitude, and Mach number.*

11.2 AEROELASTIC ANALYSIS FOUNDATIONS

11.2.1 Equations of Motion

In general, the equations of motion for the aeroelastic system may be cast in a form similar to that of the forced mass-spring-damper system discussed in Chapter 9. Writing them in matrix form

$$M\ddot{x} + C\dot{x} + Kx = F_a + F_t \qquad (11.1)$$

where M, C, and K are the matrices that represent the structural mass, damping, and stiffness, respectively, and x is the vector of coordinates which may represent displacements and rotations resulting from bending and torsional motions, for example. The vectors F_a and F_t represent loads. F_a denotes motion-dependent (that is, the "unsteady") aerodynamic loads,

$$F_a = M_A\ddot{x} + C_A\dot{x} + K_A x \qquad (11.2)$$

where M_A, C_A, and K_A depict aerodynamic inertial, damping, and stiffness contributions which depend upon flowfield conditions (Mach number, altitude, and airspeed) as well as the motion of the vehicle's structure. F_t denotes external loads such as those arising from gusts, turbulence, and buffeting, and may also represent loads induced by open-loop control (e.g., pilot-induced maneuvers). Equations 11.1 and 11.2 may be combined,

$$(M - M_A)\ddot{x} + (C - C_A)\dot{x} + (K - K_A)x = F_t \qquad (11.3)$$

This expression provides the means to understand the response to external loads, F_t. F_t is not required in the solution that characterizes system stability.

11.2.2 Solution Approaches

The equation may be solved in either the frequency domain or the time domain. For the frequency domain approach, harmonic motion is assumed so that $x = \beta e^{\lambda t}$; thus, substitution into eq. 11.3, with $F_t = 0$, gives:

$$\left(\lambda^2(M - M_A) + \lambda(C - C_A) + (K - K_A)\right)\beta = 0 \qquad (11.4)$$

Equation 11.4 is an eigenvalue problem in which the eigenvalues are of the form $\lambda = \delta + i\omega$, where δ represents the aeroelastic damping and ω represents the aeroelastic frequency. The eigenvector β is the aeroelastic mode shape associated with the frequency. The response of the structure is of the form $x = \beta e^{\pm \delta t} e^{\pm i\omega t}$, where $e^{\pm \delta t}$ represents a decaying or growing motion depending upon the sign of δ, and $e^{\pm i\omega t}$ represents a constant amplitude oscillatory motion with frequency ω. Figure 11.3 shows the response of a system to three cases: damped motion without frequency (δ < 0, ω = 0); growing motion without frequency (δ > 0, ω = 0); and undamped oscillatory motion (δ = 0, ω > 0). These eigenvalues indicate frequency and damping characteristics that are dependent upon flight conditions that include airspeed, Mach number, and altitude.

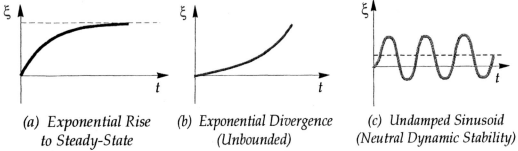

(a) Exponential Rise to Steady-State (b) Exponential Divergence (Unbounded) (c) Undamped Sinusoid (Neutral Dynamic Stability)

Fig. 11.3 Oscillatory Motion

Traditionally, in vibration analysis the natural frequency is identified as the imaginary part of the root. The real part, which represents the damping, is typically near or at zero. However, the presence of unsteady, motion-dependent aerodynamic forces creates damping and, although naturally occurring structural damping will lead to a decaying motion, aerodynamic sources may lead to loss of damping. We note that the presence of aerodynamic loads may lead to a negative damping (that is, a growing motion).

Related to eq. 11.4, three important trends are monitored during flight tests in which flight conditions are systematically modified:

(1) changes in frequency characteristics,
(2) changes in damping characteristics, and
(3) changes in characteristics in amplitude of response.

As an alternative approach to the frequency domain solution, time domain solutions (that is, numerical integration) of the governing equations (eq. 11.3) provide a simulation of the physical response to an input or disturbance. However, in such a time domain solution the task of modeling the aerodynamic loads may be more complicated, yet may address aerodynamic characteristics not fully captured by frequency domain models, such as shock - boundary layer interaction. Although physical properties associated with the structure are fixed, the flowfield characteristics such as density (altitude), velocity (airspeed, and Mach number are changed to examine various flight conditions.

For example, let us consider a wing with bending and torsion modes that are sensitive to aeroelastic instabilities. The motion for three different values of dynamic pressure is shown in Figs. 11.4-11.6. Figure 11.4 (next page) shows a decaying motion for a value of dynamic pressure below flutter conditions and illustrates that the response is, in reality, the superposition of multiple responses such as individual bending and torsion contributions. These changes are related to the motion described by the combination of $e^{-\delta t}$ (exponential decay) and $e^{\pm i\omega t}$ (sinusoidal motion with frequency ω).

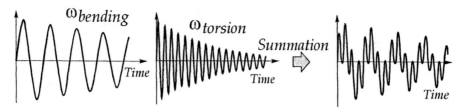

Fig. 11.4 Decaying Motion in the Time Domain

Figure 11.5 shows a growing motion for a dynamic pressure that exceeds the conditions for flutter and, as suggested by the figures, the modes of motion have coalesced to a common frequency.

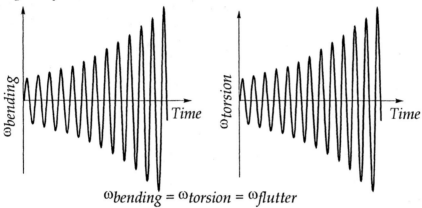

$\omega_{bending} = \omega_{torsion} = \omega_{flutter}$

Fig. 11.5 Divergent Motion in the Time Domain

Figure 11.6 illustrates a motion with constant maximum amplitude; that is, it is neutrally damped (although the motion may appear similar, this response should not be confused with limit cycle oscillations found only in nonlinear systems). This case suggests the dynamic pressure (and/or Mach number) for the aeroelastic instability has been reached. Lower values of dynamic pressure show damped motion. Higher values of dynamic pressure show growing motion. A difficulty associated with the time-domain approach is the identification of the precise point where the behavior is neutral.

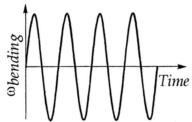

Fig. 11.6 Undamped Motion in the Time Domain

11.2.3 Aeroelastic Similarity Parameters

It is both convenient and efficient to eliminate unnecessary free variables in analyses or experiments by identifying similarity parameters. For example, dimensional analysis (introduced in Chapter 4) reveals that only two nondimensional parameters – Mach number and Reynolds number – are necessary to characterize the flow and flowfield effects on lifting surfaces. These two parameters address fluid density, viscosity, velocity, speed of sound, temperature, physical dimensions, and related variables.

In a comparable fashion, dimensional analysis for the aeroelastic equations reveals additional nondimensional terms that must be considered. The **reduced frequency**, k, is defined as

$$k \equiv \frac{b\omega}{V_\infty} \qquad (11.5)$$

where ω is the frequency of vibration, V_∞ is the freestream velocity and b is a reference length which is the semi-chord ($c/2$) of the wing. The authors recognize that the parameter b is widely used in the aerospace community as the wing span; however, in these aeroelastic discussions, we adopt the classical nomenclature prevalent in the aeroelasticity literature which uses b as the reference length. The reduced frequency is the ratio of the vibratory motion to the velocity of flow over the vehicle's surface, and represents a measure of how fast unsteady disturbances are moved downstream relative to the transverse (vibratory) motion.

The reduced velocity, $\overline{V}$, is defined as

$$\overline{V} = \frac{V_\infty}{b\omega_\alpha} \qquad (11.6)$$

where ω_α is a reference frequency of vibration, often chosen as the lowest natural frequency with a mode of motion dominated by wing torsion.

The **mass ratio**, μ, is defined as

$$\mu \equiv \frac{m}{\rho_\infty \pi b^2 s} \qquad (11.7)$$

where m is the mass of the vehicle, ρ_∞ is the density of the air, and s is the wing span. The mass ratio is a measure of the ratio of the vehicle's mass to the mass of the air that the vehicle displaces as it moves relative to the surrounding fluid.

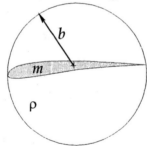

Fig. 11.7 Mass Ratio Geometry

An additional term, the **flutter speed index**, $\underline{V}$, is defined as

$$\underline{V} = \frac{V_\infty}{b\omega\sqrt{\mu}} \qquad (11.8)$$

This parameter contains both the mass ratio and reduced frequency. As the analyst compares experimental and analytical results in aeroelastic investigations, not only the geometry but also the mass distribution and structural stiffness must be similar in order for the comparisons to be valid. Results are often expressed in terms of flutter speed index and mass ratio. If two systems have identical flutter speed indices and mass ratios, then the reduced frequency is identical, since eq. 11.8 is a combination of eqs. 11.6 and 11.7. Clearly, the equality of these parameters insures dynamic similarity between the two systems, and

Chapter 11 — Aeroelasticity: Theory and Analytical Foundations

is analogous to the dimensional similarity one requires when comparing the Mach number between ground-based analysis and flight tests.

11.2.4 Static Aeroelastic Phenomena

Static aeroelastic phenomena involve the combination of aerodynamic loads and structural flexibility. They include aeroelastic divergence, control surface reversal, and lift effectiveness. An example of static aeroelasticity is the effect of aerodynamic loads on the lift distribution of a wing. The aerodynamic loads cause the wing to deform (twist and bend) and, in turn, this new shape will result in new aerodynamic loads. Naturally, one should expect a balance of forces to occur about this new deformed shape. One also observes that, as a result of deformations, the load distribution is altered and the center of pressure shifts. As a consequence, control and stability characteristics are affected.

11.2.4.1 Aeroelastic Divergence.

The possibility exists that forces will not balance since the structural deformations may continue to grow as a consequence of feedback from the deformation-dependent aerodynamic loads. These loads may give rise to great-er deformations. Eventually the surface fails if this type of feedback interaction continues.

As defined, aeroelastic divergence is a static aeroelastic instability leading to catastrophic failure of the structure, and occurs when the deformation-dependent aerodynamic forces exceed the elastic restoring forces of the structure. The instability occurs when the elastic restoring forces associated with the structural stiffness are exceeded by the aerodynamic forces. Aerodynamic loads are not constant; rather, these loads are dependent upon the displacement of the structure. It is noted that divergence is also known as "zero-frequency flutter" since the instability appears as growth without oscillation. It may be found as a subset from the dynamic equations for the case where frequency vanishes, or found directly as a static solution. Accordingly, the equations that describe static aeroelasticity may be derived from eq. 11.3

$$(K - K_A)x = 0 \tag{11.9}$$

in which velocity and acceleration forces are not considered. K_A is the aerodynamic stiffness matrix, which is composed of aerodynamic derivatives and flow properties. The dynamic pressure, q_∞, may be isolated such that the expression may be rewritten as $q_\infty \underline{K}_A$ where $q_\infty \equiv \dfrac{\rho V_\infty^2}{2}$. Therefore, eq. 11.9 may be written as

$$(K - q_\infty \underline{K}_A)x = 0 \tag{11.10}$$

Equation 11.10 is an eigenvalue problem with q_∞ the eigenvalues associated with aeroelastic divergence and x the eigenvectors associated with the shape at divergence. The term $(K - q_\infty \underline{K}_A)$ is the aeroelastic stiffness that approaches zero as q_∞ increases.

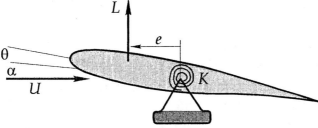

Fig. 11.8 Free-Body Diagram for Aeroelastic Divergence Example

11.2.4.1.1 *Aeroelastic Divergence – A Simple Example.*
Aeroelastic divergence is illustrated in the following example. In Fig. 11.8, a rigid wing section is mounted to an elastic support that permits one degree of freedom motion. As will be seen, loads independent of deformation such as weight, gusts, and open-loop control, will not stability. Aerodynamic loads act through the aerodynamic center.

The equation of static equilibrium for the wing deflected from the balanced system appears as follow:

$$Le - K\theta = 0 \qquad (11.11)$$

where K is the spring stiffness, θ is the rotation degree of freedom, L is the lift, and e is the distance between the elastic axis and the aerodynamic center (e is a measure of aerodynamic eccentricity). The reader is reminded that the aerodynamic center and the center of pressure are not necessarily the same locations. The aerodynamic center is a point where there is no change in moment with respect to a change in angle of attack. Furthermore, loads such as the weight of the wing are not at issue since θ is the displacement from the static equilibrium position. These parameters are all constant with respect to deformation except for the lift.

Lift is a function of dynamic pressure, which depends upon the velocity and fluid density. Furthermore, lift is a function of the total angle of attack, $\alpha + \theta$, where α is the angle of attack at equilibrium. (note: the reader is reminded to be careful in the use of terms as symbology may be confusing; for example θ has been used in previous chapters to denote the Euler pitch attitude angle. Remember that in this context, θ is the perturbation from the static equilibrium position). Thus, lift is expressed as

$$L = q_\infty S C_{L_\alpha}(\alpha + \theta) \qquad (11.12)$$

where S is the reference area, and C_{L_α} is the aerodynamic derivative defined by the slope of the lift curve. We define the aerodynamic derivative as the load per unit deflection or per unit rotation. Often, the affect of the θ term is ignored in loads analysis (the engineer might assume that θ is sufficiently small and that it may be ignored, but the reader is asked to consider what will happen in all of the following developments if the affect of θ is ignored). Substituting eq. 11.12 into eq. 11.11,

$$(K - q_\infty S e C_{L_\alpha})\theta = q_\infty S e C_{L_\alpha} \alpha \qquad (11.13)$$

which is a scalar form of the multiple degree-of-freedom system in eq. 11.9, where $K_A = q_\infty S e C_{L_\alpha}$. Solving eq. 11.13 directly for θ leads to

$$\theta = \frac{q_\infty S e C_{L_\alpha} \alpha}{K - q_\infty S e C_{L_\alpha}} \qquad (11.14)$$

Equation 11.14 provides the *response*, θ, of the wing to the angle of attack, α, and also reveals the *stability* of this aeroelastic system. The aerodynamic contribution of eq. 11.12 appears with the structural stiffness. In fact, one observes that depending upon the sign of eccentricity, e, the aerodynamic contribution is a positive or negative stiffness. As the dynamic pressure is increased, the term in the denominator approaches zero; consequently, θ approaches infinity. The terms in the denominator form the expression for the critical dynamic pressure at which aeroelastic divergence occurs. When the denominator approaches zero, the result is therefore of special interest; thus, solving for q_∞ leads to

$$q_{\infty,DIV} = \frac{K}{SeC_{L_\alpha}} \qquad (11.15)$$

One observes that the aerodynamic eccentricity, e, may be negative (that is, the aerodynamic center is behind the elastic axis) which suggests a negative divergence pressure exists. Obviously, this is not physically possible which means that divergence is not possible for this case. Thus, design considerations that minimize the effect of eccentricity will alleviate aeroelastic divergence.

It is worth noting that wing sweep has a direct correlation with aerodynamic eccentricity: e decreases with positive (aft) sweep, thus improving conditions by raising the divergence pressure; and conversely, the forward swept wing increases eccentricity thereby lowering the divergence pressure (see Fig. 11.2). Although this illustration is quite simple, the concept is essentially the same for large degree of freedom systems as described by eq. 11.9. In fact, $q_{\infty,DIV}$ is described by eq. 11.15 and is the eigenvalue of eq. 11.13.

11.2.4.1.2 *Aeroelastic Divergence – Altitude and Compressibility.*

Equation 11.15 may be expanded to illustrate altitude and compressibility effects. Since $V_\infty = M_\infty a_\infty$, the dynamic pressure may be written as $q_\infty \equiv \rho(M_\infty a_\infty)^2/2 = q_a M_\infty^2$ where q_a is a reference dynamic pressure associated with the speed of sound and density at a specific altitude. It is important to note that the aerodynamic properties in $\underline{K}_A$ depend upon Mach number. The dynamic pressure may be related to Mach number by

$$q_{\infty,DIV} = q_a M_{\infty,DIV}^2 \qquad (11.16)$$

Also, as previously stated, the aerodynamic derivatives are dependent upon flow conditions. The Prandtl-Glauert rule, described in airfoil theory, may be used to illustrate the effect of compressibility on the lift, thus

$$C_{L_\alpha} = \frac{C_{L_\alpha,0}}{\sqrt{1-M_\infty^2}} \qquad (11.17)$$

where $C_{L_\alpha,0}$ represents the aerodynamic derivative for incompressible flow. Substituting eqs. 11.16 and 11.17 into eq. 11.15 yields

$$q_a(M_{\infty,DIV})^2 = \frac{K\sqrt{1-M_{\infty,DIV}^2}}{SeC_{L_\alpha,0}} = q_{0,DIV}\sqrt{1-M_{\infty,DIV}^2} \qquad (11.18)$$

where $q_{0,DIV} = K/SeC_{L_\alpha,0}$ represents a reference divergence pressure for the incompressible assumption. Equation 11.18 may be solved for the divergence Mach number $M_{\infty,DIV}$ and compared to the reference conditions as illustrated in Fig. 11.9.

Relative to the reference conditions – namely, at sea level and incompressible -- we observe that divergence is improved with an increase in altitude and we note that compressibility effects must be included to provide the true boundary. Although not considered in this model, the aerodynamic center moves rearward as Mach number increases; thus, the aerodynamic eccentricity, e, is affected. Divergence is a static aeroelastic phenomenon, which is often confused with yield or ultimate load failures. However, we show through eq. 11.15 that aeroelastic divergence is independent of the yield or ultimate loads; rather, divergence depends upon the stiffness of the structure and the aerodynamic derivatives.

We also note that loads analysis will be adversely affected by ignoring aeroelastic affects; for example, simply consider how the deformation predicted by eq. 11.18 might be lower if the aeroelastic stiffness did not include the $qSeC_{L_\alpha}$ term in the denominator.

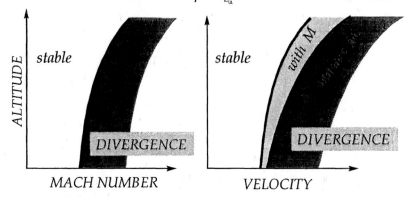

Fig. 11.9 Divergence Boundaries -- Mach Number and Altitude Effects

We could extend our development to a wing, which is stiff in bending but comparably responsive in torsion. In this case, we simplify the structural model by discretizing the wing into separate spanwise sections and we introduce the concept of influence coefficients. Influence coefficients represent the inverse of the stiffness; these coefficients relate the deformation at a specific point due to loads at all sections. For torsional behavior, these coefficients describe the torsional rigidity. We would solve the matrix of this multiple degree-of-freedom system for the critical dynamic pressure that leads to divergence.

11.2.4.2 **Lift Effectiveness and Sweep Effects**. The interaction of elastic, aerodynamic and inertial loads involves coupling between different forms of deformation; for example, coupling between wing bending and wing torsion involves geometry and structural mechanics. Our prior model included only wing torsion. However, once bending is added, torsional loads may lead to bending, and bending loads may lead to torsion.

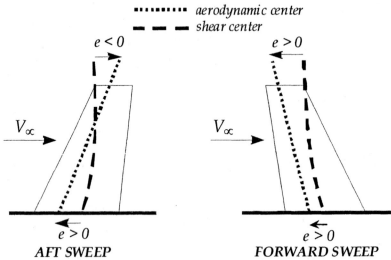

Fig. 11.10 Aerodynamic Eccentricity and Sweep

Consider the decoupling nature defined by the shear center of a structure such as a beam. The shear center is related to structural mechanics and is defined as a point at which an applied force will lead to no twist. Figure 11.10 illustrates two wings – one forward swept and one aft swept. The shear axis is the loci of shear centers, and is the point about which a force (e.g., lift) leads to no moment (that is, no twist). The aerodynamic center is related to aerodynamic loads and is defined as the (chordwise) location on the wing where there is no change in aerodynamic moment due to a change in angle-of-attack (twist). It is often related to the center of pressure.

Aerodynamic eccentricity exists when the aerodynamic center is not collocated with the elastic axis (shear center). As described by eq. 11.15, divergence conditions are made more severe by an increase in eccentricity (e) and eliminated when eccentricity is eliminated or negative. Corresponding to this trend, Fig. 11.10 shows that aft sweep improves divergence characteristics. In fact, the aft swept wing may eliminate divergence. The interaction is further complicated when one considers the effect of bending and twisting on the wing, as illustrated in Fig. 11.11 below.

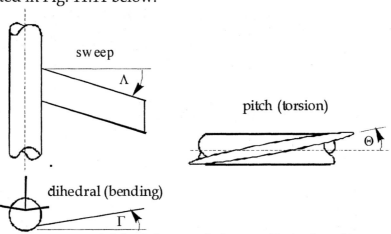

Fig. 11.11 The Wing with Sweep-Twist-Bending

For the aft swept wing shown in Figs. 11.10 and 11.11, an aerodynamic load acting through the aerodynamic center (for example, consider loads at the wing tip) will lead to an untwisting of the wing which reduces the effective angle of attack that the wing section sees. However, the opposite is true for the forward swept wing – loads lead to a growth in angle-of-attack, leading to an increase in loads, leading to a further increase in the adverse deformations, and so forth. An approximation of lift, derived assuming small deformations, is provided by eq. 11.19, The equation describes how the lift on the deforming wing is affected by sweep (Λ), twist (Θ), and bending (Γ).

$$C_L = C_{L_\alpha} \alpha_{effective} = C_{L_\alpha} \cos^2 \Lambda \, (\Theta - \Gamma \tan \Lambda) \tag{11.19}$$

Sweep is positive for the aft swept case. Thus, for a deforming wing (bending and twisting), eq. 11.19 states that lift is reduced for the aft swept wing and increased for the forward sweep wing. Lift effectiveness, which may be considered as the ratio of the flexible wing to the rigid wing,

$$\textit{Lift effectiveness} = \frac{L_{flexible}}{L_{rigid}} \tag{11.20}$$

is affected by wing sweep. Considering eq. 11.19, then $L_{flexible_{AFT\,SWEPT}} < L_{rigid} < L_{flexible_{FWD\,SWEPT}}$.

And, due to the interaction beween loads and deformations, aeroelastic divergence is affected as depicted in Fig. 11.2. Considering sweep, and the derivation of eqs. 11.15 and 11.19, a nondimensional equation for divergence is found,

$$\bar{q}_D = \frac{2\bar{e}}{\cos^2 \Lambda \, (2\bar{e} - \bar{K} \tan \Lambda)} \tag{11.21}$$

where $\bar{q}_D$ is the divergence pressure normalized by the no-sweep case, $\bar{K}$ is the ratio of torsion to bending stiffness, and $\bar{e}$ is the aerodynamic eccentricity normalized by the semichord, b. It is noted that eccentricity (and/or sweep) may be positive or negative; accordingly, $\bar{q}_D$ may be greater or less than unity. Most notably, $\bar{q}_D$ drops considerably for the forward sweep case.

11.2.4.3 **Control Reversal**. Control surface reversal occurs due to the deformations of the structure. As the structure deforms, the ability of the control surface to produce a desired reaction approaches zero. The surface is said to lose effectiveness. Further deformations result in the reversal of control surface performance; that is, deflecting the surface produces the opposite effect from that which was intended.

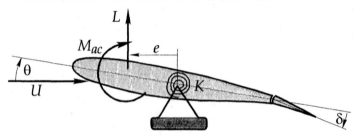

Fig. 11.12 Control Reversal

The single degree-of-freedom example is now extended to include control surface loads. An important premise for understanding control reversal is that the purpose of the control surface is to create a lift load that leads to a moment for maneuver. Thus, if the wing section represents the outboard portion of a wing, then the new lift leads to a rolling maneuver. On a vertical tail section the rudder is meant to produce a side force that causes a yawing moment. But, if no lift or side force is generated for any control input due to flexibility of the primary structure, then no maneuvering moments are created.

Summing moments about the elastic axis of the wing-control surface model,

$$\sum M_{EA} = 0 \;\Rightarrow\; Le + M_{ac} = K\theta \;\Rightarrow\; \theta = \frac{Le + M_{ac}}{K} \tag{11.22}$$

in which the aerodynamic lift is
$$L = L_{wing} + L_{control} = qSC_{L_\alpha}\theta + qSC_{L_\delta}\delta \tag{11.23}$$
and the aerodynamic moment at the aerodynamic center is
$$M_{ac} = qScC_{m_\delta}\delta \tag{11.24}$$
where δ is the control input. Substituting eqs. 11.23 and 11.24 into eq. 11.22 yields

$$\theta = \frac{Le + M_{ac}}{K} = \frac{\left(qSC_{L_\alpha}\theta + qSC_{L_\delta}\delta\right)e + qScC_{m_\delta}\delta}{K} \tag{11.25}$$

or, solving eq. 11.25 directly for θ

$$\theta = \frac{qeSC_{L_\delta}\delta + qScC_{m_\delta}\delta}{K - qeSC_{L_\alpha}} \tag{11.26}$$

The creation of lift is of primary interest, thus eq. 11.21 is substituted into eq. 11.23

Chapter 11   Aeroelasticity: Theory and Analytical Foundations   55

$$L = qS\left(C_{L_\alpha}\frac{qeSC_{L_\delta}\delta + qScC_{m_\delta}\delta}{K - qeSC_{L_\alpha}} + C_{L_\delta}\delta\right) = qS\left(\frac{KC_{L_\delta} + qScC_{m_\delta}C_{L_\alpha}}{K - qeSC_{L_\alpha}}\right)\delta \qquad (11.27)$$

Control reversal occurs when no change in lift is created regardless of the control input. Or, mathematically, we examine $\frac{dL}{d\delta} = 0$. The consequence is the numerator of eq. 11.27 from which we obtain the dynamic pressure that leads to zero (or negative) values (note: by convention, $C_{m\delta}$ is negative).

$$L \le 0 \quad \text{or} \quad q_{rev} = \frac{-K}{ScC_{m_\delta}}\frac{C_{L_\delta}}{C_{L_\alpha}} \qquad (11.28)$$

Also, divergence conditions appear through eq. 11.27 as the denominator approaches zero for increasing dynamic pressure.

$$L \to \infty \quad \text{or} \quad q_{div} = \frac{K}{eSC_{L_\alpha}} \qquad (11.29)$$

Thus, we have two flight conditions to consider – reversal and divergence – and depending upon design (see Figure 11.2), one of conditions may be limiting. Reversal is preformance limiting; divergence is loads limiting.

11.2.5 Dynamic Aeroelastic Phenomena

Dynamic aeroelastic phenomena include flutter, buzz, buffet, and dynamic response. With these phenomena, airframe motion, aerodynamic loads, and structural deformations are dependent upon one another. Flutter is a dynamic aeroelastic instability, characterized by growing oscillatory motion involving the interaction of structural modes affected by motion-dependent aerodynamic loads, that leads to a catastrophic structural failure. Flutter is an instability in which the structural response is excited by the presence of the aerodynamic loads as the structure extracts energy from the flow field. The interested reader is encourage to the works of Bisplinghoff et al. [1], and Fung[2], Dowell et al.[3], Hodges and Pierce[4], and Theodorsen and Garrick[6], as well as the descriptions provided in Ref. 7.

Types of flutter include:
- coupled bending-torsion flutter associated with high aspect ratio lifting surfaces,
- panel flutter associated with low aspect ratio lifting surfaces,
- control-surface flutter,
- flutter of plate and shell-like structures,
- propeller-whirl flutter, and
- stall flutter.

All of these flutter instabilities, with the exception of stall flutter, are characterized by an interaction (e.g., coalescence) of modes. Stall flutter occurs due to flow separation that occurs during cyclic motion. In this case, the transfer of energy from the flow to the wing does not rely on the coalescence of modes; rather, the wing is "forced" due to energy lost due to stall. Buffet is an instability that occurs due to flow separation or the wake produced from the structure. Buzz is an instability that occurs due to shock wave oscillations. In general, dynamic response of the aircraft due to gusts and atmospheric disturbances is important to characterize since, even in the absence of instabilities, it is necessary to understand how the dynamic response will affect load distributions and ride qualities. All of these behaviors are captured by eq. 11.1.

11.2.5.1 Bending-Torsion Flutter. Of these dynamic aeroelastic phenomena, the bending-torsion flutter mechanism provides the insight into the nature of these instabilities. As previously stated, flutter occurs when the structure extracts energy from the flow field. This structure may be the primary lifting surface, a control surface, the fuselage, or the entire vehicle. As the structure vibrates in a subcritical flow field, the aerodynamic forces damp the vibration of the structure; however, as the speed (or dynamic pressure) increases, a critical condition is reached in which the energy from the flow field feeds the vibration of the structure. This critical point represents the onset of flutter. Flutter is destructive in nature; but, fortunately, flutter boundaries may be identified.

Flutter results from the interaction of inertial, structural, and aerodynamic forces and, typically, flutter is the result of coupled modes of motion. The pitch and plunge motion is representative of a wing with bending and torsional response. Although these equations of motion are straightforward, they apply to more representative aerospace structures. A minimum of two degrees of freedom is used since a wing limited only to bending motion will not flutter and a wing limited only to torsional motion will exhibit only stall flutter at high angles of attack. Although unlikely, it is possible for a single degree-of-freedom to flutter in torsion for very specific center-of-gravity and elastic-axis locations (see Ref. 1). The wing in Fig. 11.13 is limited to motion in two degrees of freedom.

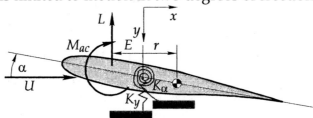

Fig. 11.13 Bending - Torsion Flutter

This wing section has a "plunging" motion, which is translational motion perpendicular to the flow, and a "pitching" motion, which is rotational motion about the elastic axis. The linearized equations of motion for the system are:

$$\begin{bmatrix} m & mr \\ mr & I \end{bmatrix} \begin{pmatrix} \ddot{y} \\ \ddot{\alpha} \end{pmatrix} + \begin{bmatrix} K_y & 0 \\ 0 & K_\alpha \end{bmatrix} \begin{pmatrix} y \\ \alpha \end{pmatrix} = \begin{pmatrix} -L \\ M \end{pmatrix} \quad (11.30)$$

where m is the mass of the wing, r is the distance between the center of mass and the elastic axis, I is the moment of inertia referenced to the elastic axis, K_y and K_α are spring constants associated with the plunge and pitch motion, respectively, L is the aerodynamic lift and M is the aerodynamic moment referenced to the elastic axis.

The form of this equation is identical to eq. 11.1. Note that in this general form, the discussion of aeroelasticity is not limited to a system of two degrees of freedom. The mass and stiffness matrices may be representative for any aeroelastic system. Structural damping may also be considered.

In real problems, these matrices may be very large. Analytically, the system is often represented through finite element methods to accurately model the structure and the associated modal characteristics. Experimentally, these matrices may be derived using measurements and employing modal methods. The modes and frequencies of such analyses are used in the solution process.

Though a wing may have a large number of vibration modes, only the lower modes are typically used, since the flutter mechanism is driven by fundamental modes. For example, the lowest ten to fifteen modes usually capture the primary behavior of the structure (the reader is referred to Ref. 7 for specific case studies). This number may increase as interaction with higher modes such as those that are associated with control surfaces becomes important. Typically, the nature of classical flutter employs the primary torsion mode with the first or second bending mode. Of importance now is to understand the content of these equations, the solution of these equations, and interpretation of results.

Examination of eq. 11.30 in closer detail reveals that the equations are coupled through the mass matrix, as evidenced by the presence of off-diagonal terms in the matrix. We could have chosen a set of coordinates in which the system is coupled through the stiffness matrix, but the stability characteristics would be identical. The choice of coordinates is for convenience in the solution. This 'inertial' coupling occurs since the center of mass does not coincide with the elastic axis; the relative location of the center of mass plays an important role in aeroelastic design as mass-balancing inherently improves stability margins.

The aerodynamic loads, L and M, are unsteady, which means these loads are motion dependent. For example, Theodorsen[6] (also see Fung[2]) provided a very rigorous development of the unsteady potential flow equation that led to the classical solution for incompressible unsteady aerodynamic loads

$$L = \pi \rho b^2 \left(\ddot{y} + V_\infty \dot{\alpha} - b a \ddot{\alpha} \right) + 2\pi \rho V_\infty b C_k \left[\dot{y} + V_\infty \alpha + b\left(\frac{1}{2} - a\right)\dot{\alpha} \right] \tag{11.31}$$

$$M = \pi \rho b^3 \left[a\ddot{y} - \left(\frac{1}{2} - a\right) V_\infty \dot{\alpha} - \left(\frac{1}{8} + a^2\right) b\ddot{\alpha} \right] + 2\pi \rho V_\infty b^2 C_k \left(\frac{1}{2} + a\right) \left[\dot{y} + V_\infty \alpha + b\left(\frac{1}{2} - a\right)\dot{\alpha} \right] \tag{11.32}$$

where a is the nondimensional distance (in terms of the semi-chord, b) measured from the mid-chord to the elastic axis (positive toward the trailing edge) and is a measure of aerodynamic eccentricity with respect to the elastic axis and C_k is Theodorsen's Function and is a complex function of the reduced frequency, k,

$$C_k \equiv \frac{H_1^{(2)}}{[H_1^{(2)} + iH_0^{(2)}]} = F(k) + iG(k) \tag{11.33}$$

where $H_1^{(2)}$ and $H_0^{(2)}$ are standard Hankel functions, and the components $F(k)$ and $G(k)$ may be approximated by

$$F(k) = \frac{k^4/2 + 0.0765k^2 + 0.000186}{k^4 + 0.0921k^2 + 0.000186} \quad \text{and} \quad G(k) = \frac{-0.108k^3 - 0.000883k}{k^4 + 0.0921k^2 + 0.000186} \tag{11.34}$$

Observe, at $k = 0$ (i.e., no frequency) $F(k) = 1$ and $G(k) = 0$.

A quasi-steady aerodynamic model, appropriate for low frequency motion ($k < 0.1$), is embedded within eqs. 11.31 and 11.32 as

$$L = 2\pi \rho V_\infty b \left[\dot{y} + V_\infty \alpha + b\left(\frac{1}{2} - a\right)\dot{\alpha} \right] \tag{11.35}$$

$$M = 2\pi \rho V_\infty b^2 \left(\frac{1}{2} + a\right)\left[\dot{y} + V_\infty \alpha + b\left(\frac{1}{2} - a\right)\dot{\alpha} \right] = b\left(\frac{1}{2} + a\right)L \tag{11.36}$$

However, a source of aerodynamic damping is present in either form.

Harmonic motion for y and α is assumed for the solution and, as a result, an eigenvalue problem is formulated as previously discussed. In the simplest form, all terms are grouped in a single matrix. The roots to the equation resulting from the expansion of the determinant of the matrix provide the stability properties. The solution is complex such that the roots contain damping and frequency information. Again, it is noted that the aerodynamic properties depend upon the airspeed, altitude, Mach number, and frequency. And, the difficulty in solution arises when one considers that the eigenvalue contains the very frequency that is needed to determine the aerodynamic loads. Thus, the solution requires an iterative process (see Hodges[4] and Hassig[8]).

11.2.5.2 Example: Two Degree-of-Freedom Aeroelasticity. Stability characteristics are obviously affected by the system parameters (mass, stiffness, geometry) contained within the matrices in the governing eq. 11.30. The designer of the airplane has some control over inertial, elastic, or aerodynamic coupling. Typically, during the analysis, the only parameters modified are associated with the flowfield (that is, velocity, dynamic pressure, and altitude) and solving these equations in the frequency domain permits one to more readily examine damping and frequency characteristics.

To further understand the flutter mechanism, an illustration based upon the two degree-of-freedom system is presented in which aerodynamic loads are modeled by the static terms in eqs. 11.35 and 11.36. Such a simplification of the aerodynamic loads may be referred to as the "quasi-static" aerodynamic model since the loads only depend upon the instantaneous angle of attack of the wing. Note that this model is not accurate but serves as a simple mechanism to describe flutter. With this approach, eq. 11.30 becomes,

$$\begin{bmatrix} m & mr \\ mr & I \end{bmatrix} \begin{Bmatrix} \ddot{y} \\ \ddot{\alpha} \end{Bmatrix} + \begin{bmatrix} K_y & 0 \\ 0 & K_\alpha \end{bmatrix} \begin{Bmatrix} y \\ \alpha \end{Bmatrix} = \begin{Bmatrix} -q_\infty S C_{L_\alpha} \alpha \\ q_\infty S C_{L_\alpha} e \alpha \end{Bmatrix} \tag{11.37}$$

or, with rearrangement

$$\begin{bmatrix} m & mr \\ mr & I \end{bmatrix} \begin{Bmatrix} \ddot{y} \\ \ddot{\alpha} \end{Bmatrix} + \begin{bmatrix} K_y & q_\infty S C_{L_\alpha} \\ 0 & K_\alpha - q_\infty S C_{L_\alpha} e \end{bmatrix} \begin{Bmatrix} y \\ \alpha \end{Bmatrix} = \begin{Bmatrix} 0 \\ 0 \end{Bmatrix} \tag{11.38}$$

Harmonic motion is assumed such that $y = Y e^{i\omega t}$ and $\alpha = A e^{i\omega t}$

$$\left\{ -\omega^2 \begin{bmatrix} m & mr \\ mr & I \end{bmatrix} + \begin{bmatrix} K_y & q_\infty S C_{L_\alpha} \\ 0 & K_\alpha - q_\infty S C_{L_\alpha} e \end{bmatrix} \right\} \begin{Bmatrix} Y \\ A \end{Bmatrix} = \begin{Bmatrix} 0 \\ 0 \end{Bmatrix} \tag{11.39}$$

These equations are simplified by dividing by mass and introducing similarity parameters previously defined. Thus,

$$\left\{ -\omega^2 \begin{bmatrix} 1 & r \\ r & r_\alpha^2 \end{bmatrix} + \begin{bmatrix} \omega_y^2 & \dfrac{bC_{L_\alpha} \bar{V}^2 \omega_\alpha^2}{\mu \pi} \\ 0 & \omega_\alpha^2 r_\alpha^2 - \dfrac{bC_{L_\alpha} \bar{V}^2 \omega_\alpha^2 e}{\mu \pi} \end{bmatrix} \right\} \begin{Bmatrix} Y \\ A \end{Bmatrix} = \begin{Bmatrix} 0 \\ 0 \end{Bmatrix} \tag{11.40}$$

where μ is the mass ratio, $\bar{V}$ is the reduced velocity, the radius of gyration is $r_\alpha = \sqrt{\dfrac{I}{m}}$, and $\omega_y = \dfrac{K_y}{m}$ and $\omega_\alpha = \dfrac{K_\alpha}{I}$ are the uncoupled (that is, $r = 0$) natural frequencies of the system.

Chapter 11 — Aeroelasticity: Theory and Analytical Foundations

To completely render the equations into a more convenient nondimensional form, the top equation is divided by the reference length and the bottom equation is divided by the square of the reference length, and both equations are divided by ω^2. Thus,

$$\left\{ -\begin{bmatrix} 1 & \bar{r} \\ \bar{r} & \bar{r}_\alpha^2 \end{bmatrix} + \begin{bmatrix} \dfrac{\omega_y^2}{\omega^2} & \dfrac{C_{L_\alpha} \bar{V}^2}{\mu \pi} \dfrac{\omega_\alpha^2}{\omega^2} \\ 0 & \dfrac{\omega_\alpha^2}{\omega^2}\bar{r}_\alpha^2 - \dfrac{C_{L_\alpha} \bar{V}^2 \bar{e}}{\mu \pi}\dfrac{\omega_\alpha^2}{\omega^2} \end{bmatrix} \right\} \begin{Bmatrix} \bar{Y} \\ A \end{Bmatrix} = \begin{Bmatrix} 0 \\ 0 \end{Bmatrix} \qquad (11.41)$$

or

$$\begin{bmatrix} \dfrac{\omega_y^2}{\omega^2}-1 & \dfrac{C_{L_\alpha} \bar{V}^2}{\mu \pi}\dfrac{\omega_\alpha^2}{\omega^2} - \bar{r} \\ -\bar{r} & \left(\dfrac{\omega_\alpha^2}{\omega^2}-1\right)\bar{r}_\alpha^2 - \dfrac{C_{L_\alpha} \bar{V}^2 \bar{e}}{\mu \pi}\dfrac{\omega_\alpha^2}{\omega^2} \end{bmatrix} \begin{Bmatrix} \bar{Y} \\ A \end{Bmatrix} = \begin{Bmatrix} 0 \\ 0 \end{Bmatrix} \qquad (11.42)$$

where the overbar signifies a nondimensional length with respect to the reference length.

We should pause to appreciate the significance of the above equation as it captures the essence of the dynamic aeroelastic system, and it illustrates the challenge to define the stability boundary (that is, where does flutter exist?). Note that the physical properties of the vehicle are defined by a few parameters (ω_y, ω_α, r_α, $\bar{e}$, and r). In an analogous manner, for routine analysis, one defines the structural system by using the vibration characteristics, namely, the natural frequencies, natural modes, and the resulting modal mass. The aerodynamics are represented by C_{L_α} in the above model only depend upon the instantaneous values of α but in reality depend upon the frequency of motion (due to $\dot{\alpha}, \ddot{\alpha}, \dot{y},$ and $\ddot{y}$) and Mach number (M_∞) due to compressibility effects. The reader should consider how the above equation might appear if all terms of eqs. 11.31 and 11.32 are included. The unknowns are μ (altitude), M_∞, $\bar{V}$ and ω.

Typically, analysis is performed at an altitude (i.e., μ) and M_∞. Then, the aerodynamic loads are determined at a frequency. For now, we will assume $C_{L_\alpha} = 2\pi$ which is the value from classical thin airfoil theory for incompressible flow (thus, $M_\infty = 0$), and for this example, these aerodynamic loads are independent of frequency. Thus, the remaining unknowns are velocity and frequency. Examination of the eigenvalues of eq. 11.42 provide the flutter velocity, $\bar{V}_F$, and the flutter frequency, ω_F, where the subscript F designates flutter. With $C_{L_\alpha} = 2\pi$, eq. 11.42 appears as:

$$\begin{bmatrix} \dfrac{\omega_y^2}{\omega_\alpha^2} - \dfrac{\omega^2}{\omega_\alpha^2} & \dfrac{2\bar{V}^2}{\mu} - \bar{r}\dfrac{\omega^2}{\omega_\alpha^2} \\ -\bar{r}\dfrac{\omega^2}{\omega_\alpha^2} & \left(1-\dfrac{\omega^2}{\omega_\alpha^2}\right)\bar{r}_\alpha^2 - \dfrac{2\bar{V}^2\bar{e}}{\mu} \end{bmatrix} \begin{Bmatrix} \bar{Y} \\ A \end{Bmatrix} = \begin{Bmatrix} 0 \\ 0 \end{Bmatrix} \qquad (11.43)$$

Several comments are made in reference to eq. 11.43:

(1) The equation has been multiplied by ω^2/ω_α^2 to avoid the singularity at $\omega = 0$.
(2) The equation is an eigenvalue form for the aeroelastic system. The determinant of the matrix forms a quadratic equation for the roots, ω^2, which represent the aero-

elastic frequencies as a function of the flutter speed index, $\bar{V}$.

(3) For the case of $\bar{V} = 0$, the roots of the equation provide the natural frequencies as a function of the mass imbalance, $\bar{r}$. The uncoupled natural frequencies are found for the case $\bar{r}=0$.

(4) For the case of $\bar{V} \neq 0$, Fig. 11.14 illustrates the characteristic behavior of the aeroelastic frequencies for increasing velocities. The flutter region is found as the two roots coalesce to a common frequency, and a complex eigenvalue appears indicating a growing motion.

(5) Aerodynamic loads introduce coupling in the system through two mechanisms. Most obvious is the fact that the loads will not usually act through the elastic axis; thus, lift leads to a moment (see $\bar{e}$ above). The second source of coupling exists since the unsteady aerodynamic loads are dependent upon displacements, velocities, and acceleration (see eqs. 11.31 and 11.32).

(6) The formulation illustrates the use of the similarity parameters.

(7) An extremely simplified formulation of the aerodynamics is used. In reality, the aerodynamics depend upon the Mach number and reduced frequency. Standard aeroelastic methodologies, such as the "k method" or "p-k method" require an iterative solution since the aerodynamic properties depend upon the reduced frequency. The aeroelastic frequency, ω, that is the eigenvalue also appears in the reduced frequency and flutter speed index.

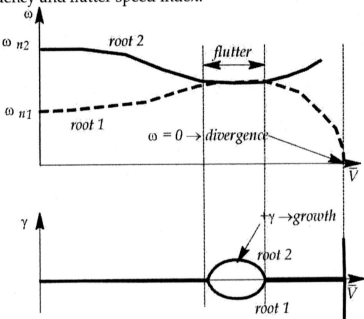

Fig. 11.14 Coupling of Flutter Modes

Solving the (flutter) determinant of eq. 11.43 yields a quadratic for the root,

$$\left(\frac{\omega^2}{\omega_\alpha^2}\right)_{1,2} = \frac{-B \pm \sqrt{B^2 - 4AC}}{2A} = \frac{-B \pm \sqrt{D}}{2A} \tag{11.44}$$

Examination of the terms in eq. 11.44 reveals that A is a system constant, and the terms B and C depend upon $\bar{V}$. The roots yield the frequencies (ω) at low values of $\bar{V}$, but may become complex conjugates (i.e., $\omega \pm i\gamma$) at higher values of $\bar{V}$, depending on the physical

Chapter 11   Aeroelasticity: Theory and Analytical Foundations   61

properties (Fig. 11.14). The appearance of the complex roots indicates flutter since a root suggests growing motion. Thus, from eq. 11.44, flutter occurs at the lowest value of $\overline{V}$ when $D = 0$. Also, aeroelastic divergence -- the "zero frequency" instability -- is found by examining the case of $C = 0$ since terms A and B are motion dependent.

The "quasi-static" assumption is an extremely simplified (and misleading) model for the aerodynamic loads. The aerodynamic terms, such as C_{L_α}, depend upon the frequency, ω, in addition to the Mach number. In general, one must know the frequency and Mach number to determine the aerodynamics, yet one must know the aerodynamics to find the frequency and Mach number. Thus, in more representative formulations of eq. 11.43, an iterative process is necessary. Nonetheless, this example serves to illustrate the concept of complex aeroelastic modes of the classical bending-torsion aeroelastic response of a high aspect ratio wing. Experiments and analysis of wings show that spanwise sections of the wing have bending and torsional components, and the modes have distinct frequencies and phase relationships that depend on airspeed. The fundamental modal characteristics are measured by ground vibration tests that serve a vital and necessary complementary role to the flight test. However, under flight conditions, unsteady aerodynamic loads contribute to these frequencies and phase relationships; and, the knowledge of these phase relationships are valuable in flutter identification. At flutter, frequencies, mode shapes, and phasing tend to coalesce and form a single aeroelastic flutter mode that involves a combination of the vibration modes (it is noted here that, although rare, there are single degree-of-freedom flutter mechanisms). Hence, the determination of the complex eigensolution (frequencies and mode shapes) is of value.

11.3 AEROELASTIC ANALYSIS TECHNIQUES

11.3.1 Modal Methods

Briefly, we review modal methods which serve as the basis for a majority of aeroelastic analyses. In words, the natural frequencies and associated mode shapes are obtained for the physical system and an aeroelastic model, including the motion-dependent unsteady aerodynamics, is developed for analysis using two powerful features of modal analysis. One, a subset of the mode shapes may accurately represent the response of the system; and two, mode shapes are orthogonal which provides a set of coordinates that decouple the system. Consider eq. 11.1 in which damping and external forces are absent,

$$M\ddot{x} + Kx = 0 \qquad (11.45)$$

This equation may represent a significantly large system. For such a system, we assume harmonic motion $x = \Theta e^{\pm i\omega t}$ where Θ represents the amplitude of motion. Substituting,

$$(-\omega^2 M + K)\Theta = 0 \qquad (11.46)$$

Given M and K from a model of the system (for example, a standard finite element model) one determines the natural frequencies, ω_i, and mode shapes, Θ_i, for $i = 1$ to n of the system, where n represents the number of degrees of freedom in the system. A subset of these mode shapes is used to represent the response of the system such that

$$x = q_i \Theta_i \text{ for } i = 1 \text{ to } m \qquad (11.47)$$

where q are the generalized (modal) coordinates and m is the number of modes (or, in other words, a reduced number of degrees of freedom) chosen to represent the response. Typically, $m \ll n$.

Generalized coordinates are time-dependent, $q = q(t)$, and the mode shapes are spatial. The response is assumed to be a combination of the mode shapes chosen, each mode weighted by the generalized coordinate. Accordingly, if the motion is dominated by a specific mode (e.g., first wing torsion mode), then the generalized coordinate associated with the mode is high relative to the others. Yet, if the mode were not used in the model, its effect was ignored (that is, filtered from the response).

Equation 11.47 may be written as

$$x = \Theta q \tag{11.48}$$

where Θ is a transformation matrix with the individual modes selected for the model from column vectors, that is, $\Theta = [\Theta_1 \ \Theta_2 ... \Theta_m]$. Substituting eq. 11.48 into eq. 11.45 yields

$$M\Theta \ddot{q} + K\Theta q = 0 \tag{11.49}$$

which must be pre-multiplied by the transpose, Θ^T, to fully complete the transform,

$$\Theta^T M\Theta \ddot{q} + \Theta^T K\Theta q = 0 \tag{11.50}$$

where the modal (or generalized) mass is

$$\Theta^T M\Theta = M_M \tag{11.51}$$

which is a diagonal matrix due to the orthogonality of the modes (note, from theory, multiplication of two modes $\Theta_i \Theta_j = 0$ if $i \neq j$). Mode shapes must be scaled (since response is proportional to external forces and no forces are present in eq. 11.45, the response is relative). If modes are scaled such that they are orthonormal with respect to mass,

$$\Theta^T M\Theta = M_M = I \tag{11.52}$$

where I is the identity matrix. Then, with these orthonormal modes, one finds

$$\Theta^T K\Theta = \Omega \tag{11.53}$$

where Ω is a diagonal matrix consisting of the natural frequencies, ω_n^2. Accordingly, the system modeled by eq. 11.45 is simplified to

$$\ddot{q}_i + \omega_{n_i}^2 q_i = 0 \quad \text{for } i = 1 \text{ to } m \tag{11.54}$$

which represents a set of uncoupled equations. It is well worth noting that a very large degree of freedom system is decoupled and reduced in size.

11.3.2 Solution – "K" Method

Now, we have an efficient and friendly basis for aeroelastic analyses. Rewriting eq. 11.3, without structural damping and external forces

$$(M - M_A)\ddot{x} - C_A \dot{x} + (K - K_A)x = 0 \tag{11.55}$$

and transforming with eq. 11.48, and assuming harmonic motion $q = \beta e^{i\omega t}$, yields the aeroelastic equations,

$$\left[I - \frac{1}{\omega^2}\Omega + \frac{1}{\omega^2}\underline{A}\right]\beta = 0 \tag{11.56}$$

where $\underline{A}$ represents the motion-dependent aerodynamic loads written in terms of the modes and ω is the eigenvalue which represents the frequency of motion.

Chapter 11   Aeroelasticity: Theory and Analytical Foundations

The aerodynamic matrix, $\underline{A}$, in eq. 11.56 depends upon Mach number, airspeed and altitude as well as the frequency. The interdependence of frequency of motion and the aerodynamics requires a solution obtained with an iterative algorithm. For example, the equations describing Theodorsen's aerodynamics (eqs. 11.31 and 11.32), using the assumption of harmonic motion, $y = Y\,e^{i\omega t}$ and $\alpha = A\,e^{i\omega t}$, may be written as

$$\left\{\begin{matrix} \dfrac{L}{qS} \\ \dfrac{M}{qSc} \end{matrix}\right\} = \pi k^2 \begin{bmatrix} -1 + 2C_k\dfrac{i}{k} & a + \dfrac{2C_k}{k^2} + \left\{1 + \left(\dfrac{1}{2} - a\right)2C_k\right\}\dfrac{i}{k} \\ -\dfrac{a}{2} + \left(\dfrac{1}{2} + a\right)C_k\dfrac{i}{k} & \dfrac{1}{16} + \dfrac{a^2}{2} + \left(\dfrac{1}{2} + a\right)\dfrac{C_k}{k^2} + \left\{\left(a - \dfrac{1}{2}\right)\dfrac{1}{2} + \left(\dfrac{1}{4} - a^2\right)C_k\right\}\dfrac{i}{k} \end{bmatrix} \left\{\begin{matrix} \overline{Y} \\ A \end{matrix}\right\} \quad (11.57)$$

The purpose of showing this equation is to demonstrate the complexity of the unsteady aerodynamics for the simple two degree-of-freedom motion. One may appreciate the escalation of difficulty in solution as one expands to the more general system in eq. 11.48. Key points include: the aerodynamics depend upon frequency which is not known but must be assumed, there is no direct solution as the terms involve nonlinear contributions of k and the C_k function, the aerodynamics are complex (literally and mathematically) which introduce phasing between loads and motion, all terms are nondimensional which follow a standard approach of similarity analysis, and the aerodynamics assume harmonic motion which is violated when there is either growing or decaying motion.

Popular methods for finding the solution to the aeroelastic equations (eq. 11.56) include the k-method and the p-k method[4] in which velocity-damping diagrams and velocity-frequency diagrams are constructed from the roots of the eigenvalue solutions (where k is the reduced frequency). In these methods, the eigenvalue term is multiplied by $(1 + ig)$ in which a fictitious damping, g, is introduced which permits the solution to the complex eigenvalue problem. Essentially, system damping is assumed to be proportional to stiffness but in phase with velocity. Frequency and damping is found for increasing velocity.

To illustrate, we return to the previous "quasi-static" example to illustrate the iterative approach. As stated, aerodynamic terms such as C_{L_α} depend upon the frequency, ω, in addition to the Mach number. In general, one must know the frequency and Mach number to determine the aerodynamics, yet one must know the aerodynamics to find the frequency and Mach number. Again, in more representative formulations of eq. 11.43 (such as eq. 11.56), an iterative process is necessary. To illustrate, we rewrite eq. 11.43, recognizing that $\overline{V} \equiv \dfrac{1}{k}\dfrac{\omega}{\omega_\alpha}$. And, with the inclusion of fictitious damping into the formulation, the eigenvalue $\dfrac{\omega_\alpha^2}{\omega^2}$ is replaced with $(1+ig)\dfrac{\omega_\alpha^2}{\omega^2}$.

$$\begin{bmatrix} (1+ig)\dfrac{\omega_\alpha^2}{\omega^2}\dfrac{\omega_y^2}{\omega_\alpha^2} - 1 & \dfrac{C_{L_\alpha}}{\mu\pi}\dfrac{1}{k^2} - \overline{r} \\ -\overline{r} & (1+ig)\dfrac{\omega_\alpha^2}{\omega^2}\overline{r}_\alpha^2 - \overline{r}_\alpha^2 - \dfrac{C_{L_\alpha}}{\mu\pi}\dfrac{\overline{e}}{k^2} \end{bmatrix} \left\{\begin{matrix} \overline{Y} \\ A \end{matrix}\right\} = \left\{\begin{matrix} 0 \\ 0 \end{matrix}\right\} \quad (11.58)$$

we recognize that the aerodynamics (in this case C_{L_α}) are not constants (as seen in eq. 11.57), but depend upon Mach number and frequency, that is, $C_{L_\alpha} = C_{L_\alpha}(k, M_\infty)$. Equation

11.56 represents a modal approach, and includes a more rigorous aerodynamic model; yet comparison of eq. 11.56 with eq. 11.58 shows strong similarities. One should observe that eqs. 11.56, 11.57 and 11.58 are written in a nondimensional manner, utilizing similarity parameters and ratios. Thus, such scaling yields solution to a broad class of systems that satisfy these same parameters. Also, one should observe that the system is desribed by a few parameters. The equations are solved in an identical manner.

The roots of eq. 11.58, $(1+ig)\frac{\omega_\alpha^2}{\omega^2}$, are complex; so they appear as $Z = Z_R + i\, Z_I$ where $Z_R = \frac{\omega_\alpha^2}{\omega^2}$ and $Z_I = g\frac{\omega_\alpha^2}{\omega^2} = gZ_R$. Likewise, the roots of eq. 11.50 are the same. The real part provides frequency and, along with Z_I, the imaginary part, provides the damping. Flutter occurs when system damping vanishes when $g = 0$. The steps to solve the aeroelastic equations – the example case in eq. 11.58 or the general case in eq. 11.56 – are identical:

(1) The physical system is defined: stiffness, mass, geometry (r_α, ω_y, ω_α, a and r). In general, stiffness and mass characteristics are captured by frequencies and modes.
(2) An altitude is set, thus defining μ.
(3) The Mach number is set, thus prescribing compressibility effects (for example, one is reminded of the Prandtl-Glauert rule $C_{L_\alpha} = 2\pi/\sqrt{1-M_\infty^2}$).
(4) Aerodynamic loads are determined for a specific reduced frequency, k. For example, in eq. 11.58 the loads are C_{L_α}/k^2. A typical range of k is $0.001 < k < 1.00$.
(5) The eigenvalues, $Z = Z_R + i\, Z_I$, are found at a specified k.
(6) Data (ω, g, and $\bar{V}$) for the velocity-frequency and velocity-damping diagrams are determined:
$$Z_R = \frac{\omega_\alpha^2}{\omega^2} \Rightarrow \omega^2 = \omega_\alpha^2/Z_R$$
$$Z_I = gZ_R \Rightarrow g = Z_I/Z_R$$
$$\bar{V} = \frac{1}{k}\frac{\omega}{\omega_\alpha} \Rightarrow \bar{V} = \frac{1}{k\sqrt{Z_R}}$$
(7) A new k is chosen, and steps (4)-(6) are repeated until $g = 0$ is found. Then at $g = 0$,
$$\bar{V} = \bar{V}_{Flutter} \text{ and } \frac{\omega_\alpha^2}{\omega^2} \Rightarrow \omega_{Flutter}$$
(8) Iterate on Mach number (step 3) if the Mach number. used does not adequately match that related to the flutter velocity and altitude.
(9) repeat step(2) as required for the range of altitudes.

This approach is referred to as the "k-method".

As illustrated in Fig. 11.15 and Fig. 11.16, the Velocity-Damping (V-g) diagram and Velocity-Frequency displays the roots of the solution (shown for a 4 degree-of-freedom example). The exact solution of the flutter conditions occurs when $g = 0$ (note: that "V-g" is a redundant terminology since the same symbology is used to describe aircraft load diagrams). Herein, $\bar{V}$ is a nondimensional velocity and g represents the overall damping of the aeroelastic system.

In summary, the roots of eq. 11.58 are examined for stability characteristics. The num-

number of roots is equal to the degrees of freedom (generalized coordinates) used to model the system. For typical aeroelastic analyses (for a wing structure or a complete aircraft) these degrees of freedom are associated with the structural mode shapes used in the analysis. These roots are traced for increasing values of velocity. The roots will naturally begin as those associated with a positively-damped system since the motion will decay with naturally occurring structural damping (consider the case of no aerodynamic effects). However, as velocity increases the path of one or more of these roots may advance into the negatively-damped regime. The crossing (sign change) represents the location of neutral stability and indicates the flutter velocity. In certain cases, one might find that for higher velocities, a root again becomes positively damped, indicating that the aeroelastic system is again stable. It is noted that the "Hump" mode illustrated in Fig. 11.15 may be eliminated with additional structural damping and, consequently, this instability may be suppressed. Regardless of structural damping, the hard crossing is present. The instability associated

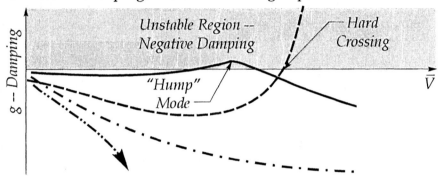

with the lowest velocity will always be of interest.

Fig. 11.15 Velocity-Damping (V-g) Diagram

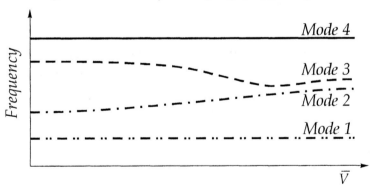

Fig. 11.16 Frequency-Velocity Diagram

In addition to examining the change in damping for increasing velocities, analysis of frequency behavior for changes in velocity as illustrated in Fig. 11.16 reveals the sensitivity of aeroelastic modal frequencies to flight conditions. The frequencies are the natural frequencies of the structure when the velocity is equal to zero. These frequencies change as the velocity increases since the aerodynamic loads couple the structural dynamics of the system. More importantly, these aeroelastic frequencies converge toward a common frequency as the velocity increases. The velocity at the point where any two (or more) frequencies coalesce is one indicator of the instability boundary for the aeroelastic system; however, damping must also be examined since the loss of system damping is the more significant measure. During flight tests both damping and frequency are monitored to identify their respective sensitivity to flight conditions -- zero damping and coalesced frequencies are of concern. In addition, changes in the amplitude of response for changes in flight conditions should be monitored.

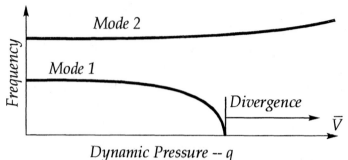

Dynamic Pressure -- q

Fig. 11.17 Frequency-Velocity Diagram (Indicating Divergence)

A root associated with system damping may also vanish as it approaches zero and, simultaneously, a frequency of the system approaches zero. This zero frequency root represents aeroelastic divergence since a characteristic of divergence is that the frequency of the diverging mode approaches zero at the divergence speed. This characteristic provides an approach to identify potential aeroelastic divergence behavior in real time during flight tests. For example, the primary frequency of a beam with an aerodynamic load on the tip is sketched in Fig. 11.17. In actuality, this solution is representative of a slender beam-like structure such as a wind tunnel sting or aircraft sensor probe with an aerodynamically sensitive device attached to the tip. The aerodynamic loading is dependent upon the displacement and rotation of the tip, as well as the velocity of the flow; consequently, the deformation-dependent loads lead to aeroelastic divergence.

11.3.3   Matched Point Analysis

It has been shown that iterative approaches are used to compute flutter conditions. For example, the sequence of steps used in the k-method showed that a V-g diagram is generated for a specific Mach number and altitude, and from the root crossing at $g = 0$ one finds the flutter velocity. Not immediately obvious here is an inherent paradox in the solution. That is, if one selects an altitude and Mach number to perform analysis, then the velocity is predetermined. Why? In the atmosphere a speed of sound is associated with each altitude (and the temperature of the air at that altitude); and, together, the Mach number and speed of sound specify a velocity,

$$V_\infty = M_\infty a_\infty \tag{11.59}$$

The iterative methods (see step 8 in the k-method) do account for an approach to converge on Mach number, but this is computationally intensive. Simply, the complexity of determining the aerodynamic loads necessitates solutions at specific altitudes and Mach numbers. Thus, the critical velocity for aeroelastic instability is found for different altitudes (mass ratios) and different Mach numbers. Is the velocity directly derived from eq. 11.59 the same as the velocity from the V-g diagram? No, not unless, the approach changes the Mach number with changes in velocity as one cycles through the iterative process (this may be done, but those in the review process are reminded to consider this paradox). Typically, the velocities are quite different except for fortuitous circumstances. Therefore, we must extend the analysis to include the "matched point" solution.

Figure 11.18 illustrates a process to find the matched point from which usable flutter boundaries may be generated from analysis. In this approach, as illustrated in the left-hand view of Fig. 11.18, the flutter velocity obtained from analysis of V-g diagrams is plotted for a series of altitudes at constant Mach number (for example, step 8 in the k-method process is ignored). The intersection of a curve through these results (the curve given by the velocity found from eq. 11.59) leads to a "matched point" in which the altitude, velocity, and Mach number are consistent. As a check, one would find that solution of the aeroelastic equations (eq. 11.58) at this altitude and Mach number would yield a velocity consistent with eq. 11.59. A new Mach number is chosen for analysis, and this process is repeated over a range of Mach numbers to complete a set of data. Mach numbers, altitudes, and velocities for which flutter occurs and from which the flutter boundaries, (middle and right views of Fig. 11.18) are generated. Flutter is primarily a low altitude, transonic instability as suggested by these figures. It is also noted that this discussion applies to vehicles in which compressibility is an issue (for example, $M_\infty > 0.5$).

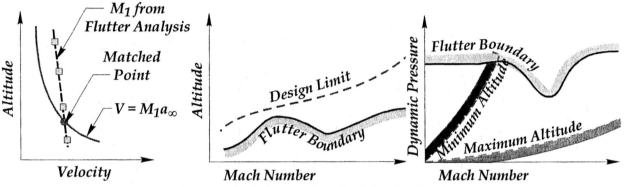

Fig. 11.18 Solution for the Matched Point

11.4 NONLINEAR AEROELASTICITY

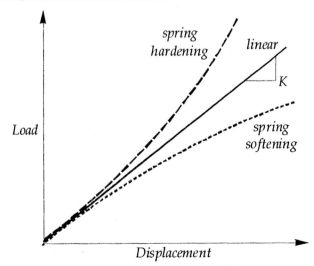

Fig. 11.19 Nonlinear Stiffness

Flight vehicles possess nonlinearities that may lead to responses not predicted by standard linear analysis. For example, one normally expects a linear relationship between load and displacement. One unit of load leads to one unit of deformation, two units of load lead to two units of deformation, and so on, but unique designs such as a wing-fold on a carrier-based aircraft may lead to a stiffening effect (the stress-strain relationship is not constant). In Figure 11.19, we illustrate spring hardening and softening effects, in addition to the traditional linear relationship with a constant stiffness, K. In these cases the stiffness may depend upon displacement, and the response will be nonlinear.

Adverse response due to nonlinearities may occur at conditions above or below that at which conventional instabilities (i.e., flutter) are predicted to occur. Such nonlinearities may lead to atypical dynamic response characteristics such as limit cycle oscillations (LCOs) which are common in the aeroelastic environment. Stall flutter and buzz are two other examples. The interested reader is referred to Dowell, Edwards and Strganac[5] for a more complete discussion; however, we present an overview of nonlinear aeroelasticity within this chapter and Chapter 12.

Physical sources of nonlinearities have been identified through mathematical models, wind-tunnel tests, and flight tests. Aerodynamic nonlinearities may include mechanisms such as flow separation due to flow viscosity, oscillating shocks due to flow compressibility, and may include both in the form of shock-boundary layer interaction. Large shock motions may lead to a nonlinear relationship between the motion of the structure and the aerodynamic loads that act on the structure. Separated flow, possibly due to the shock motion, may create a nonlinear relationship between structural motion and the aerodynamic flow field. Structural geometric nonlinearities are important and result from aerodynamic loads on the structure creating a response that is no longer linearly proportional to the load. Nonlinearities may arise from structural origins including, but not limited to, freeplay in the control linkage, joint slop in the structure, hardening or softening effects in the structural stiffness, and damping mechanisms. In addition, internal damping forces in a structure may have a nonlinear relationship to structural motion, with dry friction being an example. It is noted that structural damping is usually modeled empirically within linear aeroelastic mathematical models. The knowledge base of the fundamental mechanisms of damping and their impact aeroelastic response is sparse.

11.4.1 Divergence of the Nonlinear System

For discussion, we revisit the discussion of aeroelastic divergence. Recall, the previously developed equation for aeroelastic response,

$$(K - q_\infty S e C_{L_\alpha})\theta = q_\infty S e C_{L_\alpha} \alpha \tag{11.13}$$

The homogeneous portion of this equation, that is, $(K - q_\infty S e C_{L_\alpha})\theta = 0$ describes system stability, from which describes system stability, from which $\theta = 0$ represents an equilibrium state and from which $\frac{\partial}{\partial \theta}$ yields the aeroelastic stiffness, $(K - q_\infty S e C_{L_\alpha})$. This stiffness depends on flight conditions and the stiffness must be positive for a stable system. Divergence was provided by eq. 11.15 and occurs when the stiffness vanishes or $\frac{\partial}{\partial \theta} = 0$. Again,

$$q_{\infty, DIV} = \frac{K}{S e C_{L_\alpha}} \tag{11.15}$$

For convenience, eq. 11.60 may be rewritten as

$$(1-\bar{q})\theta = 0 \tag{11.61}$$

where a nondimensional dynamic pressure is introduced, $\bar{q} = \dfrac{q_\infty}{K/SeC_{L_\alpha}} = \dfrac{q_\infty}{q_{\infty_{DIV}}}$, and accordingly the aeroelastic stiffness is $(1-\bar{q})$ where $\bar{q} < 1$ represents a stable flight condition. Note that dynamic pressure has been normalized by the divergence pressure for the linear system.

Now, we introduce a nonlinear spring as

$$K_N = (1+a\theta^2)K \tag{11.62}$$

where $a > 0$ leads to spring hardening, $a < 0$ leads to spring softening and $a = 0$ leads to the constant stiffness, as depicted in Figure 11.19. Introducing the effect of eq. 11.62 into eq. 11.60 yields,

$$\left[K(1+a\theta^2) - q_\infty SeC_{L_\alpha}\right]\theta = 0 \tag{11.63}$$

or, in nondimensional terms,

$$(1+a\theta^2 - \bar{q})\theta = 0 \tag{11.64}$$

from which the equilibrium states are

$$\theta = 0, \pm\sqrt{\dfrac{\bar{q}-1}{a}} \tag{11.65}$$

Again, taking the derivative $\dfrac{\partial}{\partial \theta}$ of eq. 11.64 yields the aeroelastic stiffness $(1+3a\theta^2 - \bar{q})$, from which the system is stable if

$$\bar{q} < 1 + 3a\theta^2 \tag{11.66}$$

which suggests that, due to nonlinearity, divergence will be affected by displacement.

Equations 11.65 and 11.66 state that for the hardening case ($a > 0$) that multiple equilibrium states exist and divergence will occur at $\bar{q} > 1$. In other words, the dynamic pressure at which divergence occurs is higher than that for the linear system. Conversely, for the softening case ($a < 0$) multiple equilibrium states exist and divergence occurs at $\bar{q} < 1$ and, accordingly, the dynamic pressure at which divergence occurs is lowered in comparison to the linear system.

11.4.2 Aeroelastic Limit Cycle Oscillation (LCO)

An excellent illustration of a common effect of nonlinearities in the aeroelastic system is store-induced limit cycle oscillation. LCOs are not described by standard (linearized) aeroelastic analyses, and they may occur at flight predictions below those at which linear instabilities are predicted to occur. Although the amplitude of the LCO may be above structural failure limits, it is more typical that the presence of LCOs results in a reduction in vehicle performance, leads to airframe-limiting structural fatigue, and compromises the ability of pilots to perform critical mission-related tasks. Extensive and expensive flight tests for aircraft store certification are required. These LCOs have been documented[9,10] in the F-16 and F-18; and they are a characteristic of many high performance aircraft which carry external stores. AGARD[11] conducted a conference in which this behavior was the focus and, interestingly, at the time of the conference, this form of LCO response was referred to as "wings-with-store flutter" and "limited-amplitude flutter". In reality, the purist in the aeroelasticity community would agree that LCOs and flutter are closely related, and would define flutter as the simple solution of the linearized aeroelastic equations which possess many different response characteristics. Yet, the problems the store-induced LCOs present are the same today as the time of the conference – poorly modeled physics, unpredictability, performance limiting.

Another case of LCO follows the onset of flutter in plate-like structures[12], having a nonlinear stiffening due to tension induced by mid-plane stretching of the structure that arises from bending. This LCO is called panel flutter in which a very local section encounters flutter followed by LCO. It has occurred in rockets as well as on the X-15 and it is been recognized that low aspect ratio wings behave as structural plates and the entire wing may undergo plate-like flutter and LCOs. However, the phenomenon has not been clearly documented from flight, but in wind tunnel models and computational studies.

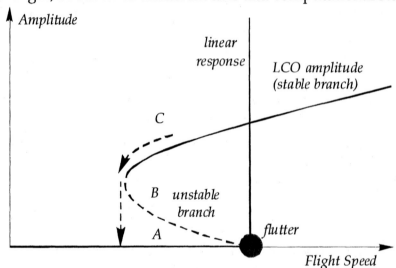

Figure 11.20 Subcritical Bifurcation

An LCO is characterized by many different features. Interestingly, although one may find them referred to as limited-amplitude flutter, suggesting a hazardous unstable situation, these LCOs are stabilizing in that the motion is bounded. Thus, the nonlinearity may eliminate catastrophic flutter. The trigger for the LCO is a sufficiently large disturbance that leads to a response of periodic, oscillatory motion with amplitude that is related to dynamic pressure (altitude, velocity and Mach no). Figures 11.20 and 11.21 illustrate two forms of LCO behavior with nonlinear characteristics. The figures present bifurcation characteristics of the system. In Fig. 11.20, we see 'subcritical' behavior in which the LCO occurs at flight conditions below flutter; this pattern is illustrative of store-induced LCOs. In both figures, the point at which flutter occurs is indicated and the vertical line emanating from that point confirms that amplitude rapidly grows to unbounded levels, as linear theory predicts. In addition, the figures also show branches associated with the amplitude of the LCO as dictated by the dynamics of the nonlinear system.

In Figure 11.20, three different levels of disturbances are shown – A, B and C:
- ◆ Disturbance level A is relatively small, and below the unstable branch (dotted line,

also referred to as the separatrix). Response to any such disturbance decays to equilibrium.

- ◆ Disturbance level B lies between the unstable branch and stable branch (solid line). Response to any such disturbance grows to the amplitude of the stable branch.
- ◆ Disturbance level C lies above the stable branch. Response to any such disturbance decays to the amplitude of the stable branch.

This case also illustrates the hysteretic behavior of subcritical behavior (see the dashed line from point B). That is, once the system is entrained in the LCO, the amplitude grows or decays with changes in flight speed, but does not return to zero amplitude (the zero line) until the flight speed drops below the lowest speed at which the stable LCO is present.

This phenomenon has been observed during flight tests. Consider an aircraft configured with wing stores in straight and level flight that does not experience LCO. However, an LCO is triggered when the vehicle is maneuvered (that is, during a windup turn) and the LCO persists when the vehicle returns to straight and level flight. Within the framework of a linear aeroelastic model, such behavior is not expected. Now, an explanation is offered based upon consideration of a nonlinear aeroelastic system. Considering the bifurcation diagram presented in the previous figure, if the disturbance is sufficiently small, LCO will not occur, but if sufficiently large, then LCO is induced. Once LCO exists, it will persist with the hysteresis even if one returns to the nominal original flight condition. Two different response states are possible at the same flight speed.

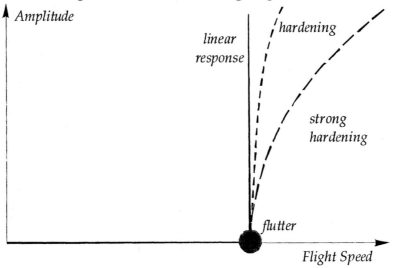

Fig. 11.21 Supercritical Bifurcation

In Figure 11.21, we see 'supercritical' behavior in which LCOs occur at flight conditions above flutter. Physical nonlinearities exist and one is asked to consider a structure that becomes stiffer with loads and/or a wing that no longer sees a growth in aerodynamic loads due to stall. These situations are illustrative of hardening effects. Two cases are illustrated to show that the amplitude of the LCO is governed by the strength of the nonlinearity. One is reminded that in linear analysis, stiffness is constant and stall is ignored. Again, flutter and the associated unbounded growth in amplitude are illustrated for this linear case. Thus, once the response of the system begins to grow, there are no physical mechanisms present to limit the growth.

The possible existence of LCOs is sensitive to changes in design parameters, and subtle differences – often within design specifications may be the difference in whether or not the LCO appears. As reported by the fight test community, there are cases in which two (nominally) identical vehicles flying the same path have different responses: one vehicle experiences LCOs but the other does not. Nonlinear analysis offers an explanation. Consider freeplay in the control linkage as one example; consider the inertial characteristics of an under-wing store as another. Although the two vehicles may have 'identical' measures of freeplay and/or 'identical' stores, there may be differences but within tolerances within regulatory standards. For one vehicle, the LCO amplitude might not be detected because it is too small. However, for the other vehicle, the LCO may be measurable. For example, from theoretical and experimental studies of freeplay,60 it is known that LCO amplitudes increase in proportion to the magnitude of the freeplay and that the magnitude of the angle of attack required to suppress LCO due to freeplay scales in proportion to the magnitude of the freeplay.

Denegri[13] suggests that the frequencies of LCO might be identified by linear flutter analysis; however, linear analysis fails to predict the oscillation amplitude or the onset velocity for LCO. No theory has been put forward to completely explain the mechanisms responsible for F-16 LCO. Denegri notes that although linear techniques have been used to predict the frequency of the LCO, linear analysis cannot consistently predict where within the flight envelope the onset of the oscillations occurs. Nonlinear analysis will be necessary.

Most of these LCOs are attributed to nonlinear aerodynamics due to shock wave motion and separated flow. However, there is the possibility that nonlinear structural effects involving stiffness, damping or freeplay are involved. An example is the wing-fold for storage of the carrier-based F-18. Much present day research and development effort is devoted to clarifying the basic mechanisms responsible for nonlinear flutter and LCO.

To summarize, both stable and unstable LCOs may exist, and may exist below the flutter boundary. In theory, a disturbance is required to trigger the nonlinear response and new steady-state motions of the system – that is, move from equilibrium toward a stable LCO and away from an unstable LCO. With stable LCOs, for any sufficiently small disturbance, the motion returns to the same LCO with time. Yet, if the disturbance is sufficiently large, may seek another LCO. Unstable LCOs are those for which any small disturbance will cause the motion to move away from the unstable LCO branch and move toward a stable LCO branch or equilibrium. There may be a hysteretic response as flight speed increases and then decreases. Beyond the flutter boundary, LCOs may arise due to some nonlinear effect, and typically the amplitude of the LCO increases as the flight speed increases beyond the flutter speed.

11.4.3 Residual Pitch Oscillation (RPO)

The residual pitch oscillation (RPO) involves an interaction of a flexible (wing) mode and the rigid-body degrees of freedom. For example, the B-2 bomber encountered a RPO during low-altitude, high-speed flight.[14] Neither the RPO nor any tendency of related response was predicted by (linear) analysis. For the B-2, the RPO involved a symmetric wing bending mode with a short-period rigid-body mode at a heavy, forward center-of-gravity configuration. To model the RPO that led to predictions similar to measurements, it was necessary to include the nonlinear response characteristic of the control surfaces. As reported, a lighter weight flight configuration did not exhibit RPO. Instead, damping increased with slight further increase in speed, typical of hump mode behavior.

11.4.4 Internal Resonance (IR)

Internal resonance (IR) occurs as a result of nonlinearities present in the system and leads to an exchange of energy between the system modes. Studies show that during IR the nonlinear modes exchanged energy over time, even in the presence of damping. It was also shown that in the presence of an external excitation, IR gives rise to coupling between the modes, leading to several motions including nonlinear periodic, almost periodic, and chaos. The amount of energy that is exchanged depends on the type of nonlinearity and the relationship of the linear natural frequencies.

IR exists when the linear natural frequencies of a system are commensurable, or nearly so, and the nonlinearities of the system provide a source of coupling. Commensurability means

$$m\omega_i \pm n\omega_j \approx 0 \tag{11.60}$$

where m and n are integers, and ω_i and ω_j are normal mode frequencies. Although an integer natural frequency ratio does not guarantee IR, it does form a necessary condition for IR. IR has been shown to exist in many systems, and its presence depends on the geometry, composition of nonlinearities, and boundary conditions.

During wind-tunnel tests by Cole[15] intended to verify the aeroelastic stability of a new wing design, an unexpected flutter-type response occurred at dynamic pressures much lower than analysis had predicted. It is noteworthy that predictive tools based on linear theory were used. For the physical structure, the natural frequency of the second bending mode of the wind-tunnel model was slightly more than twice the natural frequency of the first torsional mode. However, because frequencies in an aeroelastic system depend on the aerodynamic loads, a system's frequencies may be tuned as the velocity changes. In Cole's experiments, a resonance-type condition may have been reached before linear flutter conditions. Consequently, it was considered that the inaccurate predictions were due to the limitations presented by the use of linear theory.

Gilliatt et al.[16] and Chang et al.[17] studied the possible presence and effects of internal resonances in aeroelastic systems. Gilliatt et al. in particular, were motivated by the experimental findings of Cole. The two-degree-of-freedom model of the O'Neil et al.[18] research was a basis for the study, and the aerodynamic model was extended to include stall effects, which introduced strong cubic nonlinearities into the equations of motion. The system parameters were selected to permit the aeroelastic frequencies to pass through a 3:1, 2:1, and 1:1 ratios as the flowfield velocity was increased. Gilliatt et al. found that the presence of cubic nonlinearities in the aeroelastic system led to a 3:1 IR. Thompson and Strganac[19] explore several aspects of internal resonance as it may pertain to the aeroelastic environment, including the possibility of store-induced IR mechanisms.

In an attempt to explain the unexpected experimental results, Oh, et al.[20] developed an experiment to examine the structural dynamic behavior of Cole's experiments. These experiments were conducted in the absence of any aerodynamic loads. They determined, theoretically and experimentally, the linear natural frequencies and the mode shapes, and also experimentally showed that an antisymmetric vibration mode of a cantilever metallic plate was indirectly excited by a 2:1 IR mechanism. The IR was present because the natural frequency of the second bending mode was nearly twice the natural frequency of the first torsional mode. Their experiment consisted of a base excitation being applied to a cantilevered plate with the same aspect ratio as Cole's wing. However, in this study the second bending mode was excited by a shaker rather than by aerodynamic forces.

Using flight-test data, Stearman et al.[21] studied resonances in aeroelastic systems and showed that both combination-type and parametric resonances can occur. These resonances occurred if

$$\Omega_f \approx \frac{2\omega_n}{k} \approx \frac{|\omega_i \pm \omega_j|}{k} \tag{11.61}$$

where k is an integer, Ω_f is the frequency of external forcing function, and ω_n are normal mode frequencies. Equation 11.60 is similar to eq. 11.61, but external forcing functions are not present.

11.5 SUMMARY

In this chapter we have attempted to summarize theory and analysis considerations for one of the most complex of all aeronautical subjects. Trying to extract the appropriate level of this theory for test engineers, test planners, and test pilots is a daunting task. There are far too many issues facing flight test teams that conduct this kind of envelope expansion to cover all theory that will be needed. Our attempt to shrink the available theory into under 40 pages is likely to be incomplete for many readers and incomprehensible to others. Nonetheless, based on feedback from many students in the last 20 years of teaching this material in both standard undergraduate/graduate classes and continuing education short courses, the hope is that our pragmatic approach is sufficient background for the test community. With the theoretical considerations spelled out in this chapter, we are ready to move on to the preparations necessary to conduct aeroelastic investigations in the air.

REFERENCES

1. Bisplinghoff, R. L., Ashley, H., and Halfman, R. L., **Aeroelasticity**, Addison-Wesley, Massachusetts, 1955.
2. Fung, Y. C., **An Introduction to the Theory of Aeroelasticity**, John Wiley and Sons, New York, 1955.
3. Dowell, E. H., Clark, R., Cox, D., Curtiss, H. C. Jr., Edwards, J. W., Hall, K. C., Peters, D. A., Scanlan, R., Simiu, E., Sisto, F., and Strganac, T. W., **A Modern Course in Aeroelasticity**, Kluwer, Massachusetts, 2004.
4. Hodges, D.H. and Pierce, G.A., **Introduction to Structural Dynamics and Aeroelasticity,** Cambridge Aerospace Series, Cambridge, UK, 2002.
5. Dowell, E., Edwards, J., and Strganac, T.W., "Nonlinear Aeroelasticity", *Journal of Aircraft*, Vol. 40, No. 5, 2003, 857-874.
6. Theodorsen, T. and Garrick, I. E., "Mechanism of Flutter -- A Theoretical and Experimental Investigation of the Flutter Problem," NACA TR-685, 1940.
7. "Introduction to Flutter of Winged Aircraft," von Karman Institute for Fluid Dynamics, Lecture Series 1992-01, von Karman Institute, 1992.
8. Hassig, H. J., "An Approximate True Damping Solution of the Flutter Equation by Determinant Iteration", AIAA Journal of Aircraft, Vol. 3, No. 11, Nov. 1971, pp. 885-889.
9. Denegri, C. M., Jr., "Limit-Cycle Oscillation Flight Test Results of a Fighter with External Stores," *Journal of Aircraft*, Vol. 37, No. 5, 2000, pp. 761-769.
10. Bunton, R. W., and Denegri, C. M. Jr., "Limit-Cycle Oscillation Characteristics of Fighter Aircraft," *Journal of Aircraft*, Vol. 37, No. 5, 2000, pp. 916-918.
11. AGARD Specialists Meeting on Wings-with-Stores Flutter, 39th Meeting of the Structures and Materials Panel, CP 162, AGARD, Oct. 1974.
12. Dowell, E. H., **Aeroelasticity of Plates and Shell**, Kluwer Academic, Norwell, MA, 1975.
13. Denegri, C.M., Jr., "Correlation of Classical Flutter Analyses and Nonlinear Flutter Responses Characteristics," International Forum on Aeroelasticity and Structural Dynamics, 1997, pp. 141–148.
14. Dreim, D. R., Jacobson, S. B., and Britt, R. T., "Simulation of Non-Linear Transonic Aeroelastic Behavior on the B-2," NASA CP-1999-209136, CEAS/AIAA/ICASE/NASA Langley International Forum on Aeroelasticity and Structural Dynamics, 1999, pp. 511–521.
15. Cole, S. R., "Effects of Spoiler Surfaces on the Aeroelastic Behavior of a Low-Aspect Ratio Wing," 31st AIAA Structures, Structural Dynamics, and Materials Conference, AIAA, Washington, DC, 1990, pp. 1455-1463.
16. Gilliatt, H. C., Strganac, T.W., and Kurdila, A. J., "Nonlinear Aeroelastic Response of an Airfoil," 35th Aerospace Sciences Meeting and Exhibit, AIAA, Reston, VA, 1997, pp. 258-266.

17. Chang, J. H., Stearman, R. O., Choi, D., and Powers, E. J., "Identification of Aeroelastic Phenomenon Employing Bispectral Analysis Techniques," International Modal Analysis Conference and Exhibit, Vol. 2, 1985, pp. 956–964.

18. O'Neil, T., Gilliatt, H., and Strganac, T., "Investigation of Aeroelastic Response for a System with Continuous Structural Nonlinearities," AIAA Paper 96-1390, 1996.

19. Thompson, D. E., and Strganac, T. W., "Store-Induced Limit Cycle Oscillations and Internal Resonance in Aeroelastic Systems," AIAA Paper 2000-1413, 2000.

20. Oh, K., Nayfeh, A. H., and Mook, D. T., "Modal Interactions in the Forced Vibration of a Cantilever Metallic Plate," Nonlinear and Stochastic Dynamics, Vol. 192, 1994, pp. 237–247.

21. Stearman, R. O., Powers, E. J., Schwarts, J., and Yurkorvich, R., "Aeroelastic System Identification of Advanced Technology Aircraft Through Higher Order Signal Processing," 9th International Modal Analysis Conference, 1991, pp. 1607–1616.

Chapter 12

AEROELASTICITY: GROUND AND PRE-FLIGHT PREPARATION

The Federal Aviation Administration requires aeroelastic stability evaluations, including considerations of flutter, divergence, control reversal and any undue loss of stability and control as a result of structural deformations. Compliance must be shown by analyses, wind tunnel tests, ground vibration tests (GVT), and flight tests. Full scale flight flutter tests must be conducted for new designs and modifications to existing designs (unless the modification is proven to be insignificant). The reader is referred to Part 23 - Normal, Utility, Acrobatic, and Commuter Category Airplanes[1] and Part 25 - Transport Category Airplanes[2] for detailed discussion of the standards. In a similar manner, military specifications[3] detail design requirements to prevent static and dynamic aeroelastic instabilities.

The methods of aeroelastic analysis described in Chapter 11 provide the conditions for the onset of the instability including altitude, Mach number, velocity, and dynamic pressure at which an instability occurs, as well as frequency and damping trends that provide valuable tracking information. In this chapter, we describe the use of analytical results, ground based wind tunnel and ground vibration tests, and regulatory information to establish an effective strategy to quantify the aeroelastic characteristics of the flight vehicle.

12.1 THE FLIGHT ENVELOPE

In Fig. 12.1, a flight operating envelope is shown with an expanded margin that must be cleared. In addition, a boundary is shown which represents a potential flutter instability. In Fig. 12.2, an altitude-Mach operating envelope with a potential flutter boundary is shown. The boundaries in these two figures are indicative of the low altitude, transonic, high dynamic pressure nature of flutter. Yet, flutter and other aeroelastic instabilities are not limited to this regime. Flight test procedures to substantiate the location of potential instabilities within the flight envelope will be addressed in Chapter 13.

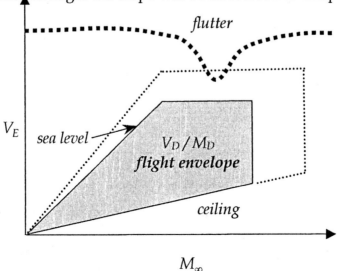

Fig. 12.1 The V_D/M_D Operating Envelope

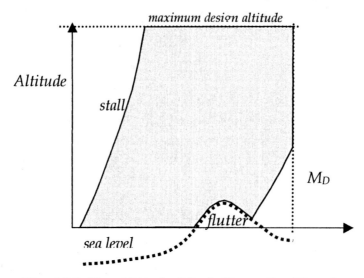

Fig. 12.2 The Altitude-Mach Operating Envelope

12.2 FAR ADVISORY CIRCULARS: PARTS 23 AND 25.

As detailed in the Advisory Circulars[1,2], velocity-damping (V-g) and velocity-frequency (V-ω) diagrams are approved approaches used to examine the results of flutter analyses. Using the approach described in Chapter 11, the complex eigenvalues (roots) are found for increasing values of reduced frequency and translated to velocity.

12.2.1 *V-g* Diagrams

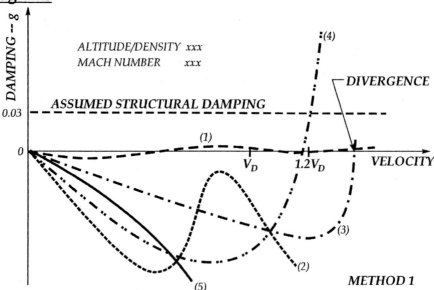

Fig. 12.3 *V-g* Diagram for a Five-Mode Analysis (adapted from AC 25.629-1A[2])

In Fig. 12.3, a velocity-damping diagram is sketched showing the damping roots for an aircraft represented by 5 structural modes. Of particular interest is the root that leads to the negatively damped system (root 4 in the sketch), indicating a growing motion. However, structural damping occurs naturally, improves the aeroelastic stability characteristics and, as a consequence, delays the onset of flutter. Two methods to address damping are provided in Reference 2.

Chapter 12 Aeroelasticity: Ground Test and Pre-flight Preparation

In Method 1, structural damping is not available from measurements. In the absence of measured damping, a structural damping factor of $g = 0.03$ is assumed to be present in the structure. It is noted that structural damping is twice the value of viscous damping, and $g = 0.03$ is a conservative assumption of structural damping. The presence of structural damping eliminates the apparent instability indicated by the soft flutter crossing ("hump mode") associated with root 1. However, as one examines the behavior of root 4 at $g = 0$ and $g = 0.03$, one sees that the velocity at the crossing is affected very little by the presence of damping. The flutter velocity is identified as the value at this crossing. However, since the velocity associate with the flutter crossing is associated with an altitude and Mach number, consideration must be given to the "matched-point" solution. As described in Chapter 11, "matched-point" analysis assures the user that the results are consistent with the flight environment within the atmosphere.

Per the FAR advisory circulars and the military specifications, *Aircraft require a 15% margin of safety on design dive speed, V_D, for flutter clearance* (Fig. 12.3). V_D is the design velocity to which a 15% margin as applied. The flutter crossing of root 4 represents $1.15V_D$. Other than the exact crossing from stable to unstable, the path of the roots should not be used to infer stability behavior as the solution strategy is only correct at zero damping (sinusoidal behavior assumed, which is only satisfied when the motion neither grows nor decays). The authors acknowledge that there are some approaches in which the V-g diagrams do provide an indication of true damping characteristics, but the user is advised to be certain of this method. Aeroelastic divergence is characterized by a root that abruptly terminates at zero damping -- a mode indicated by root 3 in Fig. 12.3. This characteristic will be substantiated by observing that the frequency of that mode approaches zero.

Method 2 uses measurements of damping made available through the GVT, but a minimum of $g = 0.02$ may be used to assure a conservative velocity. It is noted that structural damping manifests itself as a vertical shift of the velocity axis in the V-g diagram -- a consequence of the solution methodology.

12.2.2 FAR Flight Envelopes

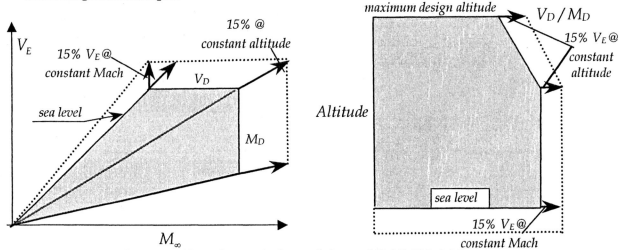

Fig. 12.4 Envelopes (adapted from AC 25.629-1A[2]).

Figure 12.4 illustrates two standard envelopes prescribed by FAR standards. A 15% margin on equivalent airspeed, V_E, is required at constant altitude and at constant Mach number. This latter case results in clearance below sea level and, although flight at this altitude is obviously not feasible, this boundary is necessitated by the low-altitude, high dynamic pressure variables that most affect flutter.

12.2.3 Velocity-Frequency Diagrams

In Fig. 12.5 a velocity-frequency (V-ω) diagram is sketched showing the frequency roots for an aircraft represented by 5 structural modes. The natural frequencies such as those measured through a GVT of the aircraft are the frequencies at zero velocity, and the affect of aerodynamic loads on these frequencies is seen as a function of velocity. Although flutter can not be determined solely from this figure, aerodynamic-sensitive modes tend to coalesce as the flutter velocity is approached. Aeroelastic divergence is noted for the mode that approaches zero frequency.

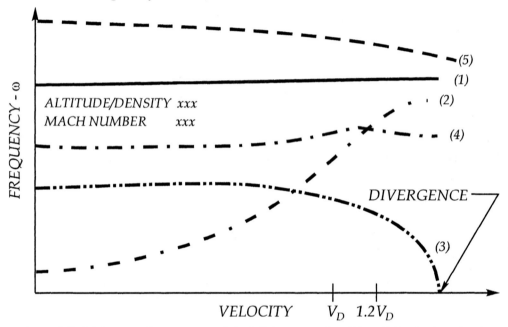

Fig. 12.5 Velocity-Frequency (V-ω) Diagram for Five-Mode Analysis. (adapted from AC 25.629-1A[2])

Accelerometers mounted to the aircraft should be used to monitor frequencies, and these frequencies may be compared with the velocity-frequency diagram for correlation during ground and flight tests.

12.3 GROUND VIBRATION TESTS (GVT)

Ground vibration tests (GVTs) are used to identify natural frequencies and mode shapes of the aircraft structure. Kehoe[4] and Norton[5] provide a thorough discussion of a GVT program in support of flight tests. Albeit a general theory of mechanical vibrations is not presented herein (the interested reader is referred to Craig[6]), a succinct overview is provided.

12.3.1 Vibration Characteristics - Frequencies and Modes

As discussed in the previous chapter, the equations of motion for the general mechanical system with mass, damping, and stiffness may be represented as

$$M\ddot{x} + C\dot{x} + Kx = F_t \tag{12.1}$$

where M, C, and K are parameters associated with the physical system, F_t is external forcing, and x is the response. If two of three components – the physical parameters (M, C, and K), the external forcing (F_t), or the response (x) -- are known, then the third component may be found.

In analysis, the physical parameters and external forcing are known, and the response is simulated. Frequencies and associated mode shapes may be directly derived from an appropriate model of the system. For example, consider eq. 12.1 in which damping and external forces are absent,

$$M\ddot{x} + Kx = 0 \quad (12.2)$$

This equation may represent a significantly large system. For such a system, we assume harmonic motion $x = \Theta e^{\pm i\omega t}$ where Θ represents the amplitude of motion. With substitution,

$$\left(-\omega^2 M + K\right)\Theta = 0 \tag{12.3}$$

Equation 12.3 may be written as

$$\left(-\omega^2 I + M^{-1}K\right)\Theta = 0 \tag{12.4}$$

where I is the identity matrix, or

$$\omega^2 \Theta = M^{-1}K\Theta = A\Theta \tag{12.5}$$

Therefore, given M and K from a model of the system such as that derived from a standard finite element model, the matrix A is formed from which the natural frequencies, ω_n, and mode shapes, Θ, are determined. In practice, although the model M, K, and A may be very large, one is interested in a much smaller set limited to the lower frequencies and associated mode shapes. Note, although this smaller set is necessitated by mathematical limitations associated with extraction of eigenvalues from A, the physics of interest is captured. Physically, the higher modes are not required since these modes do not participate in the flutter mechanism.

In Chapter 11 we discussed modal methods for aeroelastic analyses and showed that the aeroelastic system is represented by a set of mode shapes. Analytically, eq. 12.5 provides the natural frequencies, ω_{n_i}, and mode shapes, Θ_i, for $\underline{i} = 1$ to m of the system, where m represents the number of degrees of freedom chosen to model the system. And, we note this set is significantly smaller than the original finite element model.

12.3.2 The Modal Survey

A modal survey, conducted as part of the GVT, provides direct measurements of the natural frequencies, associated mode shapes, and measures of structural damping for the primary modes for the configuration. These measured modes are used to validate analysis or may be used to develop a model for predictions. It is worth noting that, as a matter of practice, one should compare GVT measurements with numerical results such as those derived from eq. 12.5 to provide verification and validation of both approaches.

In the GVT, the response to a known external forcing is measured, from which the physical parameters are derived. In a GVT, as depicted in Fig. 12.6, an accelerometer (or strain gage, if appropriately located) is used to measure response due to an input. This forcing input may be from a shaker, providing a sinusoidal sweep over a range of frequencies or a random (or pseudo-random) specta of frequency content. The forcing may be from an impact hammer, giving an impulse that captures a broad range of frequency content. The fundamental wing modes found with a modal survey are illustrated in Fig. 12.7.

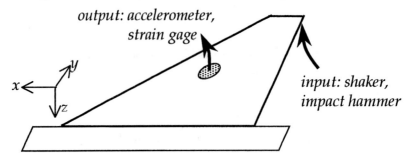

Fig. 12.6 GVT Measurements – Input and Output

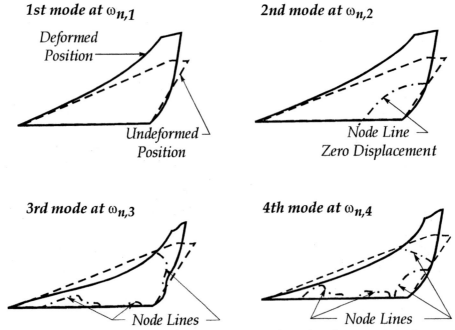

Fig. 12.7 An Illustration of the Fundamental Wing Modes

Several supporting comments are in order:

(1) A GVT may use a single accelerometer from which measurements are obtained at many locations on the structure, or the GVT may use an array of many accelerometers to provide simultaneous measurements. In either case, for a set of natural frequencies, a mode shape is derived from these measurements at each frequency.

(2) The "reciprocity" principle suggests that the source of excitation may be placed at any location leading to equivalent modal characteristics. However, node lines (as illustrated in Fig. 12.7) are lines of zero displacement and, consequently, excitation along these node lines may not excite that specific mode. Accordingly, the GVT should be conducted at more than one input location to confirm identification of all modes. For example, the second mode of Fig 12.7 may be missed if the shaker were positioned along the node line.

(3) The GVT identifies modes that should be tracked during flight tests. The GVT will identify the optimal placement of accelerometers during flight tests. Typically, a limited number of accelerometers are be used during flight tests to track modes identified during the GVT.

(4) Aircraft possess both symmetric and asymmetric mode shapes, as illustrated in Fig. 12.8. Accelerometers placed on each wing will distinguish the nature of the symmetry. Also, higher modes should be identified.

(5) Strain gages may be present from prior tests (airframe loads, etc) and may be used to measure vibratory response; however, the location may not be optimal and signal to noise ratios may be too low.

(6) The underlying theory associated with the GVT assumes linear behavior. Thus, sources of nonlinearities such as freeplay, sources of damping, and excess compliance in the structure must be identified and/or minimized. Linearity may be verified through static load-deflection tests. Preloading of the structure will eliminate freeplay. Partial deflation of tires and landing gear struts during the GVT will help distinguish rigid-body modes from flexible-body modes.

(7) Mode shapes exhibit the property of orthogonality. Mathematically speaking, measurements should behave as discussed in Chapter 11, that is, the modal mass is of the form

$$\Theta^T M \Theta = M_M \tag{11.51}$$

or, since modes may be scaled such that they are orthonormal with respect to mass,

$$\Theta^T M \Theta = M_M = I \tag{11.52}$$

where I is the identity matrix. And, with these orthonormal modes, one finds

$$\Theta^T K \Theta = \Omega \tag{11.53}$$

where Θ is a transformation matrix whose columns are comprised of measured mode shapes and Ω is a diagonal matrix consisting of the frequencies, ω_n^2.

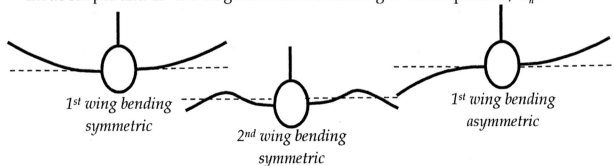

Fig. 12.8 Vibration Modes

Thus, orthogonality should be verified to validate the quality of measurements. The off-diagonal terms in the matrices resulting from the operations in eqs. 11.52 and 11.53 should be two orders of magnitude less than the terms on the diagonal.

Of primary interest for flight test applications is the use of these measured modes to observe, in real-time, changes of vibration characteristics -- that is, monitor aerodynamically sensitive modes that may lead to aeroelastic instabilities.

12.3.3 The Frequency Spectrum

The frequency of the input also affects the response of the modes of interest. At a specific test point, output data are obtained and analyzed for frequency and damping characteristics. Measurements from devices such as accelerometers are often in the form of a response time history. Typically, 60 seconds of data are obtained at a specific test point. Given an adequate sampling rate and sufficient measurements, the data are transformed (normally by Fast Fourier transforms) to the frequency domain. Figure 12.8 provides a simplified illustration of the frequency response derived from transformation of time domain measurements, and shows the amplitude of response as a function of frequency.

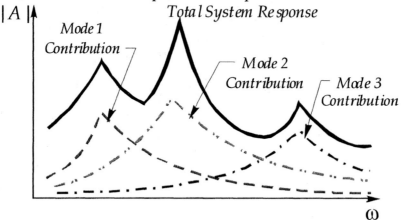

Fig. 12.9 Typical Frequency Response

Several comments are appropriate as one considers the frequency response behavior depicted in Fig. 12.9.

(1) The peaks associated with the system frequency response indicate the aeroelastic frequencies of the system, and these peaks may be observed to shift with changes in test conditions (airspeed, Mach no., altitude). For the zero velocity case, these peaks represent the natural vibration frequencies of the aircraft. As flutter is approached, two or more of these peaks will merge toward a common peak and te amplitude of response will rise. As aeroelastic divergence is approached, one of these peaks will migrate toward zero frequency.

(2) Forced excitation from any external sources (such as vortex shedding, actuator movement, oscillating shocks, and other sources with an embedded frequency content) at an aeroelastic (natural) frequency will result in large amplitude response.

(3) The response of the structure is the superposition of all modes. The magnitude of response is the additive affect of each mode -- for example, although the first mode (lowest frequency) dominates the response at that frequency, contributions from higher modes are present.

(4) During tests using mechanical excitation methods, a sinusoidal ("sine") sweep through a range of frequencies will identify aeroelastic frequencies indicated by peak amplitudes; the "sine dwell" at specific frequencies will permit examination of specific modes.

(5) The random input -- from turbulence or mechanical excitation -- will excite all modes and frequencies simultaneously. The aeroelastic frequencies will be identified by the peak response. The response from an impact hammer be similar.

(6) Structural (no-flow) and aeroelastic damping can be identified through the sharpness of the frequency response. The "half-power" method, for example, is one technique used to derive damping from the frequency response spectrum.[10]

(7) Modal characteristics must be understood (almost always based on GVTs) prior to flight tests. The presence of rigid body motion and external vibratory sources will be observed in the frequency response, and these effects must be identified.

12.3.4 Determination of Damping

Typically, models of vibration satisfy assumptions of viscous damping and exponential motion. Accordingly, the accelerometer output is used to measure damping as well as frequency. Figure 12.10 illustrates the time history of structural response, showing frequency with a rate of decay. The "log-dec" (logarithmic-decrement) and half-amplitude methods[10] are used to derive damping from the decay rate. Both methods assume exponential decay.

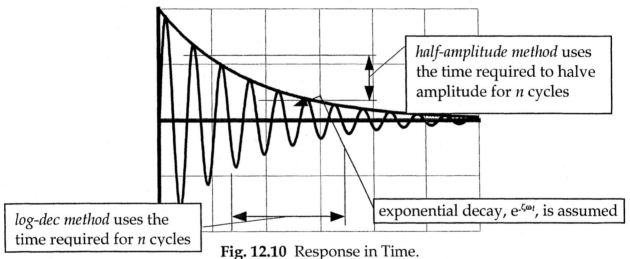

Fig. 12.10 Response in Time.

The *log-dec method* uses the time required for the system to decay an integer number of cycles to derive the damping,

$$\zeta = \frac{\delta}{2\pi n} \tag{12.6}$$

where ζ is damping factor, n is the integer number of cycles, and δ is a ratio of the amplitudes,

$$\delta = ln\left(\frac{Amp_t}{Amp_{t+n\,periods}}\right) \tag{12.7}$$

The *half-amplitude method* uses the time required for the amplitude to halve itself. Accordingly,

$$\delta = ln\left(\frac{Amp}{Amp/2}\right) = 0.693 \tag{12.8}$$

from which damping can be found

$$\zeta = \frac{\delta}{2\pi n} = \frac{0.693}{\omega t} \tag{12.9}$$

where ω is the frequency and $\omega t \equiv 2\pi n$.

For low frequency response such as that one might expect for the fundamental frequency associated with bending vibration of the wing for a transport category aircraft (on the order of 3 Hz), observations of the time response quickly provide estimates of damping. One merely needs to observe amplitudes, time, and cycles. history is often used for test points far-removed from the flutter boundary.

Alternatively, the time history from the accelerometer may be transformed into the frequency domain, from which damping may be extracted by the *half-power method*. As seen in Fig. 12.11, the vibration signal is a sinusoidal wave. The spectral density of this wave is proportional to the power within the wave per unit frequency, referred to as the mean square spectral density (*MSSD*) of the signal which is also called the power spectral density (*PSD*) in some circles. Figure 12.10 illustrates the *MSSD* derived from the accelerometer output. Figure 12.9 is representative of the *MSSD* for a multi-mode system.

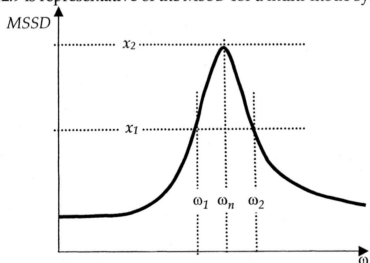

Fig. 12.11 Mean Square Spectral Density (MSSD) of Accelerometer Output

The damping is equal to the frequency bandwidth taken at the half-power point divided by the frequency of the mode,

$$\zeta = \frac{\omega_2 - \omega_1}{2\omega_n} \tag{12.10}$$

where ω_1 and ω_2 are the frequencies associated with the half-power points located at amplitude, x_1, and where $x_1 = \frac{\sqrt{2}}{2} x_2$.

12.4 WIND TUNNEL TESTS

Wind tunnel tests provide intermediate information to qualify design concepts prior to full scale testing. Such tests substantiate flutter margins by validating analyses. In Chapter 11, the requirement to satisfy similarity parameters identified through dimensional analysis was addressed. These similarity parameters capture effects of elastic and inertial prop-

Chapter 12   Aeroelasticity: Ground Test and Pre-flight Preparation

erties, geometry, and flight conditions, and included Mach number, reduced frequency, frequency ratio, mass ratio, and flutter speed index. If dynamic similitude is present, 7experimental and analytical results may be compared with a degree of confidence.

Wind tunnel test techniques share common features with flight tests in flutter identification. A major difference between wind tunnel test procedures and atmospheric flight test procedures is the control of the test environment available in the wind tunnel. The temperature, Mach number, velocity, and density may be controlled in the wind tunnel. In atmospheric tests, nature dictates the dependency between the altitude (density) and air temperature (speed of sound). This dependency creates the need for the Matched Point solution to properly interpret the analytical results.

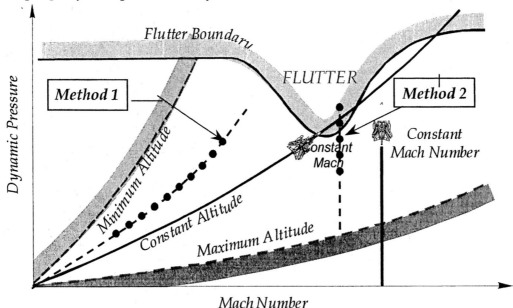

Fig. 12.12 The Flutter Boundary

Since environmental control is possible in the wind tunnel test scenario, all but one of the test parameters may be held constant when identifying the flutter boundary. Typically, two methods are employed to establish the flutter boundary – such as critical dynamic pressure versus Mach number – in the wind tunnel. The methods are illustrated in Fig. 12.12. In the first method, the dynamic pressure is increased through the use of the fan speed to increase the flow velocity for a constant static pressure (this is analogous to maintaining constant altitude while increasing aircraft velocity during flight tests). Mach number and dynamic pressure both increase with this method. In the second method, the Mach number is maintained and the total pressure is increased in the wind tunnel (this is analogous to maintaining Mach number while decreasing altitude during flight tests).

A sequence of test points is established along the path associated with each method. At each test point, conditions (velocity, pressure, Mach) are maintained and responses are measured that include frequency, damping, and magnitude. Trends are tracked and correlation is made with predictions similar to those depicted in Figure 12.5. In the next section, we describe some approaches to extrapolate this information to establish the boundary. It is pointed out that a test point along the constant Mach line can be maintained in the wind tunnel since conditions can be held constant; however, in a flight test, the constant

Mach approach requires a dive which means that density and dynamic pressure is increasing in a continual manner. Care must be taken not to monitor conditions. In the next chapter, procedures for tests that require dives are discussed.

The flutter boundary in Fig. 12.12 is shown with the classical "flutter bucket" where the minimum dynamic pressure (or "transonic dip") for flutter occurs in the transonic regime. Subsonic and supersonic regimes routinely have nearly constant flutter dynamic pressures.

An initial test is planned to identify a flutter-free path throughout the Mach number range. The speed of the flow increased well beyond the anticipated transonic dip of the flutter bucket. As stated previously, this method increases both the dynamic pressure and the Mach number, and this method resembles the constant altitude flight test. (Note: conducting a constant altitude flight test at the maximum flight altitude typically identifies this flutter-free path.) Next, the speed is decreased and the tunnel pressure (stagnation) is increased to provide a higher dynamic pressure for this same Mach number, and this method resembles a constant Mach dive maneuver to a lower altitude. Then, at this pressure (altitude) the speed of the flow is increased until the instability is detected establishing a point on the flutter boundary. This process is repeated several times to complete the description of the flutter bucket *on the subsonic side of the transonic dip*.

The flutter boundary beyond the transonic dip is found by significantly decreasing the tunnel pressure (returning to conditions associated with the original flutter-free path) and, then, increasing velocity to obtain higher Mach numbers. At this higher Mach number, the pressure is increased at constant Mach number. We observe that, at points on the flutter boundary beyond the transonic dip, **a decrease in velocity** may cause onset of flutter.

Therefore, in wind tunnel tests, instead of reducing fan speed, total pressure is dropped to avoid triggering instability by entering the transonic dip from supersonic conditions. In flight tests, this drop is analogous to an increase in altitude (requiring a climb) at constant Mach Number. Although these methods establish points on the flutter boundary, it is not desirable to actually test on the instability boundary (that is, "hit" a hard flutter point) due to the destructive nature of flutter. Therefore, subcritical response techniques are used in both wind tunnel and flight tests. As suggested by the name, these methods examine the dynamic response of the structure prior to the onset of the instability. This examination includes frequency, damping, and/or amplitude trends for increases in airspeed, and an extrapolation to actual flutter conditions. The structure or aircraft must be spared.

12.5 SUBCRITICAL RESPONSE TECHNIQUES

Wind tunnel tests and flight tests are required to establish precise flutter boundaries for the vehicle or component. Yet, since flutter is destructive by nature, testing at, or even near, the boundary is highly undesirable. Therefore, subcritical response techniques are employed to predict the onset of instabilities by extrapolating damping and frequency trends. These techniques require the use of accelerometers and/or strain gages. These instruments are mounted on the critical components so measurements of the modal characteristics can be recorded and monitored in "real time".

Modal parameters of primary interest include the frequency, damping, and amplitude characteristics. Our development of aeroelastic theory in Chapter 11 shows that shifts in frequency and damping parameters identify the presence of aeroelastic instabilities. We

Chapter 12   Aeroelasticity: Ground Test and Pre-flight Preparation   89

must take advantage of these characteristics to predict where the instability exists based upon measured response at subcritical conditions.

Two subcritical response methods are described that may be used to predict flutter characteristics: the Power Spectrum Density (PSD) method and the Peak-Hold method (see Cole[7]). Both methods use the frequency response data as shown in Fig. 12.11 as measured for the contributing modes. This measurement must be made at stabilized flight conditions -- constant dynamic pressure, altitude, and Mach number -- and measurements are repeated for several conditions (refer to the test points in Fig. 12.12).

The PSD method uses the dynamic response which has a peak for each structural mode which yields the structural damping. As shown in eq 12.11, and related discussion, structural damping is equal to the frequency bandwidth taken at the half-power point divided by the frequency of the mode. Damping is tracked for increases in flight conditions. The flutter point is determined by extrapolating the structural damping. A minimal acceptable damping of 1.5% correlates to the advisory circulars. (Note, structural damping, g, is twice that of the viscous damping, ζ, of eq. 12.11).

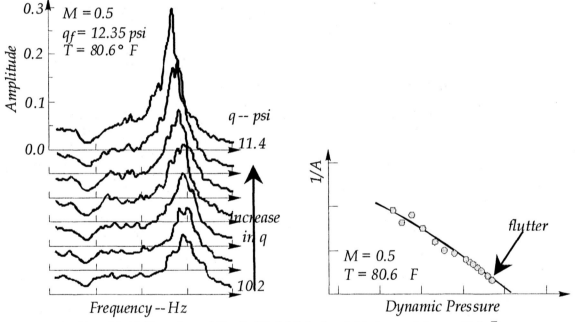

Fig. 12.13 Results of Peak-Hold Method (Adapted from Cole[7])

An inspection of the responses in Fig. 12.13 reveals a broad peak for the vibration of particular modes. This broad peak suggests well-damped modes. As the dynamic pressure is increased, the response appears as a narrow peak. The inverse of the maximum amplitude is plotted versus the dynamic pressure (right side of Fig. 12.13). As the dynamic pressure is increased towards the critical dynamic pressure, the inverse of the maximum amplitude approaches zero. This inverse is extrapolated to this zero value which corresponds to the flutter dynamic pressure. Changes in frequency as noted by the shift in the peak, and changes in damping as noted by the width of the peak, also provide indicators that may be used to predict the onset of the instability.

12.6 SUMMARY

In this chapter we attempted to connect the fundamental theory in Chapter 11 with the flight test program which will be examined in Chapter 13. The FAR standards for both general aviation and transport category airplanes are reviewed, showing the "to-be-cleared" flight envelope that provides the required margins of safety for a certified aircraft. Since aeroelastic stability is characterized by specific changes in vibrational characteristics, ground vibration tests quantify frequency and damping responses, which prove invaluable as the envelope expansion proceeds. Furthermore, prior to flutter flight tests, wind tunnel investigations are useful to identify potential instabilities. Finally, it is emphasized that, due to the explosive, destructive nature of flutter, subcritical methods are used to identify the boundary.

REFERENCES

[1] ACE-100, "Means of Compliance with Title 14, Part 23, Section 23.629, Flutter," Advisory Circular No. 23.629-1B, U. S. Department of Transportation FAA, September 2004.

[2] ANM-114, "Aeroelastic Stability Substantiation of Transport Category Airplanes," Advisory Circular No. 25.629-1A, U. S. Department of Transportation FAA, July 1998.

[3] MIL-A-8870C, Military Specification, "Airplane Strength and Rigidity Vibration, Flutter, and Divergence", March 1993.

[4] Kehoe, M. W., "Aircraft Ground Vibration Testing At The NASA Ames-Dryden Flight Research Facility," NASA TM-88272, NASA Ames Research Center, Dryden Flight Research Center, California, July 1987.

[5] Norton, W. J., "Structures Flight Test Handbook," AFFTC-TIH-98-02, Air Force Flight Test Center, Edwards AFB, California, 2nd Edition, May 2002.

[6] Craig, R. R., Jr., **Structural Dynamics - An Introduction to Computer Methods**, John Wiley and Sons, New York, 1981.

[7] Cole, S. R., "Exploratory Flutter Test in a Cryogenic Wind Tunnel," AIAA Paper No. 85-0736, Florida, American Institute of Aeronautics and Astronautics, 1985.

Chapter 13
AEROELASTICITY: FLUTTER FLIGHT TESTS

Flutter flight test holds a place of prominence among the diverse flight test disciplines. Preparing to embark on a multi-year flutter program, a flight test mentor remarked, "High-alpha and spin testing will win you a Kincheloe award. Flutter testing will get you killed[1]." Conceptually, it explores the most complex physical phenomena experienced by an air vehicle, weaving unsteady aerodynamics, structural dynamics and flight controls. Programmatically, flutter paces all the others disciplines of any large flight test effort. For many aircraft programs, flutter testing is almost synonymous with envelope expansion, with the loads, flying qualities, engine operability, performance, and weapons separation efforts waiting impatiently for the flutter team to release precious envelope in which they can then go to work. From the perspective of safety and risk management, flutter testing ranks at the top because of the likelihood for severe airframe damage (Fig. 13.1) or abrupt, catastrophic loss of both test pilot and test aircraft.

Fig. 13.1 Flutter Damage Sustained During E-6A Envelope Expansion[2]
(courtesy U.S. Navy/Boeing Company)

Part 25 of the Federal Aviation Regulations for Transport Category Airplanes defines requirements for aeroelastic certification of commercial airliners: "*Demonstration of compliance with aeroelastic stability requirements for an airplane configuration may be shown by analyses, tests, or some combination thereof*[3]." Part 25 further amplifies: "*...Full scale flight flutter testing of an airplane configuration to VDF/MDF is a necessary part of flutter substantiation. An exception may be made when aerodynamic, mass, or stiffness changes to a certified airplane are minor, and analysis or ground tests dhow a negligible effect on flutter or vibration characteristics*[3]." Similarly, a recent amendment to part 23 for Normal, Utility, Acrobatic, and Commuter Category Airplanes[4] requires certification based upon flight tests in addition to analysis or compliance with other regulatory criteria. Earlier Part 23 documents did not require both analysis and flight flutter test. Military specifications are similarly explicit: "*Aeroelastic stability flight tests shall be free of any aeroelastic instability, including sustained limit amplitude instabilities (author note, limit amplitude instabilities are also known as 'limit cycle oscillations', as described in Chapter 11), throughout the prescribed design limit speed flight envelope with not less than 3 percent total (aerodynamic plus structural) damping coefficient and no predicted occurrence of an aeroelastic instability below 1.15 times design limit speed through extrapolation of flight test data*[5]."

13.1 FLIGHT TEST METHODS

Two goals loom large during a flutter flight test effort. Ultimately, the team seeks to confirm the predicted flutter boundary so as to either clear the intended envelope, or establish new envelope restrictions. In so doing, the team does not want to directly experience the flutter boundary.

At the stage when flight tests are required, aeroelastic analyses have identified the flutter boundary and wind tunnel tests have verified this boundary. In addition, math models of the structure have predicted the vibrational characteristics and measurements have verified the actual modal characteristics. Now, we extend our theory and ground-based test experience to flight tests. Several supporting references are suggested which include the efforts of Kehoe[6], Norton[7], and van Nunen and Piazzoli[8], as well as material addressing flutter test techniques provided in[9,10]. Some content below is adapted from Niewoehner and Carriker[2].

Expansion of the flutter envelope is an iterative process, as depicted in Fig. 13.2, from inception of a new configuration to flight testing of the full scale aircraft. As the process continues and updated flutter calculations improve with new knowledge, a foundation is created for flight tests that confirm an understanding of aeroelastic frequency and damping characteristics.

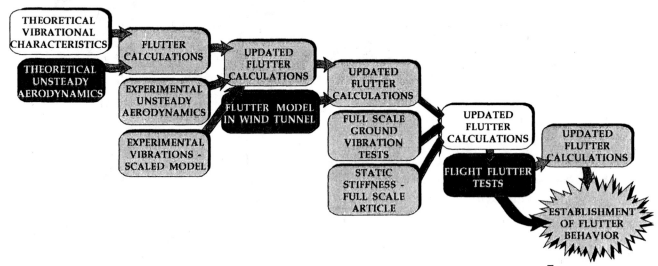

Fig. 13.2 Flutter Expansion Flow Diagram (Adapted from Norton[7])

13.2 THE FLIGHT TEST APPROACH

Figure 13.3 (next page) is presented to facilitate discussion on the basic strategy to clear a flight vehicle for compliance. For a hypothetical aircraft, the figure shows the planned flight envelope based upon design. In addition, the 1.15 limit speed expansion is shown. Normally, tests are conducted along paths of constant altitude, beginning at the maximum design altitude. The build-up philosophy – from most conservative to hazardous – proceeds from high altitude, low speed to low altitude, high speed conditions. Test points are shown where the vehicle maintains altitude and airspeed. At each test point, the test team conducts an airborne version of the ground vibration test. Data are collected (approximately, 60 seconds duration in most cases) from the aircraft instrumented with accelerometers, telemetered to the ground station, processed, and assessed.

Using the approaches described in Chapter 12; frequency, damping, and amplitude of response are tracked. If predictions are available from analysis, then correlations with such predictions are valuable in the assessment. Regardless, a rise in amplitude, a shift in frequencies, and a drop in damping are indicators that require constant monitoring. Extrapolation methods use damping and amplitude information to indicate where flutter occurs. The properly planned test project avoids flutter while clearly identifying where the instability occurs, or showing the vehicle is completely stable up to 1.15 times the limit speed. As shown in Fig. 13.3, flutter exists for this hypothetical aircraft, and one can see how the 1.15 boundary limits the vehicle to conditions below originally designed.

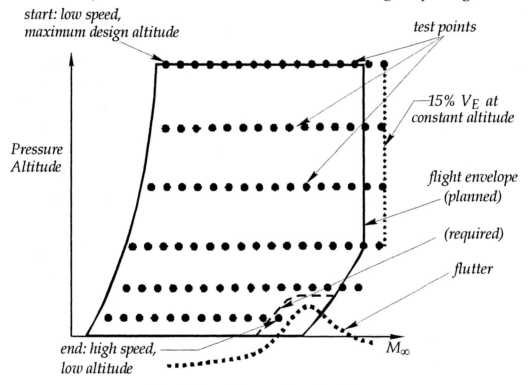

Fig. 13.3 A Flutter Test Profile

Additional comments are in order. In certain test projects (for example, the slightly modified aircraft), at the high altitude, low speed test points it is not necessary to sit on station while data are assessed. In practice, the test team quickly examines damping and frequency characteristics using approaches as simple as the "log-dec" method to count cycles and measure decay on a strip chart. The figure shows only constant altitude tests, yet in certain cases dives are required to achieve high speeds and in other cases constant Mach dives are performed to identify the boundary on the high side of the flutter bucket (see Chapter 12 for discussion).

Flutter flight test demands some of the most expensive instrumentation suites found anywhere in flight test. The principal sensors are strain gauges and accelerometers, located to capture all the dominant modes of interest/concern. Because the phenomena of interest may run past 100 Hz or more, sample rates of 1000 samples per second or more are commonplace. Because of the number of possible modes, the number of sensors can likewise run in the hundreds. This entails a massive bandwidth requirement for both onboard data capture and telemetry. For U.S. Certification trials, the applicable circulars[3,4] provide an excellent discussion of the required and desired instrumentation.

Flutter instrumentation is so expensive that it may be maintained for years after the conclusion of a flutter program, to ensure that the capability is retained to repeat any flutter work that might arise in later life of the aircraft. For example, the U. S. Navy and Grumman carefully preserved the instrumented wings from the original F-14 flutter test aircraft for over twenty years after the end of the flutter program; they would have been invaluable had any flutter anomalies arisen in the operational life of the F-14. This equipment represented a capability for which the Navy only wanted to pay once[2].

13.3 FLUTTER EXCITATION IN FLIGHT

Structural vibrations must be excited at flight conditions to examine aeroelastic sensitivity and validate predictions. As illustrated in the Fig. 13.4, several approaches are available. Techniques include:
- *natural atmospheric turbulence*
- *mechanical excitation devices (mass eccentric devices)*
- *aerodynamic excitation devices (oscillating vanes)*
- *pre-programmed loads from flight control systems*
- *pyrotechnic devices ("bonkers")*
- *pilot-induced control surface pulses ("stick raps")*

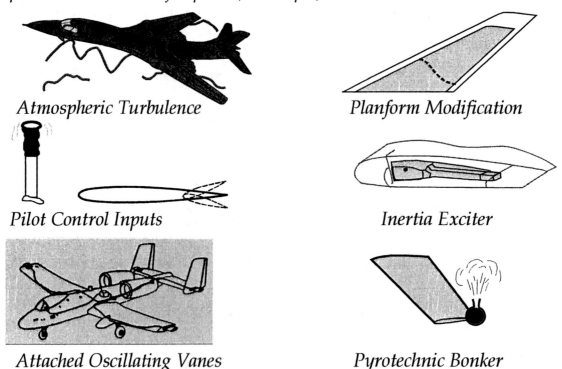

Fig. 13.4 Flutter Excitation Schemes

Inputs take several forms including: sudden impulses intended to simulate a 'step' input (stick raps and bonkers), frequency sweep and dwell loads (shakers), and random loads (turbulence). The purpose of these inputs is to excite structure at a specific frequency, or range of frequencies, and observe response characteristics -- in particular, damping and frequency – at different flight conditions. The advantages and disadvantages of each form of possible excitation are discussed briefly in the following sections.

13.3.1 Natural Atmospheric Turbulence (including wake and buffet induced vibrations):

The advantages are obvious simplicity and low hardware or aircraft modification cost. Disadvantages include: (1) excitation parameters (frequencies swept, dwell times) are not controllable; (2) data reduction requires more sophistication due to low signal-to-noise ratios; (3) quality of data collection depends on atmospheric conditions; (4) all modes are excited simultaneously and closely spaced modes may require additional data collection; and (5) excitation levels will vary with altitude.

13.3.2 Mechanical Devices (including wake and buffet induced vibrations):

Advantages include: (1) efficient use of test time; (2) precise control of frequency sweeps and dwell times; (3) short duration at test points; (4) similar excitation at all altitudes; (5) data reduction is more dependable; (6) signal-to-noise ratios are higher; and (7) specific modal frequencies can be targeted. The disadvantages of using such devices include: (1) systems may be bulky and adversely affect certain modes; (2) added complexity increases chances of test aborts; (3) external actuators may affect the aerodynamics (4) data storage requirements are increased; and (5) short duration at test points requires high quality signal processing and high data rates.

In lieu of additional flight test hardware, existing on-board flight control systems (FCS) in many aircraft configurations provides adequate inflight excitation. Pre-programmed inputs of sine-sweep, sine-dwell, and targeted frequencies are often used with automatic flight control systems. Concerns with the use of FCS actuation include notch filters or other limitations that prevent excitation at specific frequencies, or undesirable feedback between pre-programmed inputs and vehicle response/performance.

13.3.3 Pilot Control Inputs and Stick Raps

Allowing the pilot to excite the structural dynamics is a classical flight test approach. Advantages are: (1) no alteration of mass or aerodynamics since no additional hardware is used; (2) pilot exercises some control of excitation; and (3) reasonable time spent at each test point. The disadvantages of manual excitation include: (1) input frequency is limited by human response time; (2) only the lower modal frequencies (up to about 5 Hz) may be adequately excited; (3) considerable test time may be required if pulses are required in both directions and about each of the axes.

13.4 FLUTTER ENVELOPE EXPANSION

The flight envelope, including the required margin of safety, must be free of flutter. The test program must consider all store configurations, fuel management scenarios, and alternative vehicle configurations. As previously stated, flutter depends on flight conditions such as velocity, altitude, and Mach number; therefore, the test program must consider these parameters. Furthermore, an ordered priority must be used in exploring the effects of each parameter.

13.4.1 **Procedures Requiring Dives**

Dives may be required to achieve higher airspeeds and dynamic pressures for envelope clearance, beyond the maximum achievable level flight speed, V_{max}. An altitude band of approximately ±1000 feet near the targeted test altitude is usually acceptable to obtain data at an altitude (mass ratio) assumed constant. Obviously, descent rates associated with large dive angles give large altitude changes and may not allow completion of the sweep -- repeated dives may be necessary to complete a test point. Care must be taken in dive procedures to guarantee that airspeed increments are controlled.

The sweep rate available with mechanical excitation may be changed to complete the sweep within this altitude band. Often, the "burst and decay" method is used in which a targeted frequency is excited ("burst" or "dwell") for a few cycles, and the decay from this excitation measured. Frequency dwells may be conducted in a dive if a suitable forced excitation system is installed and averages of data may be needed if atmospheric excitation is used. Test points flown in a dive with control surface excitation done by the pilot are usually pulses applied about one axis at a time and several passes through the altitude band may be required for a statistically significant data sample.

13.4.2 **Case Study: the X-29**

The significant efforts of Kehoe and Rivera[12] and others (such as Freudinger[13]) provide a detailed discussion of flutter clearance flight test programs. As an illustration, results from the X-29 flutter clearance project are briefly discussed.

Table 13.1 Mode Type, Frequency, and Nature of Instability

MODE	FREQUENCY Hz	POTENTIAL MECHANISM FOR INSTABILITY
First Symmetric Wing Bending	8.6	D, ASE
First Antisymmetric Wing Bending	11.3	ASE
First Fuselage Vertical Bending	11.6	ASE
First Fuselage Lateral Bending	12.5	ASE
First Fin Bending	15.2	F
Canard Pitch	21.0	D, F
Second Fuselage Vertical Bending	24.3	ASE
Second Symmetric Wind Bending	26.3	ASE, F
Second Antisymmetric Wing Bending	26.8	ASE
Inboard Flap Rotation	31.3	F
Wing Torsion-Outboard Flap Torsion	35-37.0	F
Midflap Torsion	38.7	F
Canard Pitch-Bending	42.0	F
First Fin Bending	50.0	F
Inboard Flap Torsion	51.5	F

Adapted from Kehoe and Rivera[12]

A ground vibration test (GVT) identified frequencies and modes leading up to flight tests. Table 13-1 describes the aircraft mode, modal frequency, and potential aeroelastic instability to which the modes were predicted to contribute (according to preflight analysis). Several forms of instability were predicted for the X-29, including (1) flutter; (2) divergence -- aggravated by the forward sweep wing design of the X-29; (3) aeroservoelastic instabilities -- beyond the scope of this text, but an instability that involves an interaction of the flight control system with flexible vehicle modes; and (4) body-freedom flutter -- an instability that caused by the interaction of flexible modes with rigid body modes. In the following table, "D" indicates a potential divergence instability, "F" indicates a potential flutter instability, and "ASE" indicates a potential aeroservoelastic instability.

The GVT also identified the optimal location of accelerometers to track modes. Although the GVT likely used hundreds of accelerometers to characterize a specific mode shape, the flight test only required a few judiciously place accelerometers. Fig. 13.5 shows the relative placement of a few accelerometers on the X-29.

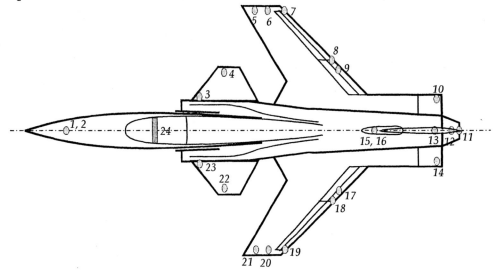

Fig. 13.5 Location of Accelerometers for the X-29 (Adapted from Kehoe and Rivera[12])

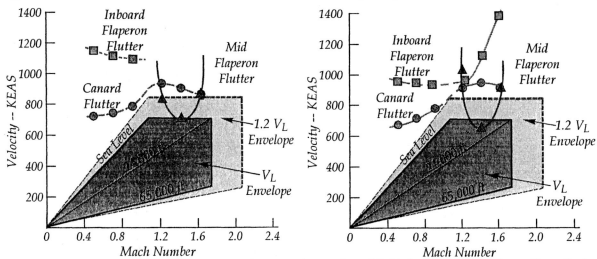

a. *Predicted X-29 Symmetric Flutter Boundaries*   b. *Predicted X-29 Antisymmetric Flutter Boundaries*

Fig. 13.6 Predicted X-29 Flutter Boundaries (Adapted from Kehoe and Rivera[12])

A single accelerometer on the wing tip permitted tracking of the bending mode. A second accelerometer on the other wing tracked symmetric versus antisymmetric bending modes. Moreover, a second accelerometer mounted on each wing -- with one placed near the leading edge and the other placed near the trailing edge – allowed tracking of torsional modes.

Figure 13.6 superimposes predicted flutter boundaries onto the flight envelope -- displayed in the standard format described previously. Symmetric modes and antisymmetric modes are shown. A 20% margin of safety on the limit velocity (V_L) is used. Several modes were predicted to approach the required clearance envelope; most notably, a midflaperon flutter mechanism that penetrated deep into the planned flight envelope. These predictions used an extremely conservative value for structural damping (0.02 - 0.03) that leads to conservative (low) predictions of flutter speed. However, flight tests were required to clear the flight envelope.

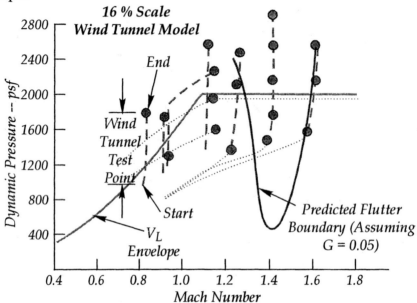

Fig. 13.7 X-29 Midflaperon Flutter Boundary Prediction
(Adapted from Kehoe and Rivera[12])

Scale model (16% dynamically scaled model) wind tunnel tests were conducted to investigate the midflaperon flutter mechanism. Several wind tunnel tests (Fig. 13.7) were conducted and then specific flight tests were flown well within this regime of predicted flutter. No instabilities were experimentally measured in flight events; differences in the analytical predictions summarized in Fig. 13.7 were attributed to greater structural damping than was assumed in the analysis. Structural damping delayed flutter onset.

Following extensive analysis, ground vibration tests, and wind tunnel investigations, a flight test program was begun, using a build-up approach. That is, flight tests were conducted in increments from lowest risk to highest risk, and proceeded from low speeds to high speeds beginning at highest altitudes and finishing at lowest altitudes. Thus, confidence was established prior to approaching potential flutter boundaries. Figure 13.8 depicts the flutter test matrix flown. No instabilities occurred within the flight envelope, and the maximum Mach number for the vehicle was established as $M_\infty = 1.4$ for reasons other than aeroelastic concerns.

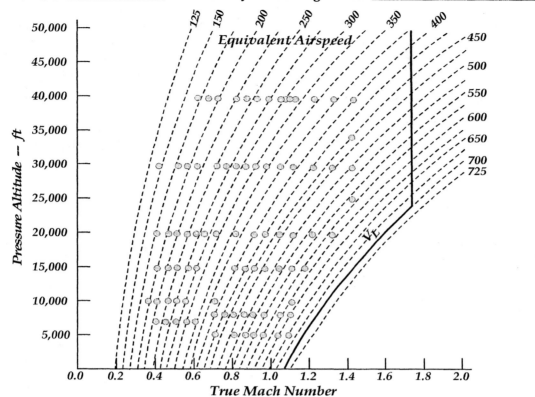

Fig. 13.8 X-29 Flutter Flight Test Matrix (Adapted from Kehoe and Rivera[12])

13.4.3 Case Study: F/A-18 E/F Super Hornet [adapted from Niewoehner and Carriker[11]]

The F/A-18 E/F Super Hornet is one of the most recent large-scale flutter campaigns conducted on a U.S. tactical aircraft, for which your third author flew as one of the team's flutter pilots during development. The flutter program for the 'clean' airplane comprised 150 flights, 259 hours, and 13 aircraft months (that is, no pylon stores, with and without wing tip AIM-9 missile). After three years of flying, the program had cleared the clean airplane and over a dozen weapons loads. Those store loads were regarded as the most significant flutter loads, and once evaluated, cleared all permutations and potential downloads.

The first aircraft of the seven aircraft in the development program was dedicated almost exclusively to flutter testing. Because flutter paced the activities of all seven development airplanes, a second airplane also had provisions installed for flutter instrumentation, in the event any mishap restricted the availability of the primary airplane. The flutter instrumentation on the primary airplane included accelerometers, strain gauges, and a tape recorder with several hours' capacity. A flutter card inserted in the flight control computers permitted up to one hundred pre-programmed sweeps or burst excitations of any flight control surface or pair of surfaces. The programs were uploaded prior to flight, but could be modified through a Multi-Function Display (MFD) and keypad. The Flutter Exciter Control Unit (FECU) required software unique to the airplane in which it was installed.

Only two of the test team's ten pilots flew flutter missions. This constraint was initially dictated by safety considerations (the FECU required some training and adaptation). Later, as confidence grew in the stiffness of the vehicle and the proficiency of the engineering teams, the flutter pilots for the team remained at two. A flutter data point costs more than any other type of data collection, with the possible exception of weapons separation. Once beyond the speeds achievable in level flight (V_{max}), a four-hour flight might yield less than five minutes of data. Since pilot technique dramatically impacted test productivity, the team only assigned as many pilots to flutter testing as could remain thoroughly proficient in the techniques.

Fifteen months were required to complete the full envelope clearance for the clean aircraft, during which this airplane flew little else but flutter. The clean airplane campaign included symmetric and asymmetric excitation of ailerons, rudders, and stabilators. The aileron excitations were performed both with and without wing tip missiles. Testing progressed in bands that correlated with equivalent airspeed, and the completion of the band resulted in a clearance being passed to the other airplanes in their test disciplines (propulsion, flying qualities, weapons separation and loads). Within a dynamic pressure band, tests were performed at as many as four altitudes. Significant test flexibility and productivity were gained by briefing all altitudes within a band of dynamic pressures for a given flight. In this way, the pilot could adapt those daily tests to the weather on the range.

Stores flutter testing required approximately 6 weeks of testing per loading. Each such series included one to two weeks of ground vibration testing (GVT) followed by four to five weeks of flying. Stores testing was restricted to asymmetric excitation of the ailerons alone; the clean testing of rudders and stabilators sufficed for all loadings, and the clean symmetric aileron testing indicated such high stiffness levels as to be cleared by analysis alone (shaving months from the planned effort).

During the clean configuration testing effort, the airplane was scheduled almost every day, six days per week. The data analysis/flight preparation cycle was approximately 48 hours, requiring two complete teams to support daily flights. On a day when Team A was flying (tip missile off), Team B was analyzing the prior day's data and planning the next day's flight (tip missile on). This approach permitted optimal use of the airplane; suitable weather became the constraint. Instrumentation deliberately placed no sensors on the missiles. They could be uploaded and downloaded without disturbing any instrumentation, enabling the alternating schedule.

Once the program moved into stores testing, instrumented wing stores could not be disturbed flight-to-flight, and the highest fly rate achieved was three sorties per week, allowing for a data analysis/planning day between flights. At this point, data analysis became the pacing activity, rather than either airplane availability or weather. Two weeks of GVT were followed by four weeks of flying, after which that configuration was cleared. The other airplanes in the program frequently flew loads, performance, propulsion, or weapons separation flights within days of the flutter team clearing a load. As mentioned, the entire test program's productivity hinged on the flutter program proceeding apace and staying out ahead of the other disciplines.

Flights were planned for 3-4 hour durations, during which the test airplane and chase expected to refuel 3-6 times from a tanker. Flights began on the ground with frequency sweeps of all surfaces and tests of the safety cut-offs. The sweeps established the integrity of the instrumentation and confirmed that no changes to the stiffness properties of the airframe or flight controls had occurred flight-to-flight (for example, the incipient failure of a stabilator actuator was spotted during one pre-flight frequency sweep, thereby avoiding the embarrassment of flying with a non-representative component).

Test points consisted of three-minute frequency sweeps in each of the desired axes, followed by a 5-10 minute analysis of the sweep and a plan for targeted dwells. These short analysis respites worked naturally into the refueling flow such that the analysis and new program plan could be drafted during the time required to refuel. After disconnecting from the tanker, the FECU programs were refined based upon information from the previous sweep. The dwells would then be performed at the test condition, injecting 2-3 second energy bursts into the test surface at the frequencies identified during the sweep.

Progress through individual bands progressed quickly to V_{max}, but then slowed dramatically as test speeds pushed beyond V_{max}. A sweep at a stabilized flight condition below V_{max} required only three minutes. Achieving the same data beyond V_{max} necessitated climbing well above the test condition, an unloaded acceleration onto a KEAS line, then a partial sweep during the 10-30 seconds passing through an specified altitude band about the test condition. Three to four such dives were typically accomplished on a single tank of fuel. Testing productivity plummeted from 3-6 sweeps per tank of fuel to 2-3 tanks of fuel per sweep. Multiple dwells were performed when diving through bands, depending on the severity of the power deficit. The further the test point was beyond V_{max}, the steeper the dive and the shorter the time window the event could remain "on condition." Consequently, testing productivity (test points per flying day) dropped considerably beyond V_{max}. This cost was justified by the value of these points; these data were essential to establishing the predicted damping for the V_{NE} (red line) dives to follow.

Once the build-up points were complete, the final test of the airplane consisted of a red-line (or V_{NE}) dive. These dives were initiated from V_{max} at the maximum ceiling, diving at full power until reaching the limit Mach number. That Mach number was then held by adjusting both power and pitch attitude, until intersecting the constant KEAS line which represented red-line (envelope boundary). This KEAS line was then held until rounding out the dive near the surface. Throughout the dive, dwells were performed at a small sample of frequencies for which the damping was expected to be closest to the flutter margin. These were set-up to repeat automatically so that the pilot could initiate a dwell every 4-5 seconds, with the dwell program cycling back through the sequence.

Several lessons are worth noting:
- *Cockpit Displays*. The flutter team's engineers spoke exclusively in KEAS, but the pilot displays read only KCAS. While this airspeed conversion is easy for stabilized points, it does require a significant amount of concentration from the pilot when tasked to hold a constant KEAS line through an altitude block, which converts to a constantly decreasing KCAS trajectory. We lost data by hitting the right speed at the top of a block but then not precisely controlling the KCAS bleed rate to maintain the desired KEAS through the altitude band. A single flutter test point costs thousands, even tens of thousands of dollars; only a few lost points costs more than the modest effort needed to display KEAS in a digital cockpit. If the test team wants pilots to fly KEAS, display KEAS in the cockpit.
- *The Unexpected*. The test team was embarrassed by an aileron buzz event for which we were unprepared. Our guard was down while rapidly sampling the damping at

a Mach number well below the predicted flutter speed and observers missed an aileron buzz, initiating the next test point prior to realizing we had experienced a significant aeroelastic event. There is no room for complacency even 100 knots short of the predicted limit speed.

- *Aerial Refueling.* The test program would have taken years rather than months without aggressive, dedicated aerial refueling support. Typically two KC-130s were drained in a 3.5-4.0 hour mission with the test aircraft refueling 6 total times, and the chase refueling 3-4 times. The tankers were actively involved in the test flow, monitoring the hot-mike communications channel and adjusting their orbit so that they were within visual sight of the test aircraft when tank-state was reached. Since the test aircraft did not have radar, a proactive chase pilot also speeded up the rendezvous by providing a vector to the rendezvous.
- *FECU programming errors.* The FECU stored 100 programs controlling surfaces, frequencies, amplitudes, durations, sweep rates, etc. These inputs were all loaded during the third shift in preparation for a dawn launch. While parameters could be adjusted in flight, the first hour of testing usually followed the planned programs. On one occasion, early in the flutter tests, a programming error for a dwell startled the whole team with a significantly larger excitation than planned, though fortunately at low dynamic pressure. A quality assurance process was then instituted to ensure that error-free programs were loaded into the aircraft.
- *Ground sweeps.* While requiring 5-10 minutes of ground turn time prior to takeoff, ground sweeps served their purpose in both ensuring data quality and promoting efficiency. The ground sweeps were responsible for detecting at least one failing actuator. While total failure was unlikely, flying with a faulty actuator would have invalidated the data and the degradation might not have been detected for several flights, perhaps invalidating a large amount of data.
- *Safety Chase.* As with other high-risk tests, a chase plane enhances test safety by providing a set of eyes on the outside of the airplane. Flutter flight tests purposely shake the airplane to see if anything shakes loose; sometimes it does (Fig 13.1), and neither the test pilot nor his ground engineers are likely to see or feel pieces depart. The safety chase is invaluable in halting a deteriorating situation, or appraising the test team of the structural state of affairs. Additionally, flutter flight test (along with performance testing) demands intensive instrument flying, because of the required precision. The test pilot can devote little attention to see-and-avoid; the chase crew must assume that role.
- *Hot-mike-to-chase.* Many programs bundle the pilot's intercom audio ("hot-mike") on the telemetry stream, so that the ground station can hear his commentary apart from a radio transmission. The result is that other airborne assets such as the chase and tankers only hear the ground-to-air half of a conversation. Providing a path for the test airplane's cockpit audio to those teammates is vital to them being able to fully contribute; otherwise, they simply cannot mentally follow events in the test airplane cockpit. This simple enhancement to the test equipment saved the program thousands of dollars. As soon as the test pilot told the data station, "point complete," a sharp chase pilot responded, "tanker is at 1 o'clock, angels 18, paralleling."

The open communications path meant the tanker knew exactly where to be to facilitate rendezvous, and the chase pilot, having found the tanker on radar minutes earlier, anticipated the test pilot's query. This sort of teamwork became routine; it was exhilarating to be part of a team operating at this level of efficiency. It cannot happen without seamless communications.

13.4.4 Case Study: CF-18 Hornet with Pylon-Mounted GBU-24s

A Canadian CF-18 experienced a divergent flutter episode in the mid-1990s that the saved from catastrophe only by a prompt pull-up deceleration. The purpose of the flight was clearance of a large wing-mounted store. The presence of a heavy store, cantilevered well forward of the wing's elastic axis, made this configuration prone to torsional oscillations. The event occurred at the targeted clearance speed after over one minute at the desired flight condition. Interestingly, the flutter erupted during a repeated, programmed excitation, which had shown no unstable properties during the first test. The airplane was saved by prompt pilot action, when he quickly decelerating out of the flutter-prone range. Peak oscillations induced +/- 1.2 g's at the pilot station. Details can be found in Girard and McIntosh[14], an important contribution to the literature since we seldom hear from pilots who experience fully developed flutter.

13.4.5 Commercial and Cargo Aircraft

The flutter campaigns for tactical airplanes are complicated by the diversity of external stores; the test programs for commercial airplanes are complicated by the diversity of fuel and cargo loadings. Permutations of wing, tail and body fuel and cargo must all be cleared either directly in flight, or by analysis. Even the weight of interior furnishing must factor into the certification process if a wide disparity exists between viable configurations. Given the significant fuel burned during flutter tests beyond V_{max}, these tests are profoundly complicated by the difficulty of keeping the fuel appropriately distributed for valid results.

Commercial airplane test programs frequently forego large ground data stations and telemetry, preferring to carry the engineering staff onboard. Due to the extended endurance, it is common for a single test flight to span a number of test disciplines. This approach is inappropriate for flutter testing, where minimum essential crew should be on board and exposed to the hazard.

13.4.6 General Aviation

General aviation may pose the greatest challenge for flutter testing, since the resources seldom exist to support extensive analysis, ground vibrational tests, instrumentation, and experienced flutter teams one finds on commercial or military airplane programs. Indeed, one European certification official cited the flutter certification of sailplanes among his thorniest challenges.

Stick and rudder pulses are commonly the baseline excitation method for light aircraft. The governing circular for U.S. light civil airplanes (AC 23.629-1B[4]) permits pilot excitation for substantiation of modes to 10 Hz, but discourages its use for higher frequencies, suggesting instead that rotational masses be used.

The instrumentation requirements should be determined as part of the GVT, though accelerometers should be installed on the tips of each aerodynamic and control surface.

Some symmetry in the instrumentation is required in order to resolve symmetric and asymmetric modes. The Advisory Circular[4] should be consulted for details.

13.4.7 Flutter Mitigation: Procedural Rules of Thumb

In most flight test endeavors, we have typically both preventive measures, intended to avoid hazards, and responsive measures, intended to lessen the consequences of hazards encountered. Flutter typically denies the test team an opportunity to respond to actual flutter events (the Canadian CF-18 GBU experience cited above is a notable exception). A flutter episode often erupts much too fast for a pilot to back away from the sensitive condition. Indeed, many episodes are likely divergent; often such an event destroys the airplane too quickly for even a safe ejection (if so equipped). Consequently, while a team should plan and drill towards escaping a flutter eruption, the program must dedicate even more effort to avoiding a flutter episode altogether during envelope expansion. This focus was emphasized in the two preceding chapters.

Flutter testing places unusual fatigue demands on an airplane. A tactical aircraft, for example, is typically built around a design life of 6,000 hours. That design life, however, presumes a characteristic use of the airplane. Designers expect less than 5% of an airplane's expected life to be flown in the lower right hand corner of the flight envelope, at high dynamic pressure, where the most fatigue damage is incurred. A flutter test article, flying flights concentrated in that very regime, should expect fatigue life to be consumed far faster that of fleet airplanes which have a less severe fatigue spectrum. For example, a low-time F-16 at Eglin used principally for weapons flutter tests, twice returned to base having lost a trailing edge flap panel. The first was considered an anomalous failure; the second provoked a detailed engineering investigation digging deeper into the true fatigue spectrum encountered by this flutter test article. In both cases, the flap hardware had completely exhausted its fatigue life in only months of test operation in these conditions, whereas a fleet airplane would not consume that life in years of service. Preflight frequency sweeps might identify impending failures of such components.

13.4.8 Flutter Mitigation: Design Rules of Thumb

In light of the previous discussion, it is appropriate to identify methods of raising the flutter velocity or possibly eliminating flutter altogether. All of these suppression approaches have one purpose: to break the coupling, whether it be aerodynamic, inertial (mass), or elastic. Furthermore, today's flight control systems may be used to suppress the onset of aeroelastic instabilities by affecting the phase relationships of the coupled system or create offsetting loads.

One commonly used approach is to mass balance the system. Mass balancing requires the addition or redistribution of mass so that the center of gravity is moved forward. Control surfaces and rotor blades often use ballast. An increase in mass lowers the natural frequency; mass positioned away from the center of mass lowers the natural frequency in torsion; and, an increase in structural stiffness raises the natural frequency.

The presence of flutter and divergence is influenced heavily by the character of the primary torsional mode. The vibration characteristics of the structure may be adjusted to eliminate or delay the onset of flutter. An increase in the spread between the participating modal frequencies delays the onset of flutter. Typically, the primary torsional mode is higher than the primary bending mode; therefore, increasing the torsional stiffness or reducing the moment of inertia (redistribution of mass) is usually advantageous.

The addition of structural damping also provides flutter alleviation. Other than the aeroelastic modes, "aerodynamic stiffening" is typical, as most system frequencies will rise slightly in the presence of increasing aerodynamic loads.

Wing sweep has a significant effect on certain aeroelastic instabilities. As presented in Chapter 11, forward sweep adversely affects the divergence boundary and aft sweep adversely affects control reversal. Relative to these other instabilities, flutter is affected little by sweep.

Angle of attack effects are minimal except for vortex-shedding phenomena. However, wind-up turns are often used at the completion of certain test points to demonstrate insensitivity to angle of attack. Control surface flutter modes are most sensitive to altitude (mass ratio) effects. Also, control surface "buzz" is eliminated with reduced free-play of the control surface actuator.

The GVT must identify all modes of interest and flutter analysis demands the use of all participating modes. A large transport can be modeled with the first 3 bending modes and first 2 torsional modes[10], since bending-torsion flutter is typical. However, higher-order modes are necessary for multi-control surface aircraft; for example, adequate analysis of the F/A-18 High Angle-of-Attack Research Vehicle (HARV)[13] and the previously cited X-29 required in excess of 25 modes.

13.5 STORE-INDUCED LIMIT CYCLE OSCILLATION

In the context of aeroelasticity, limit cycle oscillation (LCO) is considered an oscillatory response of the wing, the amplitude of which is limited, but dependent on the nature of the nonlinearity as well as flight conditions (airspeed, altitude, and Mach number). LCOs remain a persistent problem on high performance fighter aircraft with multiple store configurations. Control surface buzz and other shock-induced oscillations are other forms of LCOs.

13.5.1 Effect of Stores on LCO Potential

It is well known that the addition of external stores to an aircraft alters the dynamic characteristics and may adversely affect flutter boundaries. Much of the test experience with aircraft LCO has been documented by the Air Force Flight Test Center and most of it has been in the context of the F-16 aircraft.

Using measurements obtained from flight tests, Denegri[15] and Bunton and Denegri[16] describe LCO characteristics of the F-16 and F/A-18 aircraft. The amplitude and frequency of the LCO may vary with flight parameters, but usually the frequency is near that predicted by classical linear dynamic flutter analysis. The LCO motion is dominated by asymmetric modes. Denegri distinguishes among three types of response – typical LCO, atypical LCO and flutter - based on observations in flight, and by use of linear flutter analysis (with recognition that the problem is nonlinear). Typical LCO begins at a certain flight condition and the LCO amplitude increases with an increase in Mach number at constant altitude. Flutter occurs when the increase in LCO amplitude with change in speed is so rapid that the vehicle is in jeopardy. Atypical LCO occurs when the LCO amplitude, first increases and then decreases (and disappears in some cases) with changes in Mach number. Often changes in flight vehicle angle of attack, induced by a wind up turn, lead to such LCO responses. Denegri presents flight-test data showing a LCO at nearly constant Mach number over large variations of altitude. Because of aircraft safety concerns, many typical LCO encounters result in termination of testing due to the increase in response levels with each increase in Mach number. Some nontypical LCO encounters feature response levels that increase to a maximum, followed by a decrease with increasing Mach number.

There are many other experiences with store-induced LCOs. The 1974 AGARD meeting on Wing-with-Store Flutter[17] provided further insights into the issues; more importantly, this reference illustrates the persistent challenge this phenomena is to the aeroelastic community.

13.5.2 Other Pathologies

Residual pitch oscillation (RPO) involves interaction of a flexible (wing) mode and rigid-body degrees of freedom. The B-2 bomber encountered a RPO during low-altitude, high-speed flight[18]. Neither the RPO nor any tendency for a related response was predicted. The B-2 RPO involved a symmetric wing bending mode with a short-period rigid-body mode. Behavior of the RPO was better understood with analysis that modeled the nonlinear response of the control surfaces. This RPO was sensitive to vehicle weight.

Aircraft with freeplay in the control surfaces experience LCO. Croft[19] describes a flutter/LCO event and explores typical situations. The oscillation was of limited amplitude, and it was uncertain as to whether or not this LCO was sufficiently large to be a threat to the integrity of the aircraft structure.

13.6 SUMMARY

In this chapter we have considered flight tests for aeroelastic instabilities. The highly-destructive nature of these instabilities requires a judicious approach to identify boundaries and clear the flight envelope. Ground-based tests and studies that include ground vibration tests, wind tunnel tests, and flutter analyses are necessary to provide a strong foundation of understanding prior to flight tests. Flight tests at flutter or divergence conditions are not performed directly; rather, subcritical methods that include a continual examination of frequency, damping, and amplitude of response are used to identify instabilities via extrapolation approaches. A build-up approach in the flight test program that permits the lowest risk test points (low velocity, high altitude) to be conducted first is advised. The flight test team should be formed with an emphasis on appropriate experience and expertise, and one should proceed cautiously and "expect the unexpected".

REFERENCES

1. Fred Madenwald, McDonnell-Douglas Corporation, admonishing the third author in November, 1995.
2. Niewoehner, R. J. and Carriker, M. "Flutter", <u>Flight Test Safety Handbook</u>, Society of Experimental Test Pilots, Lancaster, CA, 2003, pp. 4-1ff.
3. ANM-114, "Aeroelastic Stability Substantiation of Transport Category Airplanes," Advisory Circular No. 25.629-1A, U. S. Department of Transportation FAA, July 1998.
4. ACE-100, "Means of Compliance with Title 14, Part 23, Section 23.629, Flutter," Advisory Circular No. 23.629-1B, U. S. Department of Transportation FAA, September 2004.
5. MIL-A-8870C, Military Specification, "Airplane Strength and Rigidity Vibration, Flutter, and Divergence", March 1993.
6. Kehoe, M. W., "Aircraft Flight Flutter Testing at the NASA Ames-Dryden Flight Research Facility," NASA Technical Memorandum 100417, NASA Ames Research Center, Dryden Flight Research Facility, California, May 1988.
7. Norton, W. J., "Structures Flight Test Handbook," AFFTC-TIH-98-02, Air Force Flight Test Center, Edwards AFB, California, 2nd Edition, May 2002.
8. van Nunen, J. W. G. and Piazzoli, G., "Aeroelastic Flight Test Techniques and Instrumentation," AGARD Flight Test Instrumentation Series, AGARD-AG-160, Vol. 9, AGARD, London, England, Feb. 1979.
9. "Flutter Testing Techniques," NASA SP-415, NASA, 1976.
10. "Flutter Flight Testing Symposium," NASA SP-385, NASA, 1958 (reprint 1975).
11. "Introduction to Flutter of Winged Aircraft," von Karman Institute for Fluid Dynamics, Lecture Series 1992-01, von Karman Institute, 1992
12. Kehoe, M. W. and Rivera, J. A. , "Flutter and Aeroservoelastic Clearance of the X-29A Forward-Swept Wing Airplane," NASA Technical Memorandum 100447, California: NASA Ames Research Center, Dryden Flight Research Facility, Sept. 1989.
13. Fruedinger, L. C., "Flutter Clearance of the F-18 High Angle-of-Attack Research Vehicle With Experimental Wingtip Instrumentation Pods," NASA Technical Memorandum 4148, NASA Ames Research Center, Dryden Flight Research Facility, California, 1989.
14. Girard, Maurice and McIntosh, Stuart, "CF-18 GBU-24B/B Flutter Flight Testing," *1996 Report to the Aerospace Profession*, Society of Experimental Test Pilots, Beverly Hills, CA, Sep 1996, p. 190.
15. Denegri, C. M., Jr., "Limit-Cycle Oscillation Flight Test Results of a Fighter with External Stores," **Journal of Aircraft**, Vol. 37, No. 5, 2000 pp. 761–769.
16. Bunton, R. W., and Denegri, C. M. Jr., "Limit-Cycle Oscillation Characteristics of Fighter Aircraft," **Journal of Aircraft**, Vol. 37, No. 5, 2000, pp. 916–918.
17. AGARD Specialists Meeting on Wings-with-Stores Flutter, 39th Meeting of the Structures and Materials Panel, CP 162, AGARD, Oct. 1974.

18 Dreim, D. R., Jacobson, S. B., and Britt, R. T., "Simulation of Non-Linear Transonic Aeroelastic Behavior on the B-2," NASA CP-1999-209136, CEAS/AIAA/ICASE/NASA Langley International Forum on Aeroelasticity and Structural Dynamics, 1999, pp. 511–521.

19 Croft, J., "Airbus Elevator Flutter: Annoying or Dangerous?," **Aviation Week and Space Technology**, Aug. 2001.

Chapter 14
STALL, POST-STALL GYRATIONS, AND SPIN TESTS

So far in this introductory survey of flight test, the motions under discussion have been well-behaved small perturbations from some equilibrium condition below the stall angle of attack. Classically, the stall angle of attack is that angle where the C_L-α curve breaks sharply, as shown in Fig. 14.1. If the trajectory proceeds to higher angles of attack, the vehicle may go out of control and experience uncommanded motions. This chapter briefly summarizes the theory for these post-stall motions and outlines some of the precautions that must be taken when tests are planned that deliberately place the vehicle outside of the normal operating envelope and into this hazardous flight regime.

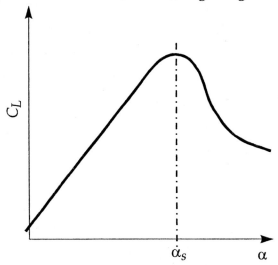

Fig. 14.1 Classical Aerodynamic Stall

The hazardous nature of post-stall flight tests underscores their importance. During the 1970s, post-stall accidents provoked a considerable amount of research aimed at preventing loss of control. Through careful tailoring of the aerodynamic configuration and skillful use of automatic control systems, designers created highly spin-resistant airplanes. The payoff for this departure-resistant design emphasis became evident during the flight test programs of the current generation of operational fighters, and the development of recent light civil airplanes. The F-14, F-15, F-16, and F-18 all have remarkably good high angle of attack characteristics compared to their predecessors. The latter three airplanes all completed rigorous high angle of attack programs without the loss of a single airplane to an out-of-control incident: an unprecedented safety record in fighter development! But it is not enough to simply design departure- and spin-resistant airplanes. Maneuverability in the post-stall arena is important for a significant number of tactical situations[1]. In the late 1980s and early 1990s there were three major research projects in flight test[2,3,4] meant to help understand and exploit any tactical advantages that come from maneuvering at angles of attack well above the stall. The YF-22 and YF-23 Advanced Tactical Fighter prototypes demonstrated significant capability in this part of the flight envelope during their competition. The Sukhoi design team in Russia also produced a design and sold it to the Indian Air Force (IAF) under the designation Su-30 MKI (see Fig. 14-2, next page) which incorporates thrust vectoring. Reportedly, the IAF has up to three squadrons of these machines in operational service at the time of this writing[5]. With this sort of thrust in design requirements (driven, of course, by perceived operational advantages) and the inherent hazards associated with high angle of attack testing, it is imperative that an introduction to flight test engineering include a discussion of post-stall testing.

Fig. 14.2 Operational IAF Su-30MKI with Thrust Vector Control[5]

14.1 THEORETICAL FOUNDATIONS

The early practitioners of aerospace engineering were keenly aware that it was dangerous to try to fly beyond the aerodynamic "boundary" called "stall". Those who pushed that boundary often lost their life; Otto Lilenienthal was one of those unfortunate "birdmen". But by World War I, the inquisitive and adventurous were deliberately crossing this boundary with regularity. Engineers and test pilots at Farnborough in England were trying to unlock the mysteries of the post-stall flight regime. A physicist named Lindemann undertook the development of a theory to explain the motions described by pilots like Parke, Courtney, and Gooden. He produced and later corroborated a mathematical theory that matched rather well some of the developed motions of the machines being tested in that era. Lindemann postulated several key assumptions that will serve our purposes in this chapter to introduce the essential elements of the most common post-stall motion, the spin. By assuming that the spin trajectory was an established equilibrium with the inertial forces and moments balanced by the aerodynamic ones, Lindemann concluded that the aircraft c.g. followed an essentially helical path. For such a dynamic motion mode, the angle of attack is almost constant (or at least periodic in behavior).

14.1.1 Post-Stall Definitions

The terminology used to discuss the post-stall flight regime is unique and an expanded vocabulary is a necessity. Our first step, then, is to understand what a stall, a departure, the incipient spin, a post-stall gyration (PSG), and a developed spin are and how they differ from one another. These definitions are the subject of this section.

14.1.1.1. **Stall**. No two sources define stall in exactly the same way. The FAA Flight Training Handbook[6] defines this term about as simply as possible:

> A **stall** occurs when the smooth airflow over the airplane's wing is disrupted and the lift degenerates rapidly.

These words describe the classical stall as depicted in Fig. 14.1. On the other hand, military specifications define the stall in subtly different terms:

Chapter 14   Stall, Post-Stall, and Spin Testing

> *The angle of attack for maximum usable lift at a given flight condition is the **stall angle of attack**[7].*

The same document explains "maximum usable lift" by spelling out conditions other than reaching the peak of the C_L-α curve that may limit available lift. Possibilities include:

- Uncontrollable pitch, roll, or yaw
- Intolerable buffet
- Excessive sink rates
- Inability to perform altitude corrections

Stall occurs when maximum usable angle of attack is reached. It is not always an uncontrollable event in itself; rather, it merely describes a critical condition reached by the lift-generating surfaces of the vehicle. Exceeding this critical condition may cause drastic changes in the trajectory of the vehicle and in its response to control inputs. Susceptibility to loss of control may increase, but a stall does not inevitably lead to loss of control.

14.1.1.2 **Departure**. If a stall does not always produce uncontrolled flight, what event does mark the transition from controlled to uncontrolled flight? The military has coined the term "departure" for that event.

> ***Departure*** *is that event in the post-stall flight regime which precipitates a PSG, spin, or deep stall*[7].

Notice two things about this definition. First, departure occurs in the post-stall flight regime; that is, the stall always precedes or is at least coincident with the departure event. It can be inferred, then, that the angle of attack for maximum usable lift is always less than or equal to the departure angle of attack. Secondly, only one of three motions may result from a departure: the airplane either enters a PSG, a spin, or a deep stall. Implicit in this definition of departure is the idea that an immediate recovery cannot be made. For example, a general aviation airplane whose stall is defined in the classical sense illustrated by Fig. 14.1, may recover immediately if the longitudinal control pressures are relaxed when the stall occurs. However, the same airplane might go out of control, or depart, if ailerons were deflected significantly at the time stall occurred. With pro-spin ailerons applied, the same docile general aviation airplane might readily spin. By the definition above, it would then have departed. This latter fact suggests that one must specify susceptibility or resistance to departure for given control positions, along with other configurational parameters.

The above definition fails to capture those departures which are precipitated by gyroscopic coupling between rotations in multiple axes. This class of departures can occur at low angles of attack, but requires the presence of high rotational rates in at least one axis coupling into another. The result, however, is the same as the departures described above: large uncommanded airplane motions.

The departure event is usually a large amplitude, uncommanded, and divergent motion. Such descriptive terms as nose slice or pitch-up are commonly used to portray departure. By uncommanded motions, it is meant that the pilot or controller did not intend the control movements to produce the actual resulting motions. The airplane may not follow the pilot's commands for a number of perfectly valid aerodynamic reasons:

- High local α may render control surfaces ineffective.
- Aerodynamic control moments lack the strength to overcome very high angular momentum due to high body rates.

◆ The pilot may be unable to position the control stick properly due to lateral or transverse accelerations at his cockpit position.

In any of these cases, the motion is uncommanded. A divergent motion is one which either continuously or periodically increases without bound. Some airplanes, like the T-33, periodically "buck" when forced past the stall; that is, the nose periodically rises and falls. However, there is no divergence unless full aft stick, ailerons, or some other pro-spin control is held. The T-38 will occasionally produce a non-divergent lateral oscillation near the stall angle of attack. Neither of these motions is usually counted as a departure, though their presence does warn of impending departure if the controls are further abused. In sum, a departure is a very difficult event to precisely describe, but there is little doubt in the pilot's mind when it has occurred!

14.1.1.3 **Post-Stall Gyration**. A post-stall gyration is an uncontrolled motion about one or more axes following departure[7]. It is also a difficult term to define concisely because it occurs in so many different forms. Frequently, the motions appear to be completely random about all three axes and no more descriptive adjective or noun can be applied than PSG. A snap roll or a tumble are motions that fit the definition of a PSG, and these more common (if less precise) words are sometimes used. The primary difficulty lies in distinguishing between a PSG and the incipient phase of a spin, especially an oscillatory spin. The chief distinguishing feature is that a PSG may involve angles of attack that are intermittently below the stall angle of attack, whereas the angle of attack must be greater than the stall angle of attack for a spin to exist. Some post-stall gyrations can be self-recovering, meaning that no deliberate recovery technique is required.

14.1.1.4 **Spin**. Finally, a spin must be defined[7].

> A **spin** is a sustained yawing and rolling rotation at angles of attack above the stall angle of attack.

Sometimes the spin is described as a yawing rotation rather than a yawing and rolling rotation, but such a description is only valid to define a perfectly flat spin with $\alpha = 90°$. Typically, the yaw rate about the z body axis is the dominating rotation. A developed spin can be either upright or inverted, the latter meaning that the airplane is stalled at negative angle of attack. The pilot of an airplane will likely consider a sustained yaw rate one of the most important visual cues indicating that he is in a spin. In fact, if the spin is steep (that is, the angle of attack is relatively low, near α_s), pilots often have difficulty recognizing the motion as a spin. Such a trajectory may be perceived as a roll rather than a spin. A steep inverted spin is another kind of spin that can be even more confusing and disorienting because the body axis roll and yaw rates are in opposite directions, and the pilot is hanging upside down in the seat. Even in an inverted spin, yaw rate dictates the direction of rotation and the control manipulations necessary to recover.

14.1.1.5 **Deep Stall**. The last out-of-control flight condition to be discussed is called the deep stall:

> A **deep stall** is an out-of-control flight condition in which the airplane is in equilibrium at an angle of attack well above α_s while experiencing negligible rotational velocities.

Chapter 14   Stall, Post-Stall, and Spin Testing

The deep stall is distinguished from a post-stall gyration by the lack of significant angular rates. Some motions with graphic names like "falling leaf" are categorized as deep stalls, but so are others that have large amplitude angle of attack oscillations. In a deep stall, there is almost always a very high rate of descent. It is fairly common for the lateral and directional controls to be ineffective and sometimes even the longitudinal control loses its authority. The latter condition, of course, is to be avoided in any design. The motion during a deep stall may not be steady. Oscillations in angle of attack may be quite large in amplitude. These oscillations sometimes are helpful in that longitudinal control inputs of the appropriate frequency have been used to dynamically generate the moments necessary to recover from the deep stall.

Table 14.1 Spin Mode Modifiers

Group 1	Group 2	Group 3
Upright	Steep	Smooth
		Mildly Oscillatory
Inverted	Flat	Oscillatory
		Highly Oscillatory
		Violently Oscillatory

14.1.2 Spin Modes

Three types of adjectives are used to describe the general characteristics of a given spin and they identify the spin mode. Note that many airplanes exhibit more than one distinguishable mode. Table 14.1 summarizes these modifiers. The sign of the angle of attack, for example, classifies a spin as either upright ($\alpha > 0$) or inverted ($\alpha < 0$). Average values of angle of attack also classify a spin as either flat or steep, meaning that flat spins occur at very high α (>60°) and steep spins occur with lower α. Finally, the peak amplitudes of the oscillations compared with the average rate of rotation spells out the oscillatory character of the spin mode. One descriptive modifier from each of these adjective groups may be used to characterize the spinning motion.

Mode identification terminology can be confusing. The most confusing facet of adjective identifier usage is associated with groups 2 and 3 in Table 14.1. The F-4E example[8] in Table 14.2 should help clarify usage of these modifiers.

Table 14.2 F-4E Spin Modes

Mode Description	Average AOA (°)	AOA Oscillations (°)	Yaw Rate (°/sec)	Roll Rate (°/sec)	Pitch Rate (°/sec)
Steep-Smooth	42	±5	40-50	50	15
Steep-Mildly Oscillatory	45-60	±10	45-60		
Steep-Oscillatory	50-60	±20	50-60 (with large oscillations)	Same as yaw rate	
Flat-Smooth	77-80	Negligible	80-90	25	7

14.1.3 Spin Evolution

A developed spin is an equilibrium dynamic state. However, this state of equilibrium is not reached by every trajectory that progresses into this nonlinear post-stall flight regime. Equilibrium for the strongly coupled, large amplitude governing equations is not necessarily a steady state solution in the classical linear sense; it is more often a periodic solution (like a limit cycle). Such a complicated trajectory does not occur instantaneously; the transient motions are quite observable and are especially important to the flight test planner (as well as to the operational pilot!). We will discuss three stages or phases (Fig. 14.3) of spin evolution; but, keep in mind that not all spins reach equilibrium. Thus, a given spin trajectory may pass through all or only one of the phases. It does not always progress through the complete evolution to the developed equilibrium state, and even deliberate attempts to force a stall may inexplicably pop out of an incipient spin to controllable, low angle of attack flight.

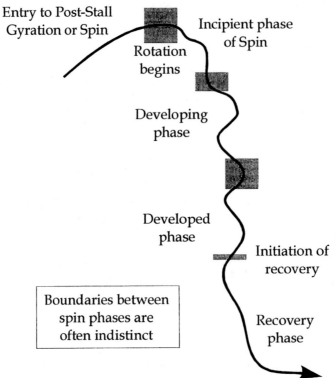

Fig. 14.3 Spin Phases

14.1.3.1 Incipient Phase.
The incipient phase of a spin is the initial, transitory part of the motion during which it is impossible to identify the spin mode. The yawing and rolling rotation (usually perceived by the pilot as yaw rate) begins as the incipient phase begins. The cue to the pilot of an incipient spin is that of a sustained, though not necessarily steady, increasing yawing rotation. The increase can be either monotonic or oscillatory, the oscillatory entry exhibiting increasingly higher peaks. Even during this incipient phase, the angle of attack is greater than the stalled angle of attack. This characteristic, stalled angle of attack in the spin is one of the distinctions between the incipient phase of the spin and any post-stall gyration that occurs after departure. The incipient phase of the spin ends when a recognizable mode can be identified. Sometimes, this mode is not recognizable in flight but must be identified by post-flight data analysis. To summarize, the in-

cipient phase of the spin is a transitory motion easily confused with a PSG, but distinct from either a PSG or the developed phase of a spin.

14.1.3.2 Developing Phase. The developing phase of a spin is that stage at which the spin mode can be ascertained. Oscillations may be present, but the mean periodic motion is discernible. The aerodynamic forces and moments may not be completely balanced by the corresponding linear and angular accelerations, but equilibrium conditions are being approached. Generally, it is evident when the developed phase is reached, though the exact point at which it began may be unclear. Motion is approaching an equilibrium state, and it is frequently advisable to initiate recovery before true equilibrium is achieved. For example, during the T-38 test program, warning lights were installed to signal a buildup of yaw rate. Recoveries were initiated when these lights came on at a specified threshold. In the flat spin mode of this basic trainer, a peak yaw rate was considerably above this threshold value. When the lights came on, the spin chute was deployed. The longitudinal acceleration (along the negative x-axis) at the pilot's station approached physiological limits, often causing temporary "red-out". Moreover, this spin could only be terminated by deployment of the spin chute[9]. Thus, the developing spin, while it may be more comfortable if it is less oscillatory than the incipient phase, can be deceptively dangerous. Careful preflight planning, continuous monitoring of the testing, and a reliable emergency recovery device is absolutely essential when intentionally allowing a spin to develop.

14.1.3.3 Developed Phase. A spin is developed when it reaches the final equilibrium state. In this stage of its evolution the motion is typically a repetitive periodic trajectory. Theoretically, all the equilibrium states are constant and all linear and angular accelerations are zero on average over a complete period. Instantaneously, at least some of the state rates of change are nonzero if the motion is oscillatory in any fashion. Practically, attaining this state may take a very long time. It may also be hard to recognize in the cockpit or from real-time telemetry data if the periodic motion is oscillatory. One very important type of fully developed spin is, however, quite recognizable: the flat spin, in which the angle of attack is theoretically 90°. The difficulty in recognizing a post-stall equilibrium condition is one of the main dangers inherent in post-stall/spin flight tests. It is one of the main reasons this kind of testing is hazardous and must be treated with special attention to detail throughout test planning and during the conduct of such tests. All systems, whether unique to the test effort or production systems, must enable the test pilot to safely recover from the worst case of dynamic equilibrium in a developed spin. Thus, the developed phase of the spin deserves considerable attention from the flight test engineer. Consequently, we will shortly develop six degree-of-freedom, fully coupled, nonlinear equations of motion for a rigid aircraft in this kind of trajectory. But before we proceed to these equations, we must anticipate the likely flight path and the underlying aerodynamic forcing functions so we can make intelligent approximations.

14.1.4 Flight Path in a Spin

A spin is a coupled motion that requires all six equations of motion for satisfactory analysis. Extreme attitudes and large angular rates are often encountered in such motions. In the general case the airplane center of gravity describes a helical path while it rotates about an axis of rotation at the center of the helix. As sketched in Fig. 14.4, the helical path may be curved and the total angular velocity vector is continually changing in both magnitude and direction. In general the wings are not level and sideslip is not zero. This general motion is very complex. However, if we make the following simplifying assumptions, some of the basic characteristics of spin trajectories can be clarified.

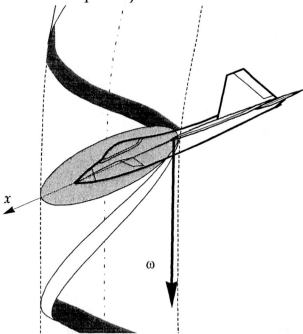

Fig. 14.4 Helical Flight Path in a Fully Developed Spin

14.1.4.1 **Assumptions**. The following assumptions simplify the analysis while retaining the more important features of spin trajectories.

- A developed spin is one in which the helical axis is vertical (ω is constant).
- No sideslip occurs, which suggests that the wings are approximately horizontal.
- The side force is negligible compared to the lift and the drag.

These assumptions mean that the angular velocity vector ω is approximately parallel to the helical flight path (and, consequently, to the relative wind).

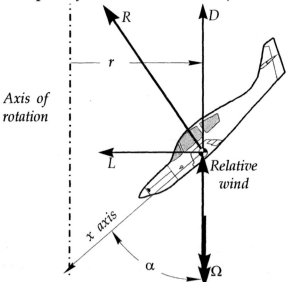

Fig. 14.5 Changes in Lift and Drag Coefficients at High Angles of Attack

14.1.4.2 **Balance of Forces**. Next, consider the balance of forces in a such a steady spin without sideslip as shown in Fig. 14.5. Because of assumption 3, the resultant aerodynamic force acts within the plane of symmetry and is approximately normal to the wing chord. Summing the vertical forces gives:

$$W = D = \frac{\rho V^2 S C_D}{2} \qquad (14.1)$$

Similarly, the forces in the horizontal direction just balance the centripetal acceleration:

$$mr\omega^2 = L = \frac{\rho V^2 S C_L}{2} \qquad (14.2)$$

The physical implications of these two seemingly trivial equations are extremely important. First, eqn. 14.1 says that, as α increases, C_D increases and, consequently, true airspeed decreases. Since the trajectory is vertical, true airspeed is approximately equal to rate of descent. In other words, as α increases, the airplane becomes more and more like a rotating bluff body in the descent and the descent rate slows in a flat spin. Furthermore, for $\alpha > \alpha_s$, C_L decreases as α increases. With these facts in mind, the left side of eqn. 14.2 must decrease as α increases. It will be shown shortly that the rotation rate ω tends to increase with increasing α. In turn, then, the radius of the helix r must decrease rapidly as α increases. These observations suggest that in a fully developed, flat spin, ω and V are nearly coincident. The angular difference between the flight path and vertical is approximated by Babister[10] as: $\tan \lambda = \frac{r\omega}{V}$. A typical variation of λ with angle of attack is about 5.5° at an α of 50° and about 1° at an α of 80°. Hence, it is justified to assume in developed spins that ω is nearly parallel to V and, for flat spins, the two vectors essentially coincide.

These assumptions are valid only if the wings are horizontal and sideslip is zero. This complexity is beyond the scope of this introductory text; both Babister[10] and Kerr[11] provide insight into these topics for the interested reader. Remember, these comments are valid only for the developed spin. The underlying physical phenomena are clarified by considering this simplified case, and hopefully, these simplifications give the student a better appreciation of the more important components of this type of motion.

14.1.5 Aerodynamic Factors in a Spin

In the post-stall flight regime aerodynamic forces acting upon the vehicle are very different than those during flight at low angle of attack, where the forces and moments vary linearly with disturbances in α, β, body rates or controls. Aerodynamic derivatives, usually described as linearized approximations with first order partial derivatives, are highly dependent on both α and β and second order derivatives become important to the analyst. Often these derivatives change sign rapidly as flows separate and reattach and vortex lift patterns break down. Hysteresis may also occur. Terms which are insignificant at low angle of attack may become very important. Stough[12,13] and his associates at Langley have concluded that, at least for straight-wing, wing-loaded airplanes (typical general aviation aircraft), the aerodynamic behavior of the wing is often the single most important factor in how the airplane itself enters and recovers from a post-stall maneuver. Though these conclusions cannot be blindly applied to all airplanes, they do suggest that the wing's behavior is a good place to start our consideration of post-stall aerodynamics.

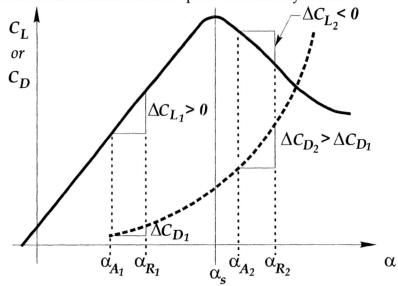

Fig. 14.6 Aerodynamic Mechanisms for Autorotative Moments

14.1.5.1 **Autorotative Couple of the Wing**. The wing is often the major contributor to the aerodynamic mechanism that initiates and sustains autorotation. If a wing is operating at a low α and a sudden local change in α occurs due to a rapid roll, there is a restoring moment developed because of the increase in lift generated by the higher α on the downgoing half of the wing and the decrease in lift on the half of the wing that is moving up (see the left side of Fig. 14.6, where ΔC_{L_1} and ΔC_{D_1} occur). This difference in lift is the mechanism for roll damping. In contrast, when a wing rolls rapidly at $\alpha > \alpha_s$ (see the right side of Fig. 14.6, where ΔC_{L_2} and ΔC_{D_2} occur), the downgoing wing experiences a loss in lift and an increase in drag. This loss of lift creates a propelling rolling moment (negative roll damping) and the increase in drag produces propelling yawing moment in the direction of the roll. This latter situation is clearly unstable and self-perpetuating; it leads to an autorotation.

A more complete understanding of the wing's aerodynamic forcing mechanism comes from Fig. 14.7 (next page), where the orientation of the resultant forces is shown, as well as the components of the resultant forces resolved along the x and z body axes. The net difference in the X forces at corresponding sections in the advancing wing panel and in the retreating wing panel produces a couple $(X_A - X_R)2y_1$. This couple is shown at the top of this plan view of the wing. The yawing moment produced by $(X_A - X_R)2y_1$ is positive for a positive yaw rate. The rolling moment produced by the couple $(Z_A - Z_R)2y_1$ is also positive. Thus, the incremental moments generated as the angular velocity increases lead to a moment contribution from the wing that propels the airplane into a spin. Of course, eventually the damping in yaw and damping in roll will become large enough to substantially oppose the autorotative motion, but it is quite clear that a fully developed spin can be sustained indefinitely. A fully developed spin is an equilibrium dynamic state and its equilibrium nature is part of what makes it dangerously hard to recover from such a motion. Aerodynamic controls are not always powerful enough to provide a large enough moment to break up the equilibrium.

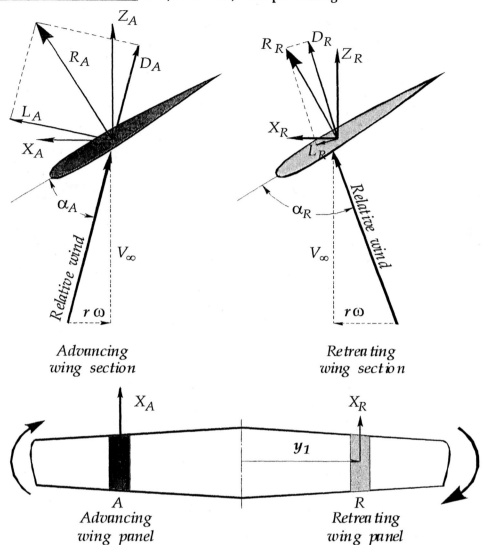

Fig. 14.7 Lift and Drag on Advancing and Retreating Wing Panels

14.1.5.2 Effect of Forebody Shape. The fuselage, especially the nose, may also contribute significant aerodynamic moments with $\alpha > \alpha_s$. The pressure field around the nose depends strongly on its cross-sectional shape. Since the flow character is highly dynamic, often dominated by vortex shedding, prediction of the aerodynamic contribution of the forward fuselage components is rather difficult. Small geometric changes in this region can make dominating changes in rolling and yawing moment coefficient in the post-stall flight regime. One possible fuselage effect is illustrated in Fig. 14.8 (next page). In these sketches we are looking at a roughly elliptical cross-section for a fuselage nose as an observer facing forward in the airplane. Naturally, this shape acts like an airfoil, producing both aerodynamic forces perpendicular (like lift) and parallel (like drag) to the local flow direction. This local flow direction is strongly influenced by the rotational rate of the airplane. As shown on the left side of Fig. 14.8, if the aircraft is rolling and yawing to the left and the net aerodynamic force vector is to the left, the local effect is to reinforce the autorotative rolling and yawing moments. The right side of Fig. 14.8 shows the same cross-section modified with a strake. With this modification the nose produces an aerodynamic moment that opposes autorotation. The spin characteristics of the T-37 trainer were considerably improved by such a modification, incorporated during developmental flight tests.

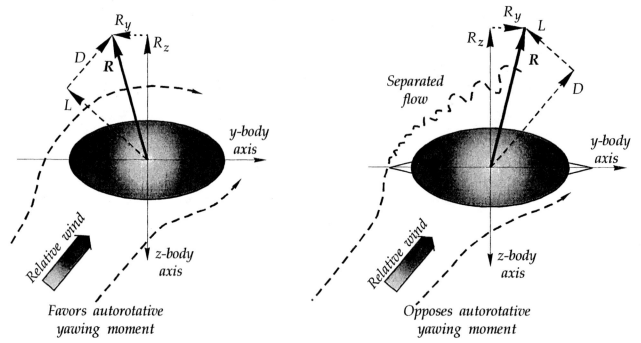

Fig. 14.8 Aerodynamic Contributions from the Forebody Shape

14.1.5.3 Effect of Damping Derivatives. Aerodynamic damping derivatives are significantly affected at high angles of attack, especially roll damping (C_{ℓ_p}). In fact, often both C_{ℓ_p} and C_{n_r} become instantaneously positive at or near α_s. These aerodynamic effects can be deduced by considering moments produced by the local lift and drag forces depicted in Figs. 14.5 and 14.7. Several authors consider this loss of damping to be one of the main aerodynamic factors resulting in autorotative behavior[10,12,13]. Obviously, a complete discussion of these aerodynamic coefficients and how they behave in the post-stall flight regime is beyond the scope of this introductory text. But, these references suggest additional reading for the interested student.

14.1.6 Effect of Mass Distribution

The moments of inertia meaningfully express the mass distribution. The inertial forces and moments, which depend directly on the mass distribution, become important in the post-stall region. While frequently neglected as insignificant in linear models of cruise flight dynamics, they may be of the same order of magnitude as the aerodynamic forces and moments. The relative sizes of the inertial forces and moments and the aerodynamic forces and moments are features peculiar to flight at post-stall angles of attack.

14.1.6.1 Principal Axes. Orthogonal body axes for which all three products of inertia are zero are called the principal axes for the body. Principal axes can be found for any rigid body. If the vehicle has a plane of symmetry, as is common for most aircraft and missiles, it is fairly easy to choose a coordinate system that approximates the principal axes. The xz plane is simply chosen to lie in the plane of mass symmetry, thus guaranteeing that I_{xy} and I_{yz} are 0. The direction of the x axis is usually chosen so it can be easily identified with configuration geometry. Thus, the term **fuselage reference line (FRL)** is often seen in airplane layout drawings. For brevity this reference set of body axes is usually called simply the body axes. (Strictly speaking, the principal axes and the stability axes used with linearized, small perturbation equations of motion are also body axes.) To simplify our later discussion and to make the large amplitude equations of motion easier to understand, we will take the body axes and the principal axes to coincide. An illustration of the difference between these two body axes is shown in Fig. 14.9.

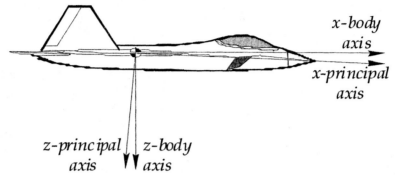

Fig. 14.9 Relationship between Body and Principal Axes

In the principal axis system, one of the moments of inertia is the maximum possible for the body. Choosing body axes that are close to principal axes means the product of inertia I_{xz} is quite small. A small value of I_{xz} reduces the cross coupling between the large amplitude moment equations that govern the symmetric (longitudinal) motion and the asymmetric (lateral-directional) motion. For the linearized small perturbation equations, this term had to be negligible in order to separate the longitudinal equations from the lateral-directional ones. That simplification gave us the luxury of solving two independent quartic characteristic equations rather than a higher order characteristic equation. Even though I_{xz} is typically small due to our choice of body axes, its coupling effects can rarely be ignored in post-stall maneuvers. We will show in a later section that pitch rate can have a significant effect on ω, even if I_{xz} can be safely ignored.

14.1.6.2 **Relative Density**. The relative density parameter μ is a measure of the ratio of the airplane's density to the density of the air its volume displaces. It is a very useful non-dimensional parameter that appears in the requisite equations.

$$\mu = \frac{2m}{\rho S b} \qquad (14.3)$$

14.1.6.3 **Center and Radius of Gyration**. Two other definitions are used in discussing mass distribution: center of gyration and radius of gyration. The center of gyration of a body with respect to an axis is that point at which the mass of the body could be concentrated and the resulting mass moment of inertia about the specified axis would be unchanged. Similarly, the radius of gyration k_i with respect to an axis is the distance from the center of gyration to that reference line. Recalling the basic definitions of the mass moments of inertia, we can write expressions for each of the three radii of gyration:

$$\int (y^2 + z^2)dm \equiv I_x = k_x^2 m$$
$$\int (x^2 + z^2)dm \equiv I_y = k_y^2 m \qquad (14.4)$$
$$\int (x^2 + y^2)dm \equiv I_x = k_x^2 m$$

14.1.6.4 Relative Magnitude of Airplane Moments of Inertia.

The relative sizes of moments of inertia for an airplane are a convenient means of classification (Fig. 14.10). The moment of inertia about the z body axis is always the largest of these inertias; this fact is assured because airplanes have little mass outside the xy plane. From a mass-distribution point of view, the vehicle is "flattened" into this plane. A wing-loaded airplane is defined as one that has more mass concentrated in the wings. The wings may be structurally heavy; fuel may be carried inside integral tanks in the wing; engines may be slung in nacelles from the wing; or stores may be loaded externally on pylons attached to the wings. Any one or all these factors cause more mass to be located in the wings. $I_z > I_x > I_y$ for wing-loaded airplanes. Figure 14.10a illustrates such an airplane. A neutrally loaded airplane is one in which the wing and the fuselage have roughly equal mass, that is, $I_x = I_y$ (Fig. 14.10b). Finally, a fuselage-loaded airplane is one in which most of the mass is concentrated in the fuselage (Fig. 14.10c). Most modern fighter airplanes are fuselage-loaded, which helps to give them very distinctive spin characteristics. We can gain some insight into just what kinds of spin modes might develop for different mass distributions (or "loadings") by careful inspection of the equations of motion.

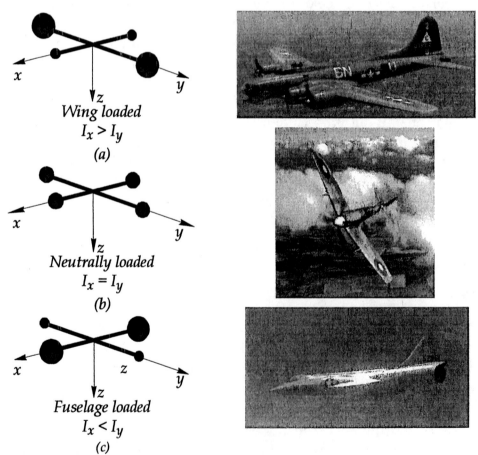

Fig. 14.10 Magnitude of Aircraft Moments of Inertia (USAF[12,13] and NASA photos[14])

14.1.7 Simplification of the Post-Stall Equations of Motion

We have been referring to the large amplitude equations of motion freely and discussing terms within these equations; it is now time to examine these equations directly. Equations of motions describing aircraft dynamics in the post-stall flight regime are more

Chapter 14   Stall, Post-Stall, and Spin Testing   123

complicated than the linearized, small perturbation equations discussed in Chapter 9 of Volume 1. The post-stall equations differ in the following ways:

- all six rigid body degrees of freedom must be considered in any solution -- pitching motions do couple into the lateral-directional response and vice versa;
- the changes in states are often quite large -- small perturbation assumptions for angular positions, rates, and accelerations are simply not warranted; and
- the aerodynamic coefficients are not constants -- stability derivatives are functions (and seldom analytic ones) of both angle of attack and sideslip angle at least.

14.1.7.1 **Simplifying Assumptions**. The basic simplification we will use is that only a developed spin with the wings horizontal is considered. This motion state is an equilibrium condition, in which the aerodynamic forces and moments and the inertial forces and moments are balanced. To further simplify the discussion, we further require that all applied moments are aerodynamic; in fact, we do not allow even propulsive moments. This latter assumption avoids a lengthy and confusing discussion of gyroscopic moments generated by rotating parts of the engines. Further, any coupling between the propulsion system and the aerodynamic controls is ignored. Summarizing these simplifications:

- Because the wings are horizontal, the angular velocity vector ω lies entirely within the xz plane. Its only components are P and R; Q is zero.
- Both V and ω are vertical for a developed spin and V = rate of descent.
- Angular velocity is constant, making $\dot{P} = \dot{Q} = \dot{R} = 0$ and time per turn constant.
- Body axes are assumed to coincide with the principal axes; that is, $I_{xz} = 0$.

The force balance takes on one of the two following forms. In an inertial frame of reference or a body frame of reference, the vector equations are:

$$F = m \frac{dV}{dt} \text{ for the inertial frame} \quad \text{or} \quad F = m\left(\frac{dV}{dt} + \omega \times V\right) \text{ for the rotating body axis frame}$$

of reference. Since we will only look at steady state conditions, $\frac{dV}{dt} = 0$ in either reference system. Also, ω and V are parallel (assumption 2), so $F = 0$ in body axis coordinates.

14.1.7.2 **Simplified Large Amplitude Moment Equations**. The summation of moments is by far the more important balance for equilibrium and it is usually only considered in the rotating frame of reference. In component form the moment equations are:

$$\frac{dP}{dt}I_x + QR(I_z - I_y) - \left(\frac{dR}{dt} - QR\right)I_{xz} = \mathcal{L}$$

$$\frac{dQ}{dt}I_y + PQ(I_x - I_z) - \left(R^2 - P^2\right)I_{xz} = M \quad (14.5)$$

$$\frac{dR}{dt}I_z + PQ(I_y - I_x) - \left(\frac{dP}{dt} - PQ\right)I_{xz} = N$$

Applying our assumptions and solving for the angular accelerations:

$$\frac{dP}{dt} = \frac{\mathcal{L}}{I_x} + \frac{QR(I_y - I_z)}{I_x}$$

$$\frac{dQ}{dt} = \frac{M}{I_x} + \frac{RP(I_z - I_x)}{I_y} \qquad (14.6)$$

$$\frac{dR}{dt} = \frac{N}{I_x} + \frac{PQ(I_x - I_y)}{I_z}$$

The first terms on the right side of eqns. 14.6 are aerodynamic contributions to angular acceleration and the second terms describe inertial contributions to pertinent accelerations. Thus, rewriting these moment equations in terms of angular accelerations shows that, for equilibrium to occur (fully developed spin; all angular accelerations vanish), the inertial term must balance the aerodynamic term in each equation. Even with our simplifying assumptions, coupling between the equations is still present. These simplified equations also suggest that, to calculate absolute angular accelerations, it is important to know the moments of inertia accurately. Obtaining them experimentally is a tedious job[17]; one that flight test personnel are often tasked to complete.

Before we expand the aerodynamic terms, it is useful to point out a difference in the usual nondimensionalization of aerodynamic coefficients for use in the coupled equations that govern post-stall trajectories. These equations differ from the small perturbation equations that spawned our original nondimensionalization of stability derivatives. They cannot be separated into uncoupled longitudinal and lateral-directional subsets. Typically, data sets must be modified from our usual concept of stability derivatives for use in modeling post-stall dynamics. Nondimensionalization of pitching moment coefficients (C_m) uses the wing mean aerodynamic chord as the reference length, while both rolling and yawing moment coefficients ($C_\mathcal{L}$ and C_n) use the wing semispan. For the data to represent the same aerodynamic moments in the large amplitude equations of motion, a common reference length must be chosen. For this kind of analysis, it is common practice to base pitching moment on wing span and designate it as C_{mb}. Numerically, the usual small-perturbation pitching moment coefficient is altered by $C_{mb} = C_m \frac{\bar{c}}{b}$. The aerodynamic terms in eqns. 14.6 can be rearranged using the radius of gyration of eqns. 14.4.

$$\frac{\mathcal{L}}{I_x} = \frac{\rho V^2 SbC_\mathcal{L}}{2k_x^2 m} = \frac{V^2 C_\mathcal{L}}{2\mu k_x^2} \quad \text{and} \quad \frac{dP}{dt} = \frac{\mathcal{L}}{I_x} + \frac{QR(I_y - I_z)}{I_x}$$

$$\frac{M}{I_y} = \frac{\rho V^2 SbC_{mb}}{2k_y^2 m} = \frac{V^2 C_{mb}}{2\mu k_y^2} \quad \text{and} \quad \frac{dQ}{dt} = \frac{M}{I_y} + \frac{RP(I_z - I_x)}{I_y} \qquad (14.7)$$

$$\frac{N}{I_z} = \frac{\rho V^2 SbC_n}{2k_z^2 m} = \frac{V^2 C_n}{2\mu k_z^2} \quad \text{and} \quad \frac{dR}{dt} = \frac{N}{I_z} + \frac{PQ(I_x - I_y)}{I_z}$$

Expressing the definitions of the aerodynamic moments in nondimensional terms,

Chapter 14 Stall, Post-Stall, and Spin Testing

$$\frac{dP}{dt} = \frac{\rho V^2 Sb C_l}{2k_x^2 m} + \frac{QR(I_y - I_z)}{I_x}$$

$$\frac{dQ}{dt} = \frac{\rho V^2 Sb C_{mb}}{2k_y^2 m} + \frac{RP(I_z - I_x)}{I_y} \quad (14.8)$$

$$\frac{dR}{dt} = \frac{\rho V^2 Sb C_n}{2k_x^2 m} + \frac{QR(I_x - I_y)}{I_z}$$

14.1.8 Aerodynamic Conditions for Dynamic Equilibrium

Combining eqns. 14.8 with the first assumption, gives equilibrium conditions.

$$0 = C_l$$

$$-\frac{\rho V^2 Sb C_{mb}}{2k_y^2 m} = \frac{RP(I_z - I_x)}{I_y} \quad (14.9)$$

$$0 = C_n$$

Remember that these equations are greatly simplified; they do not completely represent post-stall motion. But they allow us to introduce the idea that a developed spin is an equilibrium or "trim" condition, similar to straight and level flight. With no external disturbances, the airplane will remain in this dynamic state until it impacts the ground. Now, let's look at each of the component moment equations above and understand their implications for a fully developed spin. We start with the pitching moment equation, since it has the most interesting form in eqns. 14.9.

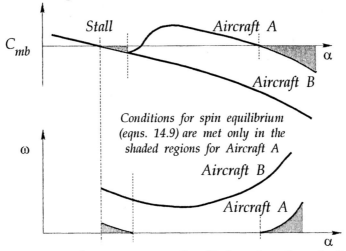

Fig. 14.11 Pitching Moment Coefficient in a Steady Spin

14.1.8.1 Pitching Moment Equation.
Consider eqn. 14.10b in conjunction with typical pitching moment curves for a fighter aircraft (Fig. 14.11). From our original definition of a spin, we're interested in α greater than α_s. For the typical highly maneuverable airplane, our simplified pitching moment equation implies that the term with C_{mb} in it must be of opposite sign to the inertial term on the right side of eqn. 14-10b. For an upright spin, this condition means that C_{mb} must be negative, since in an upright spin P and R are of the same sign (see Fig. 14.12). Combining this observation with the fact that $I_z > I_x$ for any airplane, clinches the argument that $C_{mb} < 0$.

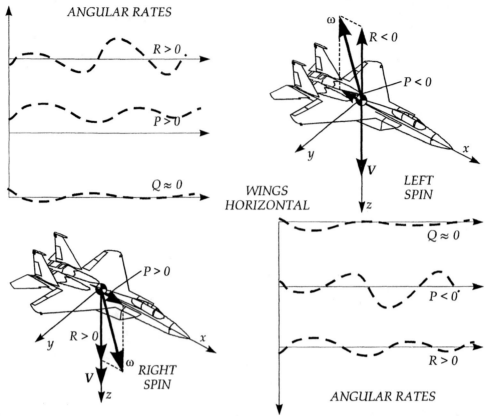

Fig. 14.12 Angular Velocity Components in a Typical Upright Spin

The slope of the C_{mb} - α curve must be negative for stability about an equilibrium condition, even in post-stall conditions. This slope must be negative for stable longitudinal equilibrium at either all angles of attack. Conversely, $\dfrac{dC_{mb}}{d\alpha} > 0$ represents an unstable condition and absent other control inputs within that region, aircraft B above would pitch up or down to reach one of the two stable trim points on either side. A nonzero angular acceleration about any axis (in our case about the y axis) violates the basic assumption of constant ω that we have used do define a developed, steady spin. Said another way, any disturbance in α produces a ΔC_{mb} that tends to restore α to its initial value only if $\dfrac{dC_{mb}}{d\alpha} < 0$.

Figure 14.11 summarizes pitching moment constraints for a developed spin by directly comparing two different aircraft, each with quite different dynamic equilibrium possibilities. Aircraft B is capable of sustaining a fully developed upright spin at any angle above α_s, insofar as the pitching moment equation is concerned. For this aircraft, C_{mb} and $\dfrac{dC_{mb}}{d\alpha}$ are always negative. Elevator setting would likely control spin rotation rate. However, airplane A can only meet the two constraints suggested above if α is in the shaded regions in the Fig. 14.11. Airplane A has two possible spin modes indicated in the sketch, while airplane B has only one.

Of course, the pitching moment is not the whole story; the rolling and yawing moment equilibrium equations must also be considered and we should give our attention to how

Chapter 14 — Stall, Post-Stall, and Spin Testing

they impact dynamical equilibrium. But first, an estimate of ω for each of the potential equilibrium conditions is useful and is easily obtained.

Figure 14.12 also emphasizes that $P = \omega\cos\alpha$ and $R = \omega\sin\alpha$ for our idealized, fully developed spin. Substituting these expressions into eqn. 14.6b gives

$$-\frac{M}{I_y} = \frac{RP(I_z - I_x)}{I_y} = \frac{(I_z - I_x)}{I_y}\omega^2 \sin\alpha\cos\alpha \text{ or, solving for } \omega,$$

$$\omega = \sqrt{\frac{-2M}{(I_z - I_x)\sin 2\alpha}} \tag{14.10}$$

Equation 14.10 suggests that a minimum rotation rate occurs at 45° angle of attack, though the nonlinearities in C_{mb} at these flight conditions may be strong enough to change the angle of attack for minimum rotation rate noticeably. But eqn. 14.10 tells only part of the story; it does not spell out all conditions for equilibrium that must be met for a fully developed spin to occur.

14.1.8.2 Rolling and Yawing Moment Equations. Equations 14.9 suggest other constraints that must be met for a fully developed spin to occur. Although we did not emphasize the point in the preceding paragraph, all the aerodynamic derivatives (even C_{mb}) are functions of α, β, and ω. Following Anglin and Scher[18], we have considered C_{mb} to be solely a function of α in spite of its dependence on other parameters. It is also convenient to take C_ℓ and C_n as functions of ω alone, even though there is little justification for these assumptions, other than that lateral-directional parameters are more directly linked to rotation rate, while the longitudinal parameters are usually more directly linked to angle of attack. These assumptions are useful to understanding, but C_{mb}, C_ℓ, and C_n are all affected by α, β, and ω.

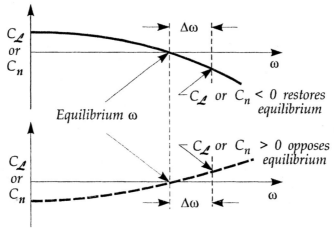

Fig. 14.13 Stabilizing and Destabilizing Slopes for C_ℓ and C_n

For a fully developed spin to occur the overall rolling moment and yawing moment coefficients, C_ℓ and C_n, must be zero and their rates of change with respect to ω must be negative. The first pair of these conditions is explicitly stated in eqns. 14.9. The second pair of conditions is a requirement for stability very much like the one that says C_{m_α} must be negative for static longitudinal stability to exist. Figure 14.13 illustrates this idea. If an increase in ω produces and increase in either or both rolling and yawing moment, then one or both of these total moments cannot remain in equilibrium for small changes in ω. That is, any change in rotation rate will cause the autorotative moments to diverge away from the nominally stable rotation rate. For a true equilibrium state to exist, any deviation from this "trim" or equilibrium ω must cause an increment in both C_l and C_n that tends to restore rolling and yawing moment to its equilibrium (non-zero) state.

These aerodynamic constraints are rarely all met exactly for an actual airplane spin. A fully developed spin is an idealization and occurs rarely in practice. It is much more common to have oscillatory spins in which these conditions are almost met or met momentarily, but true equilibrium is never fully sustained. From an engineering perspective, while it is still useful to base our concepts on idealizations like the developed spin, it is even more important to be able to estimate actual spin behavior from computed and wind tunnel aerodynamic data. Even though we expect less than perfect fulfillment of the assumptions listed, these estimates are quite important to safe and efficient flight test planning. Consequently, such estimates are often the responsibility of the flight test engineer. The following example shows how such estimates can be made.

14.1.8.3 **Estimation of Spin Equilibrium States**. Anglin and Scher[18] described in some detail how they estimated spin characteristics from both experimental and computational evidence for a model of the McDonnell F3H Demon, a carrier-based fighter of 1960s vintage. Although their original use of this empirical estimation process was intended to help them save computer time by establishing appropriate initial conditions for numerical modeling of high angle of attack trajectories, it serves as an excellent example of how to estimate post-stall behavior for flight test planning exercises. Later test planning refinements likely will be based first on the results of computer modeling of the trajectories and, when they become available, on previous post-stall flight tests. The procedure is summarized below and the interested student is referred to the original document[18] by Anglin and Scher for fuller details and a numerical example.

To use this procedure for test planning, one must have some prediction of the variation of rolling and yawing moments as functions of ω or, more commonly, the nondimensional form $\frac{\omega b}{2V}$. Using a rotary balance in an appropriately configured wind tunnel is one means of experimentally obtaining these data. It should be emphasized that each of the aerodynamic derivative variations in this data base were obtained by steadily rotating a model about an axis parallel to the relative wind direction in the vertical wind tunnel. Therefore, no oscillation in the angular rates was captured. If aerodynamic hysteresis existed (common in high angle of attack aerodynamic coefficients), these data would not account for it. This limitation (and similar shortcomings) on the data should always be kept in mind when making estimates.

Partly because the rotary balance rolling moment data for the F3H model were not "well-behaved", Anglin and Scher ignored the rolling moment data and concentrated on the yawing moment. This simplification amounted to discarding two of the limitations listed previously. The procedure then includes the following steps:

- Examine the pitching moment curve and the yawing moment curve from the rotary balance tests to locate nondimensional rotation rate(s) $\frac{\omega b}{2V}$ for which $C_n = 0$ and for which $\dfrac{dC_n}{d\left(\dfrac{\omega b}{2V}\right)} < 0$. There may be more than one of these potential equilibrium points. This step is meant to identify rotational rates for which the aerodynamic yawing moment is zero. To strictly apply the constraints of the previous section, we would have to repeat this process for rolling moment as well. Moreover, both conditions would have to be met simultaneously for a true developed spin mode to exist. Working with actual numerical data quickly convinces the practicing engineer that such idealized conditions do not exist often.

♦ Next, utilize rotary balance pitching moment data and the rotational rate (or rates if more than one exists) selected from the previous step to find a point on the pitching moment curve where inertial moment and aerodynamic moment are identical. This second step in the estimation procedure assumes that pitching moments were measured in (or can be interpolated from) the rotary balance data set for the $\dfrac{\omega b}{2V}$ at which $C_n = 0$ (as described in the preceding paragraph). Equation 14.10, rearranged to show the equality between the inertial pitching moment and the aerodynamic pitching moment, forms the basis for this second step in the procedure.

$$C_{mb} = -\dfrac{(I_z - I_x)\omega^2 \sin 2\alpha}{\rho V^2 S b} \tag{14.11}$$

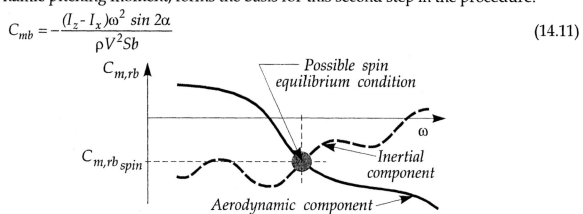

Fig. 14.14 Comparison of Aerodynamic and Inertial Pitching Moment Coefficients

C_{mb} is a function of $\dfrac{\omega b}{2V}$ and must be chosen to correspond to the candidate equilibrium value of $\dfrac{\omega b}{2V}$ that came out of step one. To check for an aerodynamic moment equal to the inertial pitching moment coefficient calculated from eqn. 14.11, a plot of the measured aerodynamic pitching moment coefficients and the calculation from this equation can be made (Figure 14.14). The intersection point represents a potential equilibrium point and, thus, a possible spin mode. Even though not illustrated on the sketch in Fig. 14.14, it is quite possible to have more than one intersection point and, consequently, more than one spin mode. The angles of attack for which a spin mode may exist are likely to occur in the vicinity of the angle of attack indicated by the point of intersection. Moreover, using this angle of attack and the value of C_{mb} indicated, ω for the estimated spin mode can be calculated. Then, individual roll and yaw rates can be estimated from this calculated ω and the estimated equilibrium angle of attack.

A comparison of the estimates that resulted from this procedure and computer simulations for the F3H Demon is shown in Table 3 (on the following page). It is unfortunate that flight test results from the spin tests of the actual airplane are not available for comparison to the computational and to the wind tunnel predictions.

Table 14.3. F3H Demon Computed Spin Modes versus Estimated Spin Modes

Computed Modes			Estimated Modes		
α (°)	ω (rad/sec)	V (ft/sec)	α (°)	ω (rad/sec)	V (ft/sec)
36.0	1.88	294	38.2	1.90	285
37.0	2.18	372	45.1	1.83	327
	Oscillated out of spin		48.2	1.89	453
51.8	2.18	619	50.5	2.18	620
80.0	4.72	494	70.0	3.50	515
36.5	2.80	380	37.4	2.69	365

Phillips[19] characterizes those attributes that commonly distinguish spin-resistant and spin-prone light airplanes for the interested student.

14.2 POST-STALL/SPIN TEST PREPARATION AND FLIGHT SAFETY

Post-stall/spin flight testing is one of the most hazardous forms of testing. By the very nature of the flight regime, the skill of the test pilot and the knowledge of the test engineer are stretched to new boundaries. After all, the test airplane and its crew are deliberately put at risk in order to make the vehicle safer for normal operations. Consequently, flight test planners must make extraordinary preparations so that exposure to dangerous conditions is minimized. A large amount of data is collected during brief events; and, while the pilot's qualitative opinion is invaluable, he cannot adequately function as a sensor in such highly dynamic tests. He can be easily misled during such maneuvers. Test equipment and test procedures specifically designed to reduce the risk of catastrophic errors must be installed, verified, and used. As the title of this section differs from the "Test Method" of earlier chapters, so do the planning and conduct of post-stall/spin tests from the usual performance or stability and control tests.

14.2.1 Pre-Test Planning and Preparation

Planning for any flight test must start several months before the vehicle rolls off the assembly line. The planning must include decisions like:

- ◆ What are the objectives of the tests? What critical questions must be answered?
- ◆ What data are required to substantiate answers to the critical questions? How can these data best be collected? What hardware is essential and what instrumentation is available? What specialized instrumentation is required?
- ◆ What legacy tests on similar airplanes might provide insight into the behavior of this airplane? (Be careful - legacy results can mislead as well as inform.)
- ◆ What special training will be required for the crew? What criteria should be used in selecting the crew?

Chapter 14   Stall, Post-Stall, and Spin Testing   131

- What safety procedures are to be followed? What special safety equipment must be installed? Spin chutes? Battery packs? What is the expected reliability of the engine under these conditions?
- How will the flight test support (chase aircraft, real-time test monitoring, video or camera coverage, communications links, and the like) be tailored? Is radar coverage required? What air traffic control coordination is needed?

In answering such questions, the test team must be guided by two competing principles:

> *Test objectives are the prime drivers behind the test; they must be met.*

> *Safety for the flight test crew, for the test resources and for the public are important (sometimes overriding) constraints to meeting test objectives.*

14.2.1.1 <u>Test Objectives</u>. Although setting clear test objectives is one of the first (both in importance and certainly in chronology) points that a flight test team must address, it is not always given the high priority it deserves. Engineers often think too much in terms of specific problems and quantified answers, without first being sure of what the purpose of their problem-solving is and determining if the problems they are solving are the ones that should be addressed. That approach to hazardous flight testing is courting disaster! Since the early 1970s, US government test agencies have been required to produce a master plan for doing test and evaluation[20], a common practice throughout the industry. There is no phase of testing where such an overview of test objectives is more important than in post-stall flight testing.

14.2.1.1.1 <u>Governing Principles</u>. But most embryo flight test engineers ask: Are these objectives not the responsibility of the design team (or the program office or the company project manager)? The answer is yes, but...! The manager ultimately responsible for the success of the design (whatever his title) is the decision-maker who sets objectives and their priorities. But you, as his flight test "experts", are his/her primary adviser on whether or not and how the critical questions can be answered by flight test. As that "expert," you will be called upon to help insure that all and only pertinent objectives are set, as well as being tasked to indicate how best to go about meeting them. To give professionally sound advice in this role means that you must understand the goals that should be set (you may even have to explain to the project manager why certain critical questions must be asked) and you must appreciate how and why the test objectives are prioritized. You must be both responsible and responsive. The governing principles in setting test objectives are:

> *Do all demonstrations that are essential to meet the requirements laid down by the customer and/or the certifying agency.*

> *For developmental testing, concentrate on those objectives that contribute to meeting the requirements. (Operational testing looks outside the requirements-driven objectives to exploit the capabilities of the new machine, but must remain within the flight envelope cleared by the developmental side of the test effort.)*

Research or exploratory objectives are relevant only when the program itself is structured for research purposes, or experimentation is the best (or only) way to credibly resolve an unknown. Operational tests often explore changes in tactics, training, and procedures that make better use of innovative features in a new weapons system. The basic guidance for OT&E test planning[21] states: "Consider organizational structure; tactics, techniques, and procedures (TTP); training; and any required supporting systems." Consequently, the two principles in the blocks often do not coincide in purpose; sometimes they even contradict! The compromises they imply must be settled **before** tests begin.

Several concrete examples should better illustrate this planning process. So, consider some representative post-stall flight test objectives for a high performance fighter built for military service and for a single engine general aviation airplane.

14.2.1.1.2 *Military Requirements*. The first step in writing such objectives is to scrutinize all applicable requirements that pertain to the post-stall flight regime, including at least the intent of MIL-HDBK-1797B[22] and MIL-S-83691 (USAF)[7] and any specific contractual paragraphs that modify or augment these high angle of attack specification documents. The excerpts (shaded below and in Table 14.4) from the latter document spell out demonstration requirements that lead to "essential" test objectives. (In Table 14.4 specific numbered paragraph and table references, as well as most of notes, are omitted for clarity.)

> *Flight Test Demonstration*. Each airplane type shall demonstrate, by flight test according to Table 14.4, the degree of compliance with the stall warning, loss-of-control warning (when required), resistance to loss of control, loss-of-control prevention, out-of-control recover, and spin recovery criteria originally specified in MIL-F-8785C (now MIL-HDBK-1797B[19]). The flight test investigation shall include an extensive evaluation of those maneuvers that may potentially result in AOA excursions to and beyond a permissible AOA flight limit, thereby subjecting the airplane to possible loss of control. Unless otherwise specified, the use of prolonged pro-spin controls to sustain a spinning condition shall not be required except for trainer-type aircraft to be cleared for intentional spins. However, reasonably delayed recovery attempts after a stall or departure, or exaggerated misapplication of controls following a stall or departure, to simulate possible incorrect pilot responses, shall be investigated under the least conservative circumstances to ascertain the degree of spin resistance/susceptibility for operational users. When spins do result as a natural consequence of testing through departures from controlled flight, a satisfactory spin recovery technique shall be demonstrated in accordance with MIL-HDBK-1797B. The results of these tests shall be used to establish the stall and limiting AOA's.
>
> *Stall/Spin Flight Test Variables*. The contractor shall establish, with the approval of the procuring activity, which of the following parameters are variables to be tested for influence on stall and post-stall flight characteristics.
>
> (a) Configuration
> (b) Gross weight
> (c) Center of gravity
> (d) Stability and control augmentation system status
> (e) External store loadings*
> (f) Stall and departure speed, altitude, and attitude
> (g) Thrust effects
> (h) Gyroscopic effects
>
> *The external store loadings should include as a minimum:
> (1) No external stores
> (2) Symmetric, fuselage-heavy [if significantly different from (1)]

(3) Symmetric, wing-heavy
(4) Asymmetric (maximum allowable asymmetry)

Table 14.4. Progressive Test Phases for Post-Stall Demonstration Maneuvers

Test Phases	Maneuver Requirements					
	Control Application	Smooth AOA Rate		Abrupt AOA Rate		Tactical
		One g	Accelerated	One g	Accelerated	
A Stalls	Pitch control applied to achieve the specified AOA rate, lateral/directional control inputs as normally required for the maneuver task. Recovery initiated after the pilot has a positive indication of: (a) A definite g-break, (b) a rapid angular divergence, or (c) the aft stick stop has been reached and AOA is not increasing.	Class: I, II, III, IV	Class: I, II, III, IV	Class: I, II, III, IV	Class: I, IV	Class: I, IV
B Stalls with Aggravated Control Inputs	Pitch control applied to achieve the specified AOA rate, lateral/directional controls as required for the maneuver task. When condition (a), (b), or (c) from above has been attained, controls briefly misapplied, intentionally or in response to unscheduled aircraft motions, before recovery attempt initiated.	Class: I, II, III, IV	Class: I, II, III, IV	Class: I, II, III, IV	Class: I, IV	Class: I, IV
C Stalls with Aggravated Control and Sustained Control Inputs	Pitch control applied to achieve the specified AOA rate, lateral/directional controls as required for the maneuver task. When condition (a), (b), or (c) has been attained, controls are misapplied, intentionally or in response to unscheduled aircraft motions and held for 3 seconds before recovery attempt is initiated.	Class: I, II, III, IV	Class: I, II, III, IV	Class: I, II, III, IV	Class: I, IV	Class: I, IV
D Spin Attempts*	Pitch control applied abruptly, lateral/directional controls as required for the maneuver task. When condition (a), (b), or (c) has been attained, controls applied in the most critical positions to attain the expected spin modes of the aircraft and held for up to 15 seconds before recovery attempt is initiated, unless the pilot definitely recognizes a spin mode.			Class: I, IV	Class: I, IV	Class: I, IV

*This phase required only for training aircraft which may be intentionally spun and for Class I and IV aircraft in which sufficient departures or spins did not result in Test Phases A, B, or C to define characteristics.

MIL-S-83691 (USAF), an older military specification[7] that succinctly spells out these demonstration requirements, lays out a progressively more and more demanding series of test phases. Where the airplane departs and/or spins determines whether or not the vehicle is departure-resistant or departure-susceptible and spin-resistant or spin-susceptible. Table 14.5 summarizes these demonstration requirements.

Table 14.5 Definitions of Departure and Spin Susceptibility and Resistance

Test Phase	Classification	
	Departures	Spins
A - Stalls	Extremely Susceptible	Extremely Susceptible
B - Stalls with Aggravated Control Inputs	Susceptible	Susceptible
C - Stalls with Aggravated and Sustained Control Inputs	Resistant	Resistant
D - Spin Attempts	Extremely Resistant	Extremely Resistant

Tables 14.4 and 14.5, along with the other excerpted paragraphs are only a very small part of the whole set of military documents that govern how a post-stall test matrix should be put together for a military airplane. But they serve our purpose, which is to introduce the student to some of the requirements and point him to those documents governing such a test. These documents and, most importantly the contractual specifications themselves, set the test objectives.

Both the F-15E Strike Eagle[23] and F/A-18 E/F Super Hornet[24,25] departure and spin programs adjusted the above sequence, and used another methodology for risk reduction. In both programs, batch simulations, using wind tunnel derived models, predicted several benign upright and inverted spin modes. The same contractor test engineer was responsible for both programs and deliberately planned spin testing to precede the aggravated input phase of testing. Aggravated inputs were expected to generate departures far more violent and disorienting than the spins, yet, after an incipient phase with fairly violent motions, were expected to progress into one of the predicted spin modes. The rationale behind the altered phase sequence was to fully explore the stable spin modes from controlled entries, permitting both the data station engineers and the pilots to gain familiarity with these modes. The intent was that, when later violent departures transitioned to familiar spin modes, the entire team would be prepared to recover from modes with which they already had experience. Both programs unfolded exactly as anticipated in this simulation-driven risk reduction planning. In the Super Hornet program, violent departures occurred several times at completely unexpected conditions, yet once the airplane settled into a spin mode the pilot recognized, recovery was quickly achieved.

14.2.1.1.3 *Civil Requirements*. For civilian certification, the following paragraphs (in the shaded and indented text) from FAR Part 23[26] are the foundations for and dictate post-stall flight test objectives for a general aviation airplane.

> **23.21 Proof of Compliance**
> (a) Each requirement of this subpart must be met at each appropriate combination of weight and center of gravity within the range of loading conditions for which certification is requested. This must be shown --
> (1) By tests upon an airplane of the type for which certification is requested, or by calculations based on, and equal in accuracy to, the results of testing; and

Chapter 14 Stall, Post-Stall, and Spin Testing 135

(2) By systematic investigation of each probable combination of weight and center of gravity location, if compliance cannot be reasonably inferred from combinations investigated.

(b) The following general tolerances are allowed during flight testing. However, greater tolerances may be allowed in particular tests:

Item	Tolerances
Weight	+5%, -10%
Critical items affected by weight	+5%, -1%
C. G.	±7% total travel

23.49 Stalling period

(a) V_{SO} and V_{S1} are the stalling speeds or the minimum steady flight speeds, in knots (CAS), at which the airplane is controllable with—

(1) For reciprocating engine-powered airplanes, the engine(s) idling, the throttle(s) closed or at not more than the power necessary for zero thrust at a speed not more than 110 percent of the stalling speed;

(2) For turbine engine-powered airplanes, the propulsive thrust not greater than zero at the stalling speed, or, if the resultant thrust has no appreciable effect on the stalling speed, with engine(s) idling and throttle(s) closed;

(3) The propeller(s) in the takeoff position;

(4) The airplane in the condition existing in the test, in which V_{SO} and V_{S1} are being used;

(5) The center of gravity in the position that results in the highest value of V_{SO} and V_{S1}; and

(6) The weight used when V_{SO} and V_{S1} are being used as a factor to determine compliance with a required performance standard.

(b) V_{SO} and V_{S1} must be determined by flight tests, using the procedure and meeting the flight characteristics specified in §23.201.

(c) Except as provided in paragraph (d) of this section, V_{SO} and V_{S1} at maximum weight must not exceed 61 knots for—

(1) Single-engine airplanes; and

(2) Multiengine airplanes of 6,000 pounds or less maximum weight that cannot meet the minimum rate of climb specified in §23.67(a) (1) with the critical engine inoperative.

(d) All single-engine airplanes, and those multiengine airplanes of 6,000 pounds or less maximum weight with a V_{SO} of more than 61 knots that do not meet the requirements of §23.67(a)(1), must comply with §23.562(d).

23.151 Acrobatic maneuvers

Each acrobatic and utility category airplane must be able to perform safely the acrobatic maneuvers for which certification is requested. Safe entry speeds for these maneuvers must be determined.

23.201 Wings level stall

(a) It must be possible to produce and to correct roll by unreversed use of the rolling control and to produce and to correct yaw by unreversed use of the directional control, up to the time the airplane stalls.

(b) The wings level stall characteristics must be demonstrated in flight as follows. Starting from a speed at least 10 knots above the stall speed, the elevator control must be pulled back so that the rate of speed reduction will not exceed one knot per second until a stall is produced, as shown by either:

(1) An uncontrollable downward pitching motion of the airplane;

(2) A downward pitching motion of the airplane that results from the activation of a stall avoidance device (for example, stick pusher); or

(3) The control reaching the stop.

(c) Normal use of elevator control for recovery is allowed after the downward pitching motion of paragraphs (b)(1) or (b)(2) of this section has unmistakably been produced, or after the control has been held against the stop for not less than the longer of two seconds or the time employed in the minimum steady slight speed determination of §23.49.

(d) During the entry into and the recovery from the maneuver, it must be possible to prevent more than 15 degrees of roll or yaw by the normal use of controls.

(e) Compliance with the requirements of this section must be shown under the following conditions:

(1) *Wing flaps.* Retracted, fully extended, and each intermediate normal operating position.

(2) *Landing gear.* Retracted and extended.

(3) *Cowl flaps.* Appropriate to configuration.

(4) *Power:*

(i) Power off; and

(ii) 75 percent of maximum continuous power. However, if the power-to-weight ratio at 75 percent of maximum continuous power result in extreme nose-up attitudes, the test may be carried out with the power required for level flight in the landing configuration at maximum landing weight and a speed of 1.4 V_{SO}, except that the power may not be less than 50 percent of maximum continuous power.

(5) *Trim.* The airplane trimmed at a speed as near 1.5 V_{S1} as practicable.

(6) *Propeller.* Full increase r.p.m. position for the power off condition.

23.203 Turning flight and accelerated stalls

Turning flight and accelerated stalls must be demonstrated in flight as follows:

(a) Establish and maintain a coordinated turn in a 30 degree bank. Reduce speed by steadily and progressively tightening the turn with the elevator until the airplane is stalled, as defined in §23.201(b). The rate of speed reduction must be constant, and—

(1) For a turning flight stall, may not exceed one knot per second; and

(2) For an accelerated turning stall, be 3 to 5 knots per second with steadily increasing normal acceleration.

(b) After the airplane has stalled, as defined in §23.201(b), it must be possible to regain wings level flight by normal use of the flight controls, but without increasing power and without—

    (1) Excessive loss of altitude;
    (2) Undue pitchup;
    (3) Uncontrollable tendency to spin;
    (4) Exceeding a bank angle of 60 degrees in the original direction of the turn or 30 degrees in the opposite direction in the case of turning flight stalls;
    (5) Exceeding a bank angle of 90 degrees in the original direction of the turn or 60 degrees in the opposite direction in the case of accelerated turning stalls; and
    (6) Exceeding the maximum permissible speed or allowable limit load factor.

(c) Compliance with the requirements of this section must be shown under the following conditions:

    (1) *Wing flaps*: Retracted, fully extended, and each intermediate normal operating position;
    (2) *Landing gear*: Retracted and extended;
    (3) *Cowl flaps*: Appropriate to configuration;
    (4) *Power*:
        (i) Power off; and
        (ii) 75 percent of maximum continuous power. However, if the power-to-weight ratio at 75 percent of maximum continuous power results in extreme nose-up attitudes, the test may be carried out with the power required for level flight in the landing configuration at maximum landing weight and a speed of 1.4 V_{SO}, except that the power may not be less than 50 percent of maximum continuous power.
    (5) *Trim*: The airplane trimmed at a speed as near 1.5 V_{S1} as practicable.
    (6) *Propeller*. Full increase rpm position for the power off condition.

23.207 Stall warning.

(a) There must be a clear and distinctive stall warning, with the flaps and landing gear in any normal position, in straight and turning flight.

(b) The stall warning may be furnished either through the inherent aerodynamic qualities of the airplane or by a device that will give clearly distinguishable indications under expected conditions of flight. However, a visual stall warning device that requires the attention of the crew within the cockpit is not acceptable by itself.

(c) During the stall tests required by §23.201(b) and §23.203(a)(1), the stall warning must begin at a speed exceeding the stalling speed by a margin of not less than 5 knots and must continue until the stall occurs.

(d) When following procedures furnished in accordance with §23.1585, the stall warning must not occur during a takeoff with all engines operating, a takeoff continued with one engine inoperative, or during an approach to landing.

(e) During the stall tests required by §23.203(a)(2), the stall warning must begin sufficiently in advance of the stall for the stall to be averted by pilot action taken after the stall warning first occurs.

(f) For acrobatic category airplanes, an artificial stall warning may be mutable, provided that it is armed automatically during takeoff and rearmed automatically in the approach configuration.

23.221 Spinning

(a) *Normal category airplanes.* A single-engine, normal category airplane must be able to recover from a one-turn spin or a three-second spin, whichever takes longer, in not more than one additional turn after initiation of the first control action for recovery, or demonstrate compliance with the optional spin resistant requirements of this section.

(1) The following apply to one turn or three second spins:

(i) For both the flaps-retracted and flaps-extended conditions, the applicable airspeed limit and positive limit maneuvering load factor must not be exceeded;

(ii) No control forces or characteristic encountered during the spin or recovery may adversely affect prompt recovery;

(iii) It must be impossible to obtain unrecoverable spins with any use of the flight or engine power controls either at the entry into or during the spin; and

(iv) For the flaps-extended condition, the flaps may be retracted during the recovery but not before rotation has ceased.

(2) At the applicant's option, the airplane may be demonstrated to be spin resistant by the following:

(i) During the stall maneuver contained in §23.201, the pitch control must be pulled back and held against the stop. Then, using ailerons and rudders in the proper direction, it must be possible to maintain wings-level flight within 15 degrees of bank and to roll the airplane from a 30 degree bank in one direction to a 30 degree bank in the other direction;

(ii) Reduce the airplane speed using pitch control at a rate of approximately one knot per second until the pitch control reaches the stop; then, with the pitch control pulled back and held against the stop, apply full rudder control in a manner to promote spin entry for a period of seven seconds or through a 360 degree heading change, whichever occurs first. If the 360 degree heading change is reached first, it must have taken no fewer than four seconds. This maneuver must be performed first with the ailerons in the neutral position, and then with the ailerons deflected opposite the direction of turn in the most adverse manner. Power and airplane configuration must be set in accordance with §23.201(e) without change during the maneuver. At the end of seven seconds or a 360 degree heading change, the airplane must respond immediately and normally to primary flight controls applied to regain coordinated, unstalled flight without reversal of control effect and without exceeding the temporary control forces specified by §23.143(c); and

(iii) Compliance with §§23.201 and 23.203 must be demonstrated with the airplane in uncoordinated flight, corresponding to one ball width displacement on a slip-skid indicator, unless one ball width displacement cannot be obtained with full rudder, in which case the demonstration must be with full rudder applied.

> (b) *Utility category airplanes.* A utility category airplane must meet the requirements of paragraph (a) of this section. In addition, the requirements of paragraph (c) of this section and §23.807(b)(7) must be met if approval for spinning is requested.
>
> (c) *Acrobatic category airplanes.* An acrobatic category airplane must meet the spin requirements of paragraph (a) of this section and §23.807(b)(6). In addition, the following requirements must be met in each configuration for which approval for spinning is requested:
>
> (1) The airplane must recover from any point in a spin up to and including six turns, or any greater number of turns for which certification is requested, in not more than one and one-half additional turns after initiation of the first control action for recovery. However, beyond three turns, the spin may be discontinued if spiral characteristics appear.
>
> (2) The applicable airspeed limits and limit maneuvering load factors must not be exceeded. For flaps-extended configurations for which approval is requested, the flaps must not be retracted during the recovery.
>
> (3) It must be impossible to obtain unrecoverable spins with any use of the flight or engine power controls either at the entry into or during the spin.
>
> (4) There must be no characteristics during the spin (such as excessive rates of rotation or extreme oscillatory motion) that might prevent a successful recovery due to disorientation or incapacitation of the pilot.

Requirements are the foundation upon which the flight test planner builds his preparation. The ones shown here are merely representative; there are other applicable documents that expand upon and explain the flight test issues. For example, FAR Part 23 certification requirements are supplemented by AC 23-8b, which clarifies some of the uncertain wording in the basic certification document. Where Part 23.221 refers to "unrecoverable spins", this advisory circular defines the term. So, the first step in laying out an overall test plan is to set these objectives down and be sure that all interested parties agree to them.

Not only must test objectives be clearly stated, they must also be prioritized. In setting down objectives and relating them to requirements, no relative priorities were suggested. Before the test team can construct an efficient test matrix and do detailed test planning, importance of each of the objectives must be established. This prioritization governs order of testing as well as allocation of test resources. Again, the flight test team and the design team must function as one entity in this phase of test preparation. There must be agreement, even though complete unanimity on the final decisions is unlikely.

With the overall test objectives stated and prioritized, the flight test team can focus on how to acquire data best and other aspects (usually safety) of test preparation.

14.2.1.1.4 *Spin Recovery.* Though not cited above in the FARs, or our excerpts from the military specifications, recommended pilot procedures for the Pilot's Operating Handbook are an important product of spin and departure testing. The test pilot has a key responsibility to all customers who fly the airplane in regular service. Specifically, he or she must outline for those users the procedures necessary for prompt recovery from any unintended motions or spins. The test program should explicitly examine variations on the recovery controls to expose any sensitivities. Some airplanes recover better with lateral stick than rudder. Some airplanes are insensitive to longitudinal stick position. Anything but perfectly centered stick position might delay recovery in still others. Should the airplane be recovered with longitudinal stick forward, aft, or neutral? Does it matter? Particularly if an airplane is to be spun regularly in a training environment, the test team needs to communicate, via the Pilot's Operating Handbook, all the appropriate subtleties. Just because it is easy for a test pilot with hundreds of spins completed to recover in a controlled environment, does not mean the disoriented student who is surprise by the unusual motions will find it easy to recover. Careful attention must be paid to providing a complete and carefully reasoned description of the best recovery procedure(s).

14.2.1.2 Data Requirements. In a sense the engineer reverts back to his profession in the next phase of preparation for post-stall testing. They lean on experience (both theirs and others) to decide what data are needed to answer the questions posed by the objectives. They seek practical and efficient ways to generate and collect these data and lay out the data presentation so that key questions are clearly answered with quantitative engineering results wherever possible. These engineers devise detailed test plans from which estimates of time, money, and people required to complete post-stall flight tests can be made. This phase of test preparation is necessarily an iterative process. Available time and resources usually shrink during the course of a project; requirements that were overlooked in the goal-setting phase often surface here. There are few general principles that apply to all classes of aircraft, but there are some guidelines to follow.

(1) Question all data requirements; "acid test" questions include:
  (a) Which objective do these data satisfy or resolve?
  (b) What is the relative priority of the objective pursued?
  (c) How will the data be used and who will use it?
(2) Is there a simpler, more cost-effective, or safer way to provide the essential data?
  (a) Did or can component testing/ground simulation provide similar data?
  (b) If so, how much flight data are needed to verify component tests?
  (c) Can simulation be used to reduce the test matrix? Can it identify the highest risk points or substitute for the highest risk points?
(3) Is the proposed data matrix an adequate basis for management decisions?
  (a) Is the planned experimental sample size statistically representative?
  (b) Do test constraints bias or skew the results?
(4) Is there a more efficient way to acquire the data in flight?
  (a) Could extended flight times (like air refueling might provide) allow coverage of more of the test matrix in each flight?
  (b) Is real-time test monitoring adequate?
  (c) How much redundancy in test support is needed?

As these questions are considered, the importance of efficient, reliable data acquisition surfaces as a relevant issue.

14.2.1.3 Instrumentation for Post-Stall Flight Tests. The instrumentation suite for post-stall testing is usually more extensive and more specialized than is used for conventional performance or stability and control testing. First, ranges of the dynamic variables when the airplane is operating at high angle of attack are usually much greater. Anables

Chapter 14   Stall, Post-Stall, and Spin Testing

when the airplane is operating at high angle of attack are usually much greater. Angle of attack variations from the trim condition may be ±90° or more, instead of ±15°, as is typical in the normal operating envelope. Angular rates may be doubled or tripled over those encountered at low angles of attack. Accelerations may be about the same in magnitude, but the acceleration directions change much more rapidly during post-stall maneuvers.

Table 14.6. Typical Sensors Needed for Stability and Control Data

Sensor	Low Angle of Attack Typical Ranges	High Angle of Attack Typical Ranges
α	-5° to 15°	±90°
β	±5°	±30°
P	±70°/sec	±100°/sec
Q	±20°/sec	±50°/sec
R	±20°/sec	±100°/sec
ϕ	±30°	±90°
θ	±10°	±90°
ψ	±20°	±360°
n_x	±0.25g	±0.5g
n_y	±0.25g	±0.5g
n_z	±4g	±4g

What kinds of sensors are needed, how do the motions encountered in the post-stall flight regime affect experimental measurements, and how must the data be handled? Conventional parameters like those shown in Table 14.6 are ordinarily what is recorded for post-stall tests. But, this list is not exhaustive; there are other measurements that are sometimes needed. Some engines don't react favorably to this environment and need to be watched closely; sometimes the data stream flags a struggling engine before the pilot notices, and a timely "knock it off" call from the ground can prevent damage or a bonafide emergency. All control surface positions, throttle position, and a number of discrete events like gear position, flap position, and speed brake position must be tracked. For the control positions, complete time histories sampled at moderately high frequencies are usually necessary. The sample rates for fast-acting surfaces should be an order of magnitude higher than the highest motion frequency of interest, even though the Nyquist criterion only requires that the sample rate be double the highest frequency of interest. Since control surface movements can be rather quick, especially when driven by high rate actuators on modern high performance aircraft, the data sample rate for control inputs is often the most important instrumentation specification for flight tests interested in the rigid body motion states. To analyze structural modes or other high frequency phenomena, other sensors likely dictate the critical sample rate. The bandwidth of the instrumentation sensors must be carefully tailored in the post-stall flight regime.

14.2.1.4 **Test Team Training**. Sometimes the test pilot is the only test team member singled out for special training during preparation for post-stall training. While the test pilot's background, training, and currency are meaningful and deserve considerable attention, the entire team needs to prepare for such hazardous tests with focus and intensity. The underlying principle is:

> *When deliberately exploring the limits and boundaries of the flight envelope,* **anticipate every eventuality** *you can.*

Even when you prepare as fully as time and resources allow, post-stall tests often reveal unexpected characteristics and behavior. But, pre-test preparation that includes both pilot (or pilots) and the entire support team does much to eliminate surprises.

First, how should the pilot be selected and trained? There is no substitute for **recent experience** in post-stall testing, though the emphasis should be on overall experience, not recency alone. The most valuable foundation is actual post-stall testing in a broad array of airplanes. A test pilot who has done at least one other post-stall test program is ideal. If the size and nature of the project permit (or even require) more than one test pilot, give this experienced pilot the responsibility for training other participants. In military programs, the senior pilot for the contractor typically does most of the initial exploratory demonstrations. He should normally be tasked with the training and preparation of every pilot who flies a post-stall test flight.

What elements should this training and preparation include? One of the most valuable considerations in preparing to conduct high angle of attack testing is a formal training course similar to those given by the USAF or USN Test Pilot Schools, the Empire Test Pilots' School, or by the French Test Pilot School. Of course, not every company or entrepreneur will have access to such personnel, but this reservoir of talent has the best basic background of formal training for post-stall testing.

Once the project pilot is chosen, preparation has only begun. It is imperative that whatever his (or their) experience level is, there should also be recent experience in an aircraft similar to the type to be tested. If possible, this "in-flight simulation" should be flown into post-stall maneuvers of the most extreme type anticipated. If there is no aircraft available for actual post-stall practice, project pilots should seek out the most realistic simulator they can find, preferably one where large amplitude motions can be experienced. The time spent in such a device, especially if it has high fidelity both in motion and in cockpit layout to the test airplane, is an ideal place for rehearsing and modifying both "normal" data collection (if data collection during post-stall maneuvers is ever normal) and emergency procedures. Realistic, repetitious practice of test profiles gives the pilots confidence, recent experience, and efficient procedures for data collection and allow them to make the best use of flight test time.

These simulations should also include as much of the test team as will be involved in real-time monitoring and quick-look data reduction. The engineers/pilots acting as test conductors and chase aircrew need to practice communication procedures and refine operational instructions. They should be just as familiar with safety rules (like minimum altitudes for initiating recovery, altitudes for bailout, expected control sequences, and the like) as the pilots who fly the test vehicle. The flight test and systems engineers who monitor key parameters (control surface positions, angular rates, engine performance, emergency recovery subsystems, ground tracking, and all essential support details) should also rehearse their roles as part of the team. The senior flight test engineer should insist that every team member participate in one or more dry runs -- including exposure to emergency situations -- if the entire team is to function efficiently. Every team member must

Chapter 14   Stall, Post-Stall, and Spin Testing   143

understand and accept his or her responsibilities, know the procedures, and appreciate the consequences if he/she fails to meet his obligations during a test flight. Above all, hazardous flight testing is an exercise in **group discipline**. The test conductor can best guard against a breakdown in discipline with realistic practice sessions, both before test flights actually begin and periodically during a program. This continuation training is particularly important if there are long breaks between test phases or a significant number of new personnel must be assimilated into the test team.

14.2.2 Safety Precautions and Special Equipment

Having discussed some of the general aspects of preparation for post-stall testing, we now turn to more specific ways to make such tests safer. Have no illusions; post-stall testing is not inherently safe; the test airplane and the crew are intentionally put at risk in either marginally controllable or out-of-control portions of the flight envelope. All flight testing should start with benign flight conditions and progress incrementally to more difficult or more dangerous ones. This principle, called the "build-up approach", is designed to preclude unwittingly pushing past boundaries beyond which there is no recovery. However, post-stall testing involves flights where nonlinear dynamical relationships and the large number of variables that are involved make it impossible to perfectly determine in advance safe increments in test parameters. The only alternatives that can reduce risks are (1) reduce the likelihood of potential failures and (2) take extraordinary precautions to mitigate the consequences of events that are still likely to occur. This section provides practical guidance for the flight test engineer planning a high angle of attack test series. It involves both procedural and hardware suggestions; and, though it would be highly presumptuous (and patently false!) to claim that these suggestions are comprehensive, they are a good starting point.

14.2.2.1 Emergency Recovery Devices.
Whenever anyone lays out a serious post-stall flight test program, inevitably these questions arise:

> *Is an emergency recovery subsystem needed? If so, what type and what kind of design requirements must be spelled out?*

The first question should almost always be answered "yes". Only airplanes with tightly limited flight envelopes or very unusual design characteristics can avoid encounters with post-stall flight. Even then, very few designers are willing to bet their reputations that no post-stall flight tests at all are necessary. Even "spin-proof" airplanes have been involved in post-stall accidents. However, answering the second question is not so easy. It cannot be answered without consideration of the following additional question:

> *What is the intended use of the design (or modification of an existing design)?*

It makes no sense to do full-blown spin tests, for example, for a large commercial airliner. Thus, the "post-stall" program for such a vehicle is almost exclusively aimed at precluding stalls -- a very limited "post-stall" (if indeed it should be called by that name) test series is in order. Most business aircraft with near centerline thrust should be treated in the same way. However, past experience has shown that even these kinds of airplanes can have post-stall catastrophes if the configuration is quite unusual or not well-known. For example, at least one early T-tail transport design[27] ran into such difficulties, as have corporate jets[28] with similar design features. There is also reason to pay careful attention to post-stall certification for canard and three-surface configurations.

> *Is it responsible to ignore post-stall testing based on handbook limitations intended for the design?*

Now this question is harder to answer and we are treading dangerously close to subjective opinion entirely. But, to offer a design to any segment of the flying public without having explored post-stall behavior is neither wise nor responsible. (It may also be legally indefensible in today's litigious social environment.) No matter what the contracting agency or certification authority requires, a professional design team owes that consideration to their customers. Please note that this professional and ethical obligation is related to vehicle characteristics and intended use, not certification requirements alone. The design process is simply not complete until predictions and engineering estimates have been validated to some margin slightly beyond the published flight envelope.

There are two types of devices usually considered for emergency recovery from spins; those which modify the ratio of aerodynamic and inertial forces and those which add additional control authority about one or more axes. Far and away the most commonly used emergency recovery scheme is some form of parachute which alters the pitching and yawing moments rather drastically. They have been used on almost every type of airplane from sailplanes to 30-ton fighters.

Design and validation of a spin recovery parachute subsystem[29] is peculiar to each configuration and thus often becomes the responsibility of the project flight test engineer. The inertia distribution matrix, likely equilibrium spin rotation rates, and an estimate of the dynamic pressure range over which the parachute must operate are key design parameters. Figure 14.15 suggests the compromises associated with designing such a subsystem and how each of these parameters are affected by one another.

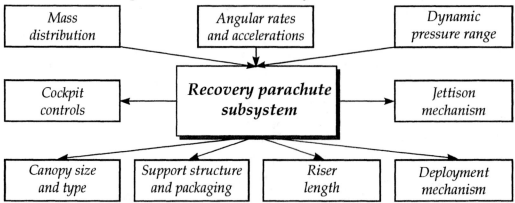

Fig. 14.15 Emergency Parachute Recovery Subsystem Design Parameters

Sizing of the canopy and its porosity are fundamental to the success or failure of this critical subsystem. Typically an off-the shelf canopy can be used and canopy characteristics are fairly completely documented[30]. Drag coefficients and opening shock load overshoots may be estimated empirically, but often ground tests are necessary to verify the assumptions. The numerical values for these parameters are strongly dependent on canopy porosity. The canopy material, its shape, and whether the canopy is solid, slotted, or vented in any way all affect the parachute drag coefficient and the opening shock loads drastically. These short duration loads are often considerably higher than the steady aerodynamic loads. Increasing the porosity and reefing the canopy so that it opens more slowly both tend to lower opening shock and thus reduce the severity of the impact loads on the risers and the attachment structure. A more porous canopy is also less likely to oscillate at the end of the risers after it is fully inflated. In fact oscillatory instability of the canopy can cause the airplane-parachute combination to be dynamically unstable after recovery from the out-of-control condition. Such dynamic behavior can be very uncomfortable (and quite unnerving!) to the pilot who is depending on this device to recover from this kind of situation. However, since a recovery parachute is typically jettisoned shortly after it performs its function, dynamic instability of the airplane-parachute combination is often of secondary importance.

The riser design and length also are quite important[31]. The risers should be long enough to guarantee the canopy opens well away from the aircraft itself. The likely worst case rotation rates and attitude combinations should be studied, as well as probable low pressure, separated flow regions, so provisions can be made to forcibly propel the parachute canopy into relatively clear air before the opening sequence begins. Canopy bags or reefing, along with drogues or ejectors are typically used to extract the parachute from its storage canister. Once the canopy opens successfully, riser length is another important factor affecting the oscillation frequency. Consideration should be given to keeping this frequency distinct from either the longitudinal short period or the lateral-directional Dutch roll frequencies of the rigid airplane.

Both the deployment and separation mechanisms for the parachute should be tested, certainly on the ground, and preferably airborne. Many programs, after lab qualification of the pyrotechnics, conduct a static ground test, high-speed taxi test, and airborne test. The most dangerous of these is likely the high-speed taxi test. The pilot should be prepared to cut-away the parachute immediately after deployment, due to the directional effect of crosswinds. Such tests have the added benefit of allowing the test pilot to physically handle the deployment controls, though he hopes never to need such actions during actual spin tests. Exercising the recovery system also provides end-to-end confirmation of the parachute, the jettison mechanism, and its instrumentation .

Rocket thrusters have been used as an alternative to spin parachutes[32]. This emergency recovery subsystem is an example of adding extra control devices to guarantee recovery from out-of-control situations. Potential advantages for such devices include:

♦ They can be installed so as to produce controlled moments about each axis.

♦ Multiple units can provide redundancy (or capability to make multiple recoveries) if the weight penalty is not too severe. Obviously, this capability could also be provided by rocket motors that can be restarted. Rarely, however, would this level of complexity be justified.

♦ Such thrusters could be used to drive the airplane into different spin modes (for research purposes), though such usage has not been reported in the literature.

Disadvantages to such devices include:

♦ Usually they involve additional complexity, both from a mechanical perspective and in terms of safety. Rocket thrusters almost always multiply safety concerns because of the larger number of failures and the potential consequences of some of those failures.

- More often than not, thrusters are pyrotechnic devices (or are susceptible to explosion at least). Again, safety issues are harder to resolve satisfactorily when multiple pyrotechnics are carried for each flight.
- It is difficult to safely verify the adequacy of recovery thrusters for their intended purpose.

Of course, both rocket thrusters and a spin recovery parachute can be installed, but rarely are. One such exception[30] was the F-100F, a version of the Super Saber that was highly susceptible to a flat spin from which aerodynamic controls could not recover the airplane. Both a spin chute on the aft fuselage and rocket thrusters on the wing tips were installed for post-stall testing. Unfortunately, when the test aircraft entered the expected flat spin mode, the recovery rockets did not stop the rotation and the parachute streamed and wrapped around the vertical tail. The test airplane was lost when both emergency recovery devices failed to perform their intended function.

Finally, the flight test team must pay careful attention to how emergency recovery devices are actuated, controlled, monitored, and terminated (jettisoned, in the case of parachutes). First, all controls for this system must be accessible to and easily identified by the pilot even under the most extreme conditions of aircraft loading and stress. The pilot must activate and deactivate (or jettison) this emergency system even when cockpit g levels are high and oscillatory. (This fact suggests one reason for additional pilot restraints, which are often used for post-stall testing.) Moreover, there must be reliable and complete warning of any malfunction in the emergency recovery subsystem. Indicator lights are one means for providing these warnings. Parachute or thruster arming, position of jettison mechanism, test limits (angular rates, minimum altitudes, and the like), flight control system status, and engine health are typical candidates for warning/caution displays during post-stall tests.

14.2.2.2 Back-up Power. Spin and departure programs on an airframe are also a test of the engine integration and the engine's ability to continue to run at unusual attitudes and with distorted inlet flow conditions. The immediate challenge is not providing thrust, but rather the ancillary power required by other aircraft subsystems. A loss of engine power can also mean the loss of both electrical and hydraulic systems, as well as pressurization. Loss of electrical power may mean loss of telemetry and ground communication. The test team has to answer 'what-if' questions surrounding the potential loss of engine and subsystem power during a departure/spin event. Preparing an airplane for spin testing often necessitates the installation of auxiliary batteries to provide instrumentation, avionics, and perhaps even hydraulic power long enough to restart the engine.

14.2.2.3 Chase Aircraft. In addition to a reliable emergency recovery subsystem, post-stall test teams usually rely on chase aircraft. Typically, these support aircraft are flown by backup pilots for the project and they often carry aerial photographers. The functions of this supporting aircraft and crew include:

- Visually checking the test aircraft for configuration; for loss of access covers, gear doors, etc.; and for obvious structural or system failures during a maneuver.

Chapter 14   Stall, Post-Stall, and Spin Testing   147

- Clearing the airspace for the test vehicle.
- Calling out critical events -- minimum altitudes, engine failure, and emergency procedures -- to the test aircrew.
- Advising the test pilot and the test conductor of any anomalies observed.
- Recording photographic data of the entire post-stall event.

The chase pilot is a critical supporting player. His piloting task is very demanding and he needs to have powers of observation every bit as keen as those of the test pilot. He must be completely familiar with all planned mission profiles, overall test objectives, as well as detailed test restrictions. To meet these responsibilities, project pilots who fly such chase sorties should practice for them -- just as the rest of the test team should rehearse. For both test pilot and supporting chase aircrew, this practice time is mandatory.

14.2.2.4 **Procedural Precautions**. The whole of section 14.2 is built around the premise of test team discipline. Any engineer or pilot who is charged with the overall responsibility for this class of flight testing must be a stickler for discipline. Insisting that comprehensive, written procedures are developed, followed, and modified constructively is a mark of professionalism. It is absolutely necessary for the successful post-stall test team leader to demand discipline. Practical suggestions include:

- Establish an independent review process to evaluate planning and procedures. The scope of this review varies from a single experienced consultant to a full-blown, structured "red-blue" team approach. The effort expended on this review is largely dictated by project size and economy, but it should **never** be omitted.

- Test procedures, especially for post-stall tests, evolve from safety constraints wrapped around clearly stated test objectives. But safety is overriding.

- Be sure procedures are communicated. Keep them simple, straightforward, and to the point. Practice and repeat the good ones; revise the bad ones.

- Keep the command lines during a test crisp and clear. The test conductor is the maestro (he carries the responsibility); the test pilot is the soloist; and the chase pilot is the harmony. But everyone else on the test team plays an essential background instrument. Nobody talks to the pilots on the radio except the test conductor, or at the test conductor's request. Supporting test engineers relay pertinent information to the test conductor, who decides what is essential to the conduct of the test. Flying a productive test sortie is always a demanding task; flying a productive post-stall mission can be downright hectic. It is the test conductor's job to see that mission objectives are not destroyed by confusion and indecision. Figure 14.16 illustrates this decision-making net for a typical post-stall flight test.

- Guarantee that each team member knows the envelope limits and the reasons for them. The senior flight test engineer takes the lead in this training; every one must be completely aware of each flight's objectives -- both primary and secondary -- and the steps planned to reach them.

- **Never** go beyond a test plan. This rule should be iron-clad; you simply must not "shoot-from-the-hip" in this environment. True, the detailed flight profiles should always be planned for more than can be accomplished; but alternatives and relative priorities are spelled out in the briefing on the ground **before** takeoff.

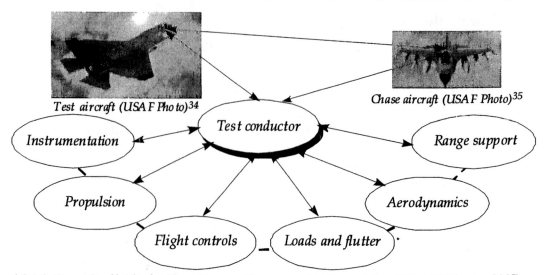

Fig. 14.16 Post-Stall Flight Test Team Communications Net (USAF Photos[34,35])

14.2.2.5 Real-Time Monitoring of Post-Stall Tests. The engineering process that needs to be so tightly disciplined must still react efficiently. Adequate information fed back to the test support engineers during post-stall maneuvers is essential to the process. Larger and adequately funded post-stall programs may have all relevant dynamic time histories telemetered back to the control room during a flight. These fortunate engineers may even have the computer power and software sophistication available to cross-plot and partially analyze these real-time data. They might, for example, superimpose dynamic states on α-β axes, rather than just time axes. Graphic images of the airplane's rotation and inertial attitude are useful. Plots of predicted boundaries and actual motions can be overlaid. For such a rich data environment the possibilities for real-time monitoring are almost infinite. At the other extreme, where there are only radio communications between the test pilot and the chase pilot, real-time analysis and advice may be sparse and subjective. But that fact does not make it unimportant. Any outside observer's observations can be useful information (data) in this often confusing test environment. Recording a running commentary from both the test pilot and the chase pilot is wise, an inexpensive "real-time" tool for post-stall tests. The point is: real-time information -- at any level between the extremes described in this paragraph -- is valuable and must not be ignored.

14.3 SUMMARY

In this chapter we have introduced another advanced topic in flight test. Most of the principles that apply to post-stall testing also apply to other forms of hazardous test: flutter envelope expansion, weapon separation tests, low level avionics system (like terrain-following) evaluations, and similar activities. The major themes of careful planning, thorough preparation, and unyielding discipline carry over into these other forms of flight testing as well. Of course, the engineering tasks, measurement techniques, and support required differ significantly. But the basic principles are still valid. In that sense, this chapter serves to draw attention to the three most common themes of all kinds of flight testing, even as it serves to introduce this topic. And those themes -- planning, preparation, and discipline – are pervasive throughout flight test activities.

REFERENCES

1. Herbst, W. B., "Future Fighter Technologies," **AIAA Journal of Aircraft**, Vol. 17, No. 8, Aug. 1980, pp. 561-566.
2. Gilbert, W. P. and Gatlin, D. H., "Review of NASA High-Alpha Technology Program," Paper at NASA High-Angle-of-Attack Technology Conference, NASA Langley Research Center, Oct. 30-Nov. 1, 1990.
3. Nix, J., "X-31 EFM: The Experiment Continues," Paper at NASA High-Angle-of-Attack Technology Conference, NASA Langley Research Center, Oct. 30-Nov. 1, 1990.
4. Trippensee, G. A. and Ishmael, S., "Overview of the X-29 High-Angle-of-Attack Program," Paper at NASA High-Angle-of-Attack Technology Conference, NASA Langley Research Center, Oct. 30-Nov. 1, 1990.
5. VayuSena, "Irkutsk/HAL Su-30MKI," http://vayu-sena.tripod.com/info-su30mki.html, ver. 2.6, May 6, 2006.
6. Flight Training Handbook, AC-61-21A, Federal Aviation Administration, 1980.
7. Military Specification, MIL-S-83691(USAF), "Stall/Post-Stall/Spin Flight Test Demonstration Requirements for Airplanes," Apr. 1972.
8. McElroy, C. E. and Sharp, P. S., "Stall/ Near Stall Investigation of the F-4E Aircraft," FTC-SD-70-20, Air Force Flight Test Center, Edwards Air Force Base, California, Oct. 1970.
9. Lusby, W. A., Jr. and Hanks, N. J., "T-38A Category II Stability and Control Tests," AFFTC-TR-61-15, Air Force Flight Test Center, Edwards Air Force Base, California, Aug. 1961.
10. Babister, A. W., **Aircraft Stability and Control**, Macmillan Company, New York: 1961.
11. Kerr, T. H., "General Principles of Spinning," **AGARD Flight Test Manual**, Volume II, Chapter 8, Pergamon Press, New York: 1962.
12. Chambers, J. R. and Stough, H. P., III, "Summary of NASA Stall/Spin Research for General Aviation Configurations," AIAA Paper 86-2597, Oct. 1986.
13. Stough, H. P., III, DiCarlo, D. J., and Steward, E. C., "Wing Modification for Increased Spin Resistance," SAE Paper 830720, Apr. 1983.
13. Stough, H. P., III, DiCarlo, D. J., and Steward, E. C., "Wing Modification for Increased Spin Resistance," SAE Paper 830720, Apr. 1983.
14. http://www.af.mil/shared/media/photodb/photos/060512-f-0558k-002.jpg.
15. http://www.maxwell.af.mil/au/afhra/photo_galleries/merhar/Photos/01097628_031.jpg.
16. http://www.dfrc.nasa.gov/Gallery/Photo/F-104/Medium/ECN-785.jpg.
17. Turner, H. L., "Measurement of the Moments of Inertia of an Airplane by a Simplified Method," NACA TN 2201, Ames Aeronautical Laboratory, Moffett Field, California, Oct. 1950.
18. Anglin, E.L. and Scher, S. H., "Analytical Study Aircraft-Developed spins and Determination of Moments Required for Satisfactory Spin Recovery," NASA TN D-2181, Feb. 1964.

19 Phillips, W.F., **Mechanics of Flight**, Wiley, Hoboken, NJ, 2004, pp 577-590.

20 Defense Acquisition Guidebook, Section 9.10, "Test and Evaluation Master Plan Recommended Format," Department of Defense," http://akss.dau.mil/dag/DoD5000.asp?view=document&rf=GuideBook\IG_c9.10.asp.

21 Defense Acquisition Guidebook, Section 9.8.2, "OT&E Best Practices," Department of Defense, http://akss.dau.mil/dag/DoD5000.asp?view=document&doc=2.

22 Military Handbook, MIL-HDBK-1797B, "Flying Qualities of Piloted Airplanes," Dec. 1997.

23 Felock, J., Habig, C., and Hancock, R., "KEEP EAGLE: F-15E High Angle of Attack Testing," **1997 Report to the Aerospace Profession, Proceeding of the 39th Symposium**, Society of Experimental Test Pilots, Lancaster CA, pp. 180-191.

24 Heller, M., Niewoehner, R. J. and Lawson, S., "High Angle of Attack Control Law Development and Testing of the F/A-18 E/F Super Hornet," **AIAA Journal of Aircraft**, vol. 38, No.5, pp. 841-847, September-October 2001.

25 Madenwald, F., Niewoehner, R. J., and Heller, M., "F/A-18 E/F High AOA Testing Update," **1997 Report to the Aerospace Profession, Proceeding of the 41st Symposium**, Society of Experimental Test Pilots, Lancaster CA, pp 5-13.

26 Federal Aviation Regulations, Part 23, Airworthiness Standards: Normal, Utility, Aerobatic and Commuter Category Airplanes, Federal Aviation Administration, June 26, 2006. (from http://www.gpoaccess.gov/ecfr/)

27 "British Aircraft Corp. Still Aims at Initial BAC 111 Delivery Dates," **Aviation Week and Space Technology**, Nov. 11, 1963, p.45.

28 "Prototype Canadair Challenger Lost," **Flight International**, Apr. 19, 1980, p. 1174.

29 Burk, S. M., "Review of Airplane Spin-Recovery Parachute Systems Technology," Paper at the AFFDL/ASD Stall/Post-Stall/Spin Symposium, Wright-Patterson AFB, Ohio, Dec. 1971.

30 Bixby, H. W., Ewing, E. G., and Knacke, T. W., "Recovery Systems Design Guide," AFFDL-TR-78-151, Air Force Flight Dynamics Laboratory, Wright-Patterson AFB, Ohio, Dec. 1978.

31 Knacke, T. W., "Recovery System Design Manual," NWC TP 6575, Naval Weapons Center, China Lake, California, Jul. 1985.

32 Farmer, G. H., "Flight Test Investigation to Determine the Effectiveness of a Rocket as an Emergency Spin Recovery Device," WADC Technical Report 53-396 (AD 76422), Wright Air Development Center, Ohio, Aug. 1958.

33 Wykes, J. H., "An Analytical Study of the Dynamics of Spinning Aircraft, Part III, Calculated and Flight Test Spin Characteristics of an F-100F with Strakes," WADC Technical Report 58-381 (AD 243600), Wright Air Development Center, Ohio, Feb. 1960.

34 USAF photo, http://www.af.mil/shared/media/photodb/web/web_020926-O-9999G-006.jpg.

35 USAF Photo (Staff Sergeant Brian Ferguson), http://www.af.mil/shared/media/photodb/photos/060507-F-2185F-575.jpg.

Chapter 15
STORES CERTIFICATION TESTING

15.1 INTRODUCTION

Not long after the invention of the airplane aircrew started dropping things from their new-found vantage point, or attaching ancillary equipment to perform some task other than pure transportation. As former military test pilots, two of the authors are inclined to associate stores compatibility with weapons carriage and separation; those are the flight tests we conducted while actively flying. The stores carriage compatibility and separation problem is much broader than weapons carriage and delivery, and should receive attention from other flight test practitioners. Logistics air-drop, water bombing of fires, externally carried remote surveillance or sensing packages, modifications to carry special equipment, and Scaled Composites' SpaceShipOne all fit within the scope of "stores compatibility" in this broader context.

The military weapons separation enterprise is massive and perpetually active with new software, new equipment, or modifications to existing airframes or weapons. Expertise in this specialized activity is developed over years of exposure to diverse projects for both pilots and engineers, under the supervision of mentors whose experience often spans decades. For the junior engineer working on such programs, this chapter provides an introduction to vocabulary and concerns, as well as highlights of lessons learned. Demonstration requirements and procedures are clearly identified in standards documents, with which the stores separation engineer must become familiar. For such readers, the chapter that follows is a simple primer.

The chapter may be of most interest and benefit to those outside the military test enterprises, simply because the lessons from military experiences are similar to fuel tank carriage and jettison for commercial aircraft. Such equipment is often added to the exterior of in-service civil aircraft to meet specific needs, perhaps for a solitary customer or application. Yet, the small aircraft company or after-market vendor may only infrequently deal with and prepare for flight testing such equipment and the engineering and flight operations staff may not possess broad experience in the special challenges and hazards of carriage or separating stores in flight. For such readers, this chapter is only a primer, but we hope awaken sensitivities to hazards that might not otherwise be considered. Moreover, the bibliography points to important reference material for more details.

15.1.1 The Key Questions

Stores carriage and release problems span the full domain of flight test. When any object is suspended externally from the airplane or ejected from an internal bay, the flutter, loads, performance, and low and high-angle of attack flying qualities are affected, usually adversely. Similarly, if the original external shape is altered or if any equipment is added that changes the mass distribution, then the same results occur. If the modification or store draws fuel, electrical or hydraulic power, and requires cooling, then the mechanical and electrical systems testing becomes more complicated. If the store or modification communicates with host airplane avionics, avionics evaluations become more complex and the electrical power system must be reexamined. Modification and stores compatibility challenge all domains of flight test.

15.1.2 Historical Background

The history of aircraft armament (indeed of aviation in general) is quite short compared to most historical topics, but to misunderstand what has gone before in a technology is the epitome of professional arrogance. Unfortunately, few historians have written about aircraft stores carriage and/or airframe modifications. Woodman[1] wrote one of the more important air armament histories on World War I in 1989 and updated it in 2003. As his title suggests, Woodman's work is only valid up through the end of World War I. Woodman points out that his sources were very "scrappy" (his term), but in his addendum in 2003 he notes that after the original publication of the book, a number of readers offered new material and photographs to broaden the factual basis of the book. Later history covering aircraft armament testing is likewise rather sparse.

15.1.2.1 Early Days. When the airplane was invented, virtually none of the designs could carry a practical payload; they were taxed to their limits to simply carry the pilot or pilots aloft safely. In 1911, Marshal Ferdinand Foch is reputed to have summarized his view of the airplane[2]: "Airplanes are interesting toys but of no military value." But as these flying machines matured, designers began giving them capabilities to carry useful payloads of substantial weight. The military uses of the airplane in World War I are the best example of explosive growth in capability and in tactics for employing armament on these largely wood and fabric machines. By the time of the Armistice all sides had advanced considerably the uses of the airplane far beyond the original notion of a reconnaissance and observation platform (or perhaps even lesser mission concepts, as suggested by Marshal Foch's ill-advised comment).

Fig. 15.1 Vickers FB.5 Gunbus with Gunner's Station in Forward Cockpit[2]

Guns of considerable size and firepower were mounted on airframes for both offensive and defensive purposes. One of the compatibility issues was how and where to mount these guns so they would be most effective. One of the first approaches to adding firepower to the airplane of World War I was to simply use a pusher propeller and mount a machine gun in a forward-facing gunner's station. The British design shown in Fig. 15.1 illustrates this concept. The performance and agility of such "fighters" left a lot to be desired and the pusher configuration soon gave way to "tractor designs" with a propeller in the nose of the airplane. But this tractor configuration posed problems for mounting the

Chapter 15 **Stores Certification Testing** 153

gun in the most favorable position for both sighting and practical use. These early guns often had to be cleared by the crew and consequently were mounted so the crew could perform this maintenance action in flight. Figure 15.2 shows such a machine gun mount on a RAF S.E.5a with a Foster mount. This mount made it easier to clear the gun in flight when necessary and also allowed firing the gun vertically. The weapon itself is a Lewis machine gun designed in 1911 by U. S. Army Colonel Isaac Newton Lewis and was used by all Allied armed forces during World War I.

Fig. 15.2 S.E.5a with Lewis Machine Gun on Top of Upper Wing (Foster Mount)[2]

Mounting a gun above the wing produces some obvious disadvantages in sighting and Roland Garros, the French aviator, who was labeled an "ace" before the term had been defined as we now use it, concluded that a fixed, forward-firing gun was the proper answer for air-to-air combat. He teamed with Raymond Saulnier, the French designer, in 1915 to find a solution. Before war broke out Saulnier had experimented with an interrupter gear meant to halt the fire of a Hotchkiss machine gun and allow propeller passage. The erratic rate of fire of that particular gun is blamed with Saulnier's initial failure. The Frenchmen chose to attach a steel plate to the propeller instead to deflect those rounds that hit the blades. The deflector plate was shaped to turn the bullet trajectory to the side, not directly aft toward the pilot. Garros quickly put the modification into combat and shot down three airplanes in less than twenty days.

Fig. 15.3 Fokker E.III with Forward-Firing, Interrupter-Equipped Machine Gun[3]

However, Garros was shot down by ground fire on the day after his third kill (April 19, 1915) and his deflector-equipped airplane was confiscated and turned over to Anthony Fokker, along with an urgent request that Fokker produce something comparable for German airmen. Fokker, unimpressed with the deflector concept, came back to the German authorities within forty-eight hours with his own interrupter design installed in an Eindecker (Fig. 15.3). The German decision makers were skeptical of Fokker's invention and demanded that he demonstrate that his interrupter mechanism worked by shooting down an airplane in actual combat. (Now, that is carrying the principle of operational effectiveness testing to the ultimate!) Though Fokker flew some sorties to attempt to prove his device, real "operational tests" on it were completed by Lieutenant Oswald Boelcke. These monoplanes, with Fokker's gun synchronization scheme, helped Boelcke score forty aerial victories before he was killed in 1916 in a mid-air collision with one his squadron mates. Boelcke's successes with Fokker's interrupter mechanism led to the beginning of the "Fokker Scourge" period of the World War I air war. Later in the Great War both Allied and German fighters were equipped with forward-firing guns using the principles of Fokker's interrupter mechanism. The most reliable of these interrupter, or synchronization, mechanisms was a hydraulically operated C.C-gear (so named after its inventors Constantinesco and Colley[4]). This device was "interoperable" (in today's terminology); it could be used with any engine and any machine gun, unlike the designs done originally by Saulnier and then by Fokker.

Boelcke went on to lead the famous squadron *Jagdstaffel* Nr. 2 (or Jasta 2). (The word *Jagdstaffel* literally means "hunting squadron"[4].) He and Max Immelmann were the first German fighter pilots to become aces (with the term defined to mean one who had at least 5 confirmed kills). Other aces from Jasta 2 included Manfred von Richthofen, Erwin Böhme and Hans Reimann. Boelcke selected many of these fighter pilots for Jasta 2 and was their mentor and teacher. He wrote the first widely used air tactics rules, "The Dicta Boelcke", which are still useful to fighter pilots[5,6]. Oswald Boelcke and Jasta 2 scored most of their other kills in early versions of the Albatros, which was armed with two forward-firing LMG 08/15 machine guns (Fig. 15.4a – a young Herman Goering is the pilot shown). This arrangement became the accepted way to mount guns on fighter aircraft by the latter stages of the Great War. Two Vickers guns, shown mounted in a Sopwith Camel (Fig. 15.4b), are typical of similar installations in British aircraft. Notice the loading handles in the Camel cockpit.

Fig. 15.4 Forward-Firing Machine Guns on Goering's Albatros D.III and Sopwith Camel[3]

Guns were not the only armament added to the airborne arsenal during the Great War; military planners were also intrigued by other capabilities of the airplane. Delivering high explosives, rockets or bombs, to distant enemy ground targets, taking out enemy observation balloons, or supporting ground troops in close combat were all roles suggested and attempted in these days of learning by experimenting (Fig. 15.5 illustrates).

Fig. 15.5a "Balloon Buster" Rockets on French Nieuport 11[7,8]

Fig. 15.5b Stick Grenades Being Loaded on German Halberstadt CL.II[3]

The German air unit structure heavily emphasized the bombing roles[8]. At first, almost any type of plane could be and was used in the crude and often improvised bombardment roles that evolved. The armament itself displayed these same improvisational characteristics. Runarsson[8] also states that the "... idea of dropping a bomb on the enemy from an airplane was first put into practice by a group of Italian army aviators. This remarkable event took place during a campaign in North Africa in 1911 where these intrepid pioneers risked life, limb and a horrible death to drop a few small bombs on native rebels form a Bleriot monoplane. Doubtless they scared the living daylights out of their adversaries who had never even seen an airplane before let alone one that dropped high explosives on them. Remarkable as this attack was, it probably did considerably more psychological than physical damage." Both sides followed this rather erratic development path during World War I; Fig. 15.6 shows both Allied and German observers trying their hand at "manual" bombing.

Fig. 15.6 Handheld Bombs Released by Observers[9]

Though the first two years of the Great War were largely devoted to experimentation as far as the ground attack use of the airplane, the combatants and their supporting munitions industries recognized the need for further specialization and development. The Carbonit bomb (Fig. 15.7) were used by the Germans during 1914-1916[8]. The series came in 50 kg, 20 kg, 10kg, 4.5 kg versions and was fused with a propeller at the front of the tear-drop shape. The tail ring served to stabilize the weapon. These bombs were typically carried in a magazine hanging from the tail rings, either internal to the fuselage or on an external rack. They were fairly high drag stores (leading to long times of flight and errors due to wind drift) and they oscillated rather severely after they were released.

Fig. 15.7 Carbonit Bomb Used by Germans in Early Years of World War I[9]

Fig. 15.8a French Bomb[9]   **Fig. 15.8b** British 1650-lb[9]   **Fig. 15.8c** German PuW Bomb[9]

The Allied munitions brain trust also improvised freely. The French converted or mimicked the shapes of artillery shells (Fig. 15.8a). The British moved rapidly toward heavy munitions; Fig. 15.8b shows the largest bomb dropped by the Allies during World War I. It weighed about 1650 pounds and was carried on the Handley Page O/400 bomber, one of the largest aircraft built during this era. The Germans established a research center that specialized in munitions and optics for sights – the *Prüfanstalt und Werft der Fliegertruppe or Test Establishment and Workshop of the Flying Corps* – PuW Institute in collaboration with the German optics firm of Göerz und Friedenau. This collaboration brought out the first truly modern series of bombs called the PuW bombs, varying in weight from 12.5 kilograms (kg) up to 1000 kg. This set of five weapons also included a 12.5 kg incendiary munition. The shape was quite aerodynamic and allowed these bombs to retain their kinetic energy longer after release, making them less susceptible to wind drift errors. Moreover, as Fig. 15.8c shows, the PuW bombs had fins that gave the weapons spin stabilization as they fell. In fact this munition is very similar in appearance to the Mark 82 bombs that U. S. pilots dropped in Vietnam, except these later bombs did had less complicated fins. The PuW bomb casing was made of high strength steel instead of cast iron like the earlier Carbonite bomb casings. Fuzing was generated by the spin rotation of the munition. From about 1916 on, the PuW series of bombs became the standard bomb load for the Germans.

Fig. 15.9 French Caudron G.4 Bomber[3]

Both German and Allied military planners recognized the necessity to design bomb-carrying capabilities into airplanes – to design for specific purposes. This understanding on both sides led to the development of very large airplanes with significant payload capability. The German bombers later in the Great War usually had two engines. Figure 15.9 is a version of the French Caudron G.4, a typical twin engine airplane that started out as a trainer and evolved into other missions. By the time the Caudron brothers, Gaston and Rene', produced their G-series, the machines had grown to the G.4 version, powered by either 80 horsepower Le Rhône or 100 horsepower Anzani radial engines. The G.4IB was had armor protection for the pilot and the G.4B2 (bomber version) could carry up to 100 kg of bombs. This rather paltry bomb load was typical of designs converted from other uses.

The British aircraft company founded by Frederick Handley Page in 1909 (the firm became the first publicly traded aircraft manufacturer in Great Britain) had a grander vision. The British Admiralty became very concerned with the maritime threat of the German Zeppelin bombers and issued a call for a "bloody paralyzer" by Commander Samson[11] to help repel the Germans at Antwerp after Britain had entered the war in 1914. Handley Page responded to the specification that Samson's plea generated for a long range patrol aircraft in December 1914 (Fig. 15.10). The machine proposed was gigantic, with a wing span of a 100 feet (supposedly how the second part of the designation "O/100" was derived)[10]. The 0/100 was underpowered, even though it went into service in squadron strength and carried out bombing raids in 1917. The O/400 variant (HP model 12) utilized two Rolls Royce Eagle engines, each producing 360 hp, which allowed the O/400, (Fig. 15.11) to carry up to 2000 pounds of bombs in its internal bay and on external racks. The machine cruised at 97 mph and had a range of about 700 miles. A total of 46 of the O/100 version were built but over 500 of the O/400 were produced and they participated in bombing raids with up to 40 aircraft participating[10].

Fig. 15.10 Preparation of a Handley Page O/100 for a Bombing Mission[3]

Fig. 15.11 Size Comparison: Handley Page O/100 Bomber and Spad Fighter[3]

Fig. 15.12a Gotha G.IV Ground Handling[3] **Fig. 15.12b** Loading Bombs on a Gotha G.IV[3]

There were similar large bombers specifically designed by the Germans and the Italians. The Gotha G-series bombers[11] were smaller than the Handley Page machines, having wingspans of about 77 feet. The engines, even on the G.IV and G.V, produced no more than about 260 hp and drove pusher propellers as Fig. 15.12 shows. The cruising speed was only about 87 mph. These machines had several design flaws that made them less than popular; the G.IV had fuel tanks in the nacelles. Whenever, a crash took place the fuel usually reached hot engine parts and ignited readily. The poor lateral stability and handling made such events all too common. Fixes were made to the later versions, but the G.Vb with an improved tail and other did not reach production status in time for combat. Virtually all 80 of these latter types were turned over to the Allied special commission and destroyed in accordance with the terms of the Armistice Treaty of Versailles.

Fig. 15.13 Russian Ilya Mourometz[3]

Last in this series of very large World War I machines is a derivative of the world's first four-engine airplane, the *Russkij Vitjaz* (Russian Knight). Designed by Igor Sikorsky, it first flew in May 1913 with only two 100 hp Argus engines from Germany. However, the configuration lacked altitude performance and was redesigned with four engines, ultimately changing to four Sunbeam engines of 150 hp each. Sikorsky renamed this modified long range machine after a Russian folk hero, Ilya Mourometand. The prototype was configured as a luxurious transport airplane, with a cabin seating for 16 passengers that large windows on each side, a glazed cabin floor, cabin heat from the inboard engines' exhaust pipes, electrical power for lighting from an air-driven generator, toilets for passenger convenience, and wicker chairs for crew and passenger comfort. The airplane weighed over 10,000 pounds and cruised at about 75 mph with a wingspan of just under 98 feet (Fig. 15.13). It served as a bomber for the Russians and was perhaps the most heavily armed of all the World War I "strategic" bombers, carrying a crew of up to 7, along with 7 machine guns (though the combat units usually opted for fewer to leave more room for bombs) or 1150 pounds of bombs. Russian Third Army chief of staff Major General Romanovsky expressed this confidence very clearly: "Give me just three Mourometsy and take away all the light aeroplanes - and I will be satisfied."[12]

As this section on air armament developments during World War I shows, the period was largely a time of improvisation and "learning by trial and error". Before we turn our attention to later time periods, here are some of the major lessons to be learned:

- Aircraft armament must be carefully tailored to the airborne environment; weight, reliability, and easy loading are important.
- To be most effective as a weapon, the airplane itself should be designed with an intended use clearly in mind.
- Larger aircraft are quite feasible but they have special problems (controllability with engines inoperative, for example) that must be addressed.

15.1.2.2. World War II.

The Second World War was also a period of almost feverish development of innovations in aircraft and armament. At the beginning of this conflict the airplane had become a productive part of the transportation system and the loads that could be carried and the range of the machines led to enormous bomb-carrying capability and serious challenges in providing reliable and safe ways to utilize that capability. We start with the gun ordnance.

Lighter, more reliable, and higher rates of fire are the three key improvements of World War II machine guns. Gustin[4] has written a fairly comprehensive comparison of the guns used starting with the Boeing P-26A Peashooter and continuing to many of the late 20th century fighter aircraft. His Fighter Gun Table is quite complete and illustrates rather clearly the progress in this set of armament. We summarize Gustin's conclusions based on the four phases of growth that he postulates.

Phase I, perhaps best described as the "rifle-calibre machine gun" era, was that period when aircraft carried little armor and lacked self-sealing fuel tanks. Light guns, like the eight .303 Brownings in early Spitfires or the four MG 17s in FockeWolf 190 could knock down an airplane if the pilot hit a critical component – the opposing crew, the fuel tanks, or the engines. Even fighter airplanes like the Messerschitt Bf 109G (Fig. 15.14), the early North American P-51, and the Mitsubishi A6M2 Zero (official Allied code name "Zeke") sporting a combination of machine guns and one or more cannons, lacked the firepower needed to compete against later designs. Before the United States was attacked by the Japanese, many fighters lacked these features, but as Gustin says, "...by 1941 a fighter without them was no longer considered suitable for combat."

Fig. 15.14 Messerschmitt Bf 109G (USAF Photo)[13]

Phase II of the World War II gun armament began about the time the United States declared war on the Axis powers and again there were two basic options of gun configurations adopted. Most of the United States fighters were fitted with either six or eight 50 caliber medium machine guns. The P-40, the P-47, the P-51D, the F4U-1 Corsair, and the F6F-3 Hellcat were all so equipped. These medium machine guns (often Brownings) were now belt-fed and more reliable but they still lacked the power to directly damage armor-protected aircraft structure. The second option typically included slower firing cannon, which had projectiles with enough energy that could take out aircraft structure, not just "hole" the skin. This approach to arming fighters with guns was used by both sides during Phase II of World War II. The P-38J Lightning (Fig. 15.15) is an example of U. S. armament of this type. Early models of this airplane carried a single 37mm cannon, along with four 50 caliber medium machine guns. Later ones carried a 20mm cannon and four 50 caliber machine guns. The machine guns were necessary because the weight and recoil of the cannon meant no more than one or two could be carried. The Bf 109E-3[5] had a typical Phase II German gun configuration, two MG-FF cannons (one in each wing) and two MG-17 machine guns in the cowl on top of the fuselage firing through the propeller.

Fig. 15.15 Lockheed P-38J (USAF Photo)[14]

Fig. 15.16 Kawanishi N1K2-J (code name "George" – USAF Photo)[15]

Phase III of the gun armament evolution did not end in 1945 but continued well beyond the end of the war, even though the principles used to decide on gun selections were quite clear by the end of hostilities. The concept driving the use of one of two options was the "kill" force was increased by the use of either 20 mm cannon or larger cannon (especially some of the German guns intended to knock down bombers). The later Spitfire, the Typhoon and Tempest, the Russian La-7, and the Japanese N1K-2J all used 20 mm cannon, usually with four of these weapons installed. This suite of weapons lasted well into the post-war era, except for U.S. designs. Four 20 mm Ho155 cannon were installed on the Kawanishi N1K2-J[15] "George" (Fig. 15.16) which appeared near the end of World War II.

Fig. 15.17 Messerschmitt Me 262A (*Schwalbe* or Swallow – USAF Photo)[16]

"Bomber killer" weaponry included 30 mm or larger guns like the four MK 108 cannon on the Me 262 (Fig. 15.17). An exaggeration of this concept involved installing larger cannon in the ME-262; Boyne[17] points out an exaggerated application of this concept occurred when "Swallows" were fitted with a single Mauser MK 214A 50mm cannon. "Happy Hunter II" (Fig. 15.18) was flown by Colonel (later Major General) Harold E. Watson and it was to have been one of the aircraft tested by "Watson's Whizzers" in their evaluation of advanced German designs immediately after World War II. Unfortunately, Happy Hunter II crashed on the way from Melun to Cherbourg after it developed engine vibrations. This cannon, could fire "…only one round at a time, due to the smoke resulting from the firing…" according to Boyne. Nonetheless, the installation illustrates the wide variety of airborne armament carried by fighter and fighter-bomber aircraft by 1945.

Fig. 15.18 Messerschmitt Me 262A -1a "Happy Hunter II (USAF Photo)[17]

Of interest for this chapter is the progress made during World War II in stores (largely bombs) carried on and in aircraft. We saw how military tacticians and strategists began to clamor for larger configurations even as early as the end of World War I. Between wars proponents of air power like Billy Mitchell and Hap Arnold in the United States raised the need for strategic (longer range) bombers. First, let us consider how the fighter-bomber or the tactical bomber evolved from the late thirties as both Axis and Allied powers began their military buildup through the end of hostilities in 1946. Allied aircraft used predominately in the fighter bomber role included the P-47 Thunderbolt and the German Stuka dive bomber was widely used as a close air support airplane (though the term "close air support" was not widely used at that time). These two aircraft are shown in Fig. 15.19. Various versions of the Thunderbolt (or "Jug", its vernacular nickname) grew the armament from eight 50 caliber machine guns in the wing – plus considerable armor plating to

Fig. 15.19a Republic P-47 Thunderbolt (USAF Photo)[19]

Fig. 15.19b Junkers Ju-87 Stuka (USAF Photo)[20]

protect the pilot – to a total of over 2500 pounds in external stores in the later versions of the airplane[18]. These external stores included drop tanks (like the centerline tanks shown in Fig. 15.19a), bombs on three different stations, and rocket pods (up to 5-inch high velocity aerial rockets or HVAR) could be carried on the venerable "Jug".

The Junkers Ju-87 is one of the best-known of German aircraft, largely based on its successes during the early campaigns of the war against Poland and France. The name is a shortened form of the German word for "diving combat aircraft". As Fig. 5.19b suggests, the Stuka was a rugged aircraft meant to operate from semi-prepared airstrips and to provide quick response to ground forces needs. Early versions of the Stuka carried a single 250 kg (551 lbs) bomb with a full crew complement; they were also armed with two 7.92 mm (17MG) machine guns in the wings and a single MG15 in the rear cockpit. By leaving the rear gunner out, these underpowered aircraft could carry a single 500 kg (1102 lbs) bomb on the centerline rack. Many early models were also fitted with a ram-air driven siren on the fixed landing gear; these sirens, called "Jericho trumpets", gave the characteristic wail of a diving Stuka as it made its bomb run. Later versions, notably the Ju-87D, could carry an 1800 kg bomb on the centerline station and on the Eastern front the Ju-87G (the tank-killer, Gustav) had considerable success with the two 37 mm cannon providing the punch to take out tanks and armored vehicles after 1943. These cannon were hung on under-wing stations just outboard of the main gear[21].

The Douglas SBD "Dauntless" had a similar mission to the Stuka; it served as the primary dive bomber for fleet operations during the first three years of World War II. During the naval battle at Midway, Dauntless crews sunk four of the Japanese fleet carriers. The SBD's could only carry a relatively small bomb load and were not as well known as the Stuka's were in Europe, but there were instrumental in the defeat of the Japanese Imperial Navy in the Pacific. The largest number of Dauntlesses in service were the SBD-5 (Fig. 15.20) which could carry up to 2250 pounds of bombs on the two wing racks and the centerline pylon. Two forward firing 50 caliber machine guns were mounted in the wing and one or two 30 caliber machine guns were mounted at the rear cockpit.

Fig. 15.20 Douglas SBD "Dauntless" (US Navy Photo)[22]

Next, we turn to the strategic bombers, a class of aircraft where some would say the most rapid progress was made during World War II. Virtually all of the combatants had famous (or notorious, depending on your point of view) machines that excelled at carrying large bomb loads for long distances. Both Allied and Axis powers utilized the long-range bomber effectively during this conflagration. Both sides also bombed targets that caused large loss of civilian life. Of course, the B-29 delivery of two United States atomic weapons at Hiroshima and Nagasaki are perhaps the culmination of this type of bombing attack that ultimately brought about the surrender of Japan without the heavy human cost of an invasion. Nonetheless, these strategic bombing raids, as well as the German raids on London, the British and U.S. pounding of the German heartland, and the incendiary raids on Japan, generated many ethical questions about the conduct of this part of the war. Ethics aside, these machines were technical marvels. One needs only look back to pages 150-152 to see how advanced these bombers were compared to those just 25 years younger.

We start with a discussion of one of the less well known Axis bombers, the Italian Piaggio P.108, of which only about 160 were built. This design was heavily armed (two 7.7 mm waist machine guns, a 12.7 mm machine gun in the lower turret, another 12.7 mm machine gun in the nose turret, and two remotely controlled twin gun turrets in the outboard engine nacelles). The four engines were 1350 hp Piaggio P.XII radial engines and the 7,700 pound (3500 kg) bomb load was carried in the internal bomb bay. The airplane carried a crew of 6. The Boeing B-29 had similar armament but flew four years after the prototype P.108B first flew in October 1939[23]. Bruno Mussolini, the Italian dictator's son, was killed in a P.108B accident in 1941[24].

The Avro Lancaster, along with the Handley Page Halifax, was a mainstay in RAF Bomber Command's long range striking force (Fig. 15.21). The Lancaster remained in various versions until 1952 and served in 7 different air forces worldwide. Some 7377 of them were produced in World War II. It was well-armed (most carried eight 30 caliber machine guns in a nose turret, a dorsal turret, and the tail, though later on the nose and dorsal turret were fitted with twin 50 caliber machine guns) and it had a 33-foot long bomb bay and carried bombs (and bomb loads) up to the 22,000 pound Grand Slam after a modification[25].

Fig. 15.21 Lancaster Dropping Incendiary Bombs (left),
and a 4000-lb "Cookie" plus J-Incendiaries (right)[25]

Germany did not successfully develop a strategic bombing capability during World War II and the Heinkel 177 story epitomizes the German approach to strategic bombing. This aircraft specification for this machine demanded a strategic, long-range bomber that could also carry anti-shipping glide bombs (like the Henschel 293) and which could also carry out dive bombing (evidently, the German Air Ministry was enamored with the early success of dive bombers). The result was a 68,000-pound, four-engined airplane with a maximum crew of six and a range of slightly over 3000 miles. The engines were housed in two nacelles and each pair of engines drove a single propeller through a gearbox. Typical armament included one 7.92 mm MG 81J in the nose, one MG 131 in the forward dorsal turret and one in the tail turrent and two MG 81 or MG 131 manually aimed from the rear gondola (and on some aircraft a 20mm cannon for piercing tank armor). Maximum bomb load was 13,200 pounds internally and an assortment of guided missiles or bombs like the Henschel 293[26], mines, or torpedoes carried externally. The He 177 never satisfactorily met the requirement to be able to dive bomb and the weight added by this requirement limited

Chapter 15 — Stores Certification Testing — 165

its speed, range, and carrying capacity. Crews derisively called it the *Reichsfeuerzeug* (Reich lighter) because the engines caught fire so often, in early versions. The engine nacelles, each with a paired engine) necessitated by the Air Ministry's insistence on the dive bombing requirement, were at least partially responsible for the fire hazard because oil leakage onto the hot manifolds on the two middle cylinder banks. Even after a new nacelle and new paired engines were installed, on one twenty-aircraft bombing sortie, 2 He 177s failed to take off, 10 returned with burning or damaged engines, and 2 of the remaining 8 aircraft were destroyed by Allied defenders. Bain also suggests that a He 177 in Czechoslovakia was identified to be the carrier of the German atomic bomb, which, thankfully, was not finished in time for the German war effort[27,28].

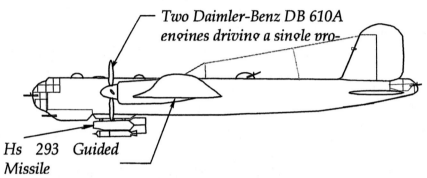

Fig. 15.22 Sketch of Heinkel 177 "Grief" Carrying Hs 123 (Adapted from[26,27,28])

Fig. 15.23 B-29 Enola Gay Landing after Dropping Atomic Bomb on Hiroshima[29]

The important fact to carry away from this brief discussion of heavy bombers and some of their more unique capabilities of World War II is that all combatants were pursuing improved technology at a furious rate and that compatibility of the weapons with the carrying aircraft was a major and continuing concern. Of course, the United States effort is epitomized in the Manhattan Project and the B-29 Superfortress that carried the first atomic weapons to Hiroshima and Nagasaki to end World War II without an invasion of the Japanese home islands (Fig. 15.23). This airplane is a large airplane: span of 141 feet, length of 99 feet, and a nominal loaded weight of 120,000 pounds (maximum of 133,500). Armament was twelve 50 caliber machine guns and one 20 mm cannon in the tail with a bomb load of 20,000 pounds. It was used by the U.S. Air Force, the U.S. Navy, the RAF, Royal Australian

Air Force, the Soviet Air Force (Tupolev re-engineered it into the Tu-4), and the People's Liberation Air Force (China). Variants (including the B-50 and its various models) remained in service through the early 1970s.

This B-29 strategic bomber also serves as a transition to the discussion of weapon and store compatibility as demonstrated during the Cold War. Nuclear weapons (heavy ordnance) and, more recently, guided weapons became dominant. But the B-29 and its derivative the B-50 (once called the B-29D) moved on to roles other than just carrying huge ordnance loads. The development of low observables since the end of the Cold War has led to more emphasis on internal carriage of weaponry, though external stores compatibility is still makes up the bulk of stores compatibility flight testing.

15.1.2.3. **Cold War Onward**. The Cold War is generally thought of as a time (roughly from the mid-1940s through 1991, at the dissolution of the Soviet Union) of tension and confrontation between the United States and the Soviet Union, with an arms race between these two superpowers and their allies fueling a rapid increase in air armament technology very similar to that shown during both World Wars. Even though this confrontation between East and West ended without direct warfare between the two major powers, the Korean and Vietnam Wars were fought in this period at a cost of over 100,000 lives lost by United States armed forces plus a cost of over $8 trillion[30] in U.S. military materiel and even larger numbers of lives lost and national treasure expended by the Soviet Union, Communist China, and their surrogates. The general trends during this period and in the decade and half after it, led to much more capable but increasingly complex systems; stealth or low observables technology came to fruition late in this period with the development of the F-117, the B-1, the B-2, and the F-22 in the United States.

The major change in aircraft immediately after World War II was the introduction of the jet engine as the primary propulsive system for military aircraft. The early jets like the Me 262 (Figs. 15.17 and 15.18) and their companion projects in England and the United States prompted an expansion of the flight envelope up to Mach 2 and beyond by the 1950s and 1960s. The Century Series fighters (F-100, F-101, F102, F-104, and the F-105) for the Unites States Air Force and similar aircraft for the U.S. Navy (F11F – later the F-11A – Tiger, the F8U Crusader, and the F4H Phantom II) all were supersonic fighters. But they carried bomb loads and other armament that made them into tactical fighters and attack aircraft for their respective services. Nuclear weapons and tactical missiles, as well as conventional bombs were carried on these tactical fighters and others of the Cold War and onward. Even heavy bombers became transonic during this period with the introduction of the "V" bombers (Vickers Valiant, Handley Page Victor, and Avro Vulcan) in England, the B-47 and B-52 in the United States, and the Tu-16 Badger in Russia. A few designs like the B-58 Hustler, the B-1 Lancer, the Soviet Tu-22 Backfire, and the Tu-160 Blackjack were even supersonic bombers. Perhaps most important most of these bomber designs relied heavily on cruise missiles, anti-shipping missiles, or other air-to-ground missiles for much of their offensive punch. It is imperative that we include some of the weapons themselves if we are to illustrate the evolution of airborne armament and the compatibility issues that come with these kinds of more modern machines. More recently, the B-2 bomber, though built in small numbers, adds low observables capability to long-range bomber technology.

Fig. 15.24 (a) F-100C with Bombs (b) F-100D Firing a Bullpup (USAF Photos)[31,32]

The North American F-100 (Fig. 15.24) was the first operational supersonic fighter for the USAF. It was designed as a "day fighter" (it was certainly not all-weather operational since it had no airborne radar) but it evolved into a workhorse multi-role airplane that could deliver both conventional weapons in support of ground troops or in an interdiction role and nuclear weapons on strategic targets. The broad array of munitions is illustrated in Fig. 15.24a except for the nuclear capability. Your senior author sat nuclear alert with 10 kiloton-equivalent nuclear weapons carried on the centerline pylon of the F-100. One otherpersonal note on the GAM-84 Bullpup shown being launched in Fig. 15.24b: the 429th Tactical Fighter Squadron (the "employer" of your senior author at the time) was the first squadron to qualify with this manually guided missile. The pilot of the single-seat F-100D flew the airplane with his right hand and guided the missile with his left hand. The Bullpup had a tracking flare on the tail fins that was visible until impact. Though it was a "stand-off" weapon, since it was launched a few thousand yards from the target, the pilot/weapon controller still had to maintain a relatively straight, steady inbound course until missile impact. Most of us fighter pilots believed we were "sitting ducks" during that tracking interval and did not relish making such a delivery in combat. Reliability of the missile was also far from 100%; on the first launch witnessed from a chase position by your senior author, the Bullpup pulled up sharply at about the point where this missile is in this picture and completed a loop behind both the firing airplane and his safety chase (me!). We both did a "hard break" to be sure the Bullpup did not take out either of us.

Fig. 15.25 F-4E Dropping Bombs (USAF Photo)[33]

The F-4 Phantom II was designed as a fleet air defense interceptor but became one of the most widely used multi-purpose aircraft ever built. Over 5,000 Phantom IIs were manufactured from 1958 through 1979 in the United States. Japan also produced over 130 Phantom IIs. The Phantom II was operated by both the U.S. Navy and the U.S. Air Force and could carry over 18,500 pounds of ordnance on nine external hard points. Though it was designed when most designers and planners thought internal guns were not needed, experience in Vietnam showed that a gun was still need and a 20 mm Vulcan (essentially, the same gun that was carried on the F-100) was integral to the armament on the Air Force's F-4E. Given the significant change in role, from fleet air defense to ground attack and interdiction, it is quite likely that the F-4 received more attention from stores compatibility testers than almost any other fighter since World War II. Both the U.S. Navy and the USAF had a stake in such testing and both pursued these capabilities vigorously.

Fig. 15.26 Sukhoi Su-25 "Frogfoot" (Copyright released Max Bryansky),[34]

The Russian Sukhoi Su-25 (Su-28) "Frogfoot" (Fig. 15.26), which first flew in 1975, is a single-purpose design differing from either the F-100 or the F-4, each of which evolved into a multi-purpose machine. Like the USAF A-10 "Thunderbolt II" (more commonly called the "Warthog") which has a similar role, the smaller Sukhoi Su-25 close air support airplane focuses on doing one thing well: carrying a large array of bombs and delivering them accurately in close proximity to engaged ground forces. The "Frogfoot" could carry over 9,500 pounds of bombs (with an empty aircraft weight of just over 20,000 pounds) along with enough ammunition for its 30 mm cannon to destroy armored vehicles. The Su-25 also has a very good reputation for operating from unprepared surfaces. It is somewhat lighter than the A-10 (which tips the scales at just under 25,000 pounds empty weight) and faster than the USAF close air support special purpose machine. The A-10 can also carry a considerably larger load of bombs, up to approximately 16,000 pounds and its GAU-8 30 mm cannon is a devastating tank-killer as it demonstrated during the First Gulf War in 1991. The Frogfoot saw extensive combat in the Afghan war, with 22 lost in that conflict[35].

The Grumman A-6 Intruder is a carrier-based attack airplane used to provide both electronic jamming (EA-6B) and deep interdiction for the fleet. Like the Sukhoi and the A-10, the Intruder carried a substantial and diverse set of stores and required careful assessment of compatibility for all the ordnance combinations it could carry. First flight was in April 1960 and the aircraft went into service in 1963. and was extended well beyond its intended life by refitting and refurbishment, notably for the electronic warfare role and after its intended successor the A-12 was canceled in January 1991, though litigation of the termination drug on for several years[33]. The A-6 remained in service until 1997 when the F-14 with a Lantirn pod took over its role. Now the F/A-18 E/F Super Hornet carries out this role for the U. S. Navy. However, the newer aircraft cannot match the Intruder in terms of useful load (maximum of just under 35,000 pounds compared to an empty weight of 25,630 pounds; the nominal armament load is 18,000 pounds distributed evenly across the five external hard points). Over 25 different types of bombs and missiles were certified for carriage and release by the A-6. Other roles for variants of the venerable Intruder included buddy refueling and the electronic warfare role (EA-6B).

Chapter 15　　　　　　　　　　Stores Certification Testing　　　　　　　　　　　　169

Fig. 15.27 Grumman A-6 Intruder (US Navy Photo by Alan Warner)[36]

The Boeing B-52 is another design, for the USAF in this case, that has lived on well past its intended life time and has been updated and modernized repeatedly. The YB-52 first flew in 1956 and the USAF still has operational squadrons of the B-52H at this writing (June 2007). Though much of its design philosophy and its operational service hinged on the nuclear deterrence of General Curtis Lemay's Strategic Air Command, the B-52 also demonstrated its usefulness as a conventional bombing platform (Fig. 15.28a) during the Vietnam war, carrying up to 30,000 pounds of conventional bombs in its internal bays and on hard points on the wings. The airplane can carry up 60,000 pounds of ordnacne. This venerable old workhorse has also outlived three generations of missiles, first carrying the Short Range Attack Missile (SRAM) in a rotatable launcher with 8 missiles per launcher, later a similar load of subsonic AGM 86A Air Launched Cruise Missiles (ALCM) from similar multiple launchers modified for the B-52 weapons bays, and more recently the stealthy AGM-129 Advanced Cruise Missile (Figs. 15.28b and c). This latter nuclear cruise missile also embodies stealth principles in its configuration. This airframe has undergone six significant upgrades to its avionics and electronic countermeasures system. The B-52 and its assorted weaponry required a continuing compatibility/release test effort as these weapons were fitted to the B-52 and each of its variants.

Fig. 15.28 B-52H (a) with　　(b) Loaded with AGM-129s[38]　　　　(c) AGM-129[38]
Conventional Bombs[37]　　(USAF Photos – MSgt Lance Cheung (a))

As suggested in the introductory paragraphs of this section, stealth technology was introduced during the latter years of the post-Cold War era and any discussion of stealth technology has to include the F-117A Nighthawk (Fig. 15.9), the first generation of such aircraft from the Lockheed Skunk Works. Shrouded in secrecy, the Nighthawk first flew in 1981 1981 and most of its developmental testing was done at night to maintain that secrecy. When it was unveiled to the public for the first time in the early 1990s, the F-117A was a veteran of the first Irag War and the Bosnian conflict. From a stores compatibility standpoint, low observables principles demand that weapons be carried internally and that, during releases, bomb bay doors open time be brief. One F-117A that was shot down over Kosovo during the second Iraqi conflict reportedly had a malfunction of the weapons bay doors, rendering it vulnerable to detection. Dropping stores from an internal bay at speeds and g-loads common today and ascertaining that low observables characteristics remain intact for as much of the mission profile as possible complicates stores compatibility testing, especially for airplanes (like the F-35 Lightning II) that are likely to have a very large complement of internally carried armament.

Fig. 15.29 Lockheed F-117A Nighthawk (USAF Photo – SSgt. Derrick Goode)[39]

This brief historical background to stores compatibility, tracing the evolution of airborne armament and weaponry for all types of aircraft, inevitably leaves out important features and interesting innovations. But hopefully it illustrates the complexity and historical evolution of this type of flight testing.

15.1.3 Acknowledgements

The authors are indebted to many specialists in stores compatibility testing who contributed their time to both offer both valuable advice in organizing this material and in reviewing the drafts we produced. A special note of thanks goes to the Seek Eagle team at Eglin AFB; Eddie Roberts and Stephen Petrillo were especially helpful in providing background material and their unique perspectives on these topics. We thank them for the time they spent is discussing their specialty knowledge with us and in reviewing the manuscript. The examples they provided and the cogent comments they made are an essential ingredient in this material more useful. It is our fervent hope that these notes will be of assistance to the Seek Eagle effort and to other groups worldwide engaged in stores certification activities.

15.2 STORES CERTIFICATION

Stores certification is a broad subject that demands careful attention to detail and disciplined test planning/preparation. The process time line chart (Fig. 15.30), adapted from Fig. 2 in MIL-HDBK-1763[40], illustrates the subprocesses and the amount of time typically required to certify a new store on a new aircraft. In many projects both the time and the effort can and should be reduced by tailoring the plan to suit the specific objectives and the concept of operations for which the store, its ancillary equipment, and the aircraft were developed.

15.2.1 Terminology, Organization, and Scope

Let us now define terms a bit more precisely and explain the organization and scope of this chapter. First, the title suggests and the subprocesses listed in Fig. 15.30 indicate, the nature of stores certification activities. Largely they fall into two broad categories: compatibility and separation issues – terms to be defined shortly – and this chapter is essentially organized around flight tests addressing these two topics. The following definitions come from the current (always check to be sure you are using the latest version of such reference documents) MIL-HDBK-1763[40] and current Federal Aviation Regulations.

Aircraft: Any vehicle designed to be supported by air, being borne up either by the dynamic action of the air upon the surfaces of the vehicle, or by its own buoyancy. The term includes fixed and movable wing airplanes, helicopters, gliders, and airships, but excludes air-launched missiles, target drones, and flying bombs.

Aircraft-stores: Any device intended for internal or external carriage and mounted on aircraft suspension and release equipment, whether or not the item not the item is intended to be separated from the aircraft in flight or not. Stores are often classified as expendable or non-expendable.

Aircraft-stores compatibility: The ability of an aircraft, stores, stores management systems, and related suspension equipment to coexist without unacceptable effects of one of the aerodynamic, structural, electrical, or functional characteristics of the others under all flight and ground conditions expected to be experienced by an aircraft-store combination. A particular store may be compatible with an aircraft in a specific configuration, although not necessarily so with all pylons (or stations) under all conditions.

All-up-round (AUR): Any completely assembled store, both mechanically and electrically, and ready for installation, carriage, and employment on or in an aircraft for a specific mission. An AUR has all mission-necessary sub-assemblies (such as guidance and control units, fins, fairings, and fuzes), associated hardware, and electrical cables installed and serviceable, as well as necessary pre-flight safety devices and any adaptation equipment normally fixed to store. An AUR does not include suspension equipment (such as bomb racks or missile rails), externally mounted electrical cables, or other items not separated with the store.

Ballistics: The science that deals with the motion, behavior, appearance, or modification of missiles or other vehicles acted upon by propellants, wind, gravity, temperature, or any other modifying substance, condition, or force.

Free-stream ballistics: A model of the weapon flight path from the time the weapon reaches steady state flight after release from the aircraft.

Separation effects: A model of weapon motion from the moment it is released until oscillations caused by the aircraft flow field are damp out. They are currently modeled as a function of release variables such as Mach number, normal acceleration, angle of attack, and dynamic pressure. These coefficients compensate for separation effects and may be incorporated into the ballistic tables and/or into a separation effects algorithm in the aircraft ballistic operational flight program (OFP). The coefficients used in the separation effect algorithm may result in aircraft velocity adjustments used in the air-to-surface trajectory calculations or may incorporate changes in the mode of trajectory calculation.

Ballistics accuracy evaluation and verification: The flight test process used to determine the accuracy of the ballistic portion of the aircraft OFP.

Ballistic trajectory: The trajectory traced after any propulsive force is terminated and the body is acted upon only by gravity and aerodynamic drag.

Carriage. The conveying of a store by an aircraft under all flight and ground conditions including taxi, take-off, and landing. The store may be located either external or internal to the aircraft. Carriage should include time in flight up to the point of complete separation of the store from the aircraft.

Symmetrical carriage: An arrangement (loading) of identical stores on either side of a dividing line or plane (usually the longitudinal axis) as related to a given aircraft, suspension equipment, or weapons bay.

Asymmetrical carriage. This term applies to carriage of stores which can be unlike in shape, physical properties, or number with reference to the plane of symmetry. 'Asymmetrical' applies to the arrangement or loading, of stores on an aircraft or suspension equipment and 'unsymmetrical' applies to an aircraft maneuver with aerodynamic loading unequally distributed on each side of the aircraft plane of symmetry, as in a roll.

Conformal (or tangential) carriage: The packaging of stores to conform as closely as practical to the external aircraft lines to reduce drag and obtain the best overall aerodynamic shape. Stores are generally carried in arrays, mounted tangentially to some portion of the aircraft, usually the bottom of the fuselage. It includes those arrangements made possible by weapon shapes configured for this purpose.

Multiple carriage: Carriage of more than one store on any given piece of suspension *equipment*, such as bombs carried on a triple ejector rack (TER) or multiple ejector rack (MER).

Single carriage: Carriage of only one store on any given station or pylon.

Tandem carriage: Carriage of more than one store on any given piece of suspension equipment such that one store is behind the other.

Authorized download: Any configuration resulting from the downloading of weapons in normal employment sequence from a configuration authorized in the aircraft Flight/Tactical Manual and which can also be authorized for flight.

Compatibility engineering data package (CEDP): A CEDP for a store is the primary data package used by the USAF to ensure that stores are physically, mechanically, electromagnetically, environmentally, structurally, and aerodynamically compatible with aircraft systems, so that supporting technical orders can be written.

Chapter 15   Stores Certification Testing   173

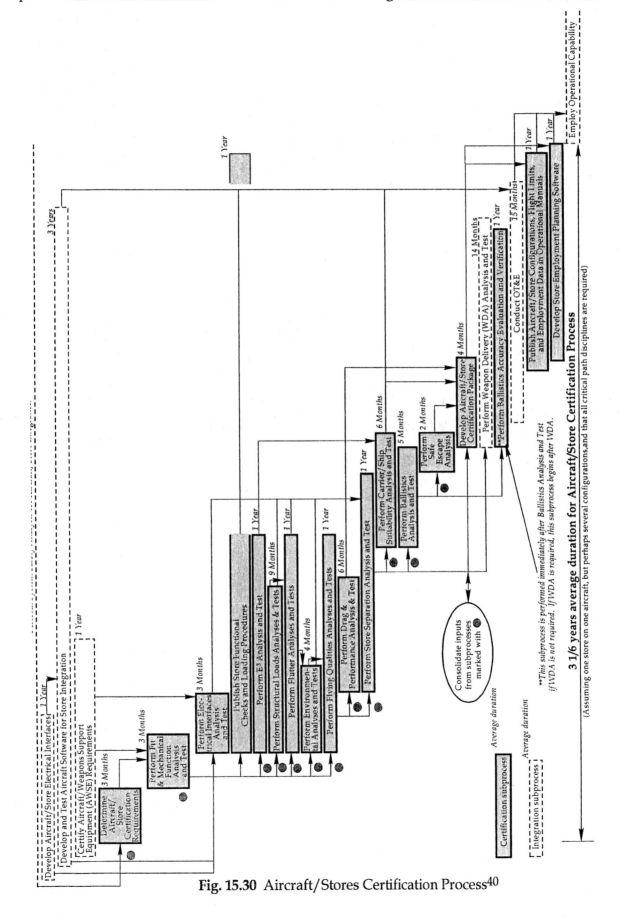

Fig. 15.30 Aircraft/Stores Certification Process[40]

Circular error probable (CEP): A measure of accuracy equal to the radius of a circle centered on the target or mean point of impact that contains 50 percent of the population impact points. CEP is measured in the normal plane for aircraft level and dive deliveries and in the ground plane for loft deliveries and is usually given in mils.

Critical conditions: A combination of operational parameters expected to be encountered by an aircraft, store, or combinations thereof; the design or operational limits of the aircraft, stores, or portions thereof are based upon these parameters.

Dispersion: A scattered pattern of hits around the mean point of impact of bombs and projectiles dropped or fired under identical conditions.

Aircraft dispersion: Errors contributing to the ballistic error budget like sensor errors, on-board avionics errors, timing delays, fire control, or variation in rack ejection forces.

Ballistic dispersion: Weapon-to-weapon variation in the free-stream ballistic flight path attributed to manufacturing tolerances such as mass and physical properties, and accidental misalignments occurring during assembly and handling of the weapon.

System dispersion: The total dispersion due to weapon and aircraft.

Electromagnetic compatibility (EMC): The capability of electrical and electronic systems, equipment, and devices to operate in their intended electromagnetic environment within a defined margin of safety, and at design levels of performance, without suffering or causing unacceptable degradation as a result of electromagnetic interference.

Electromagnetic environment effects (E^3): The impact of electromagnetic environment upon operational capability of electronic or electrical systems, equipment, or devices. It encompasses all electromagnetic disciplines, including electromagnetic compatibility; electromagnetic interference; electromagnetic vulnerability; electromagnetic pulse; electronic countermeasures; hazard of electromagnetic radiation to ordnance and volatile materials; and natural phenomena effects of lightning and precipitation static (p-static).

Electromagnetic interference (EMI): Any electromagnetic disturbance, whether intentional or not, that interrupts, obstructs, or otherwise degrades or limits the effective performance of electronic or electrical equipment.

Employment: Use of a store for purposes and in the manner for which it was designed: releasing a bomb, launching a missile, firing a gun, or dispensing a submunition.

Release: Intentional separation of a free-fall store, such as an "iron bomb," from its suspension equipment, for purposes of employment of the store.

Launch: Intentional separation of a self-propelled store (a missile, rocket, or target-drone) for purposes of employment of the store.

Fire: Operation of a gun, gun pod, or similar weapon, causing a bullet or projectile to leave through the barrel.

Dispense: Intentional separation from an airborne dispenser of devices, weapons, submunitions, liquids, gases, or other matter, for purposes of employment of the items being dispensed.

Failure mode: Malfunction of weapon components (autopilot actuation or fin deployment) which must operate normally to ensure acceptable separation.

Flight clearance/clearance recommendation: Authorization for flight, after appropriate engineering analysis, that an aircraft-store combination does not pose an unacceptable risk for a specific, limited, purpose such as DT&E or IOT&E. The flight clearance specifies flight limits and remarks about operation of the loading configuration on a specific aircraft, or group of aircraft, and is valid only for a specified finite period of time for a specific user or group of users.

Free flight (of a store): Motion of a store, either powered or unpowered, through the air after separation from an aircraft.

G-jump: Change in normal load factor resulting from store release, due to combined effects of ejection force, dynamic response, and instantaneous aircraft gross weight decrease. The g-jump effect is most apparent when several large stores are released simultaneously or a large number of smaller stores are released with a very small ripple interval. The amount of g-jump can be large enough to overstress the aircraft or other stores not being released. In dive-toss delivery g-jumps of up to 3.5g (in addition to the maneuver load factor) are common.

Hung store or hangfire (for Army): Any store (or stores) which does not separate from the aircraft or launcher when actuated for employment or jettison.

Interval: The elapsed time between the separation of a store and the separation of the next store. The minimum release interval is the shortest allowable or usable interval between successively released stores that will allow safe separation of the stores from the aircraft.

Mean point of impact (MPI): A point which has as its range/deflection coordinates the arithmetic mean of the range and deflection coordinates of the impact points.

Mixed load: Simultaneous carriage or loading of two or more unlike stores on an aircraft.

Operating limitation: Flight carriage, employment, and jettison envelopes detailing acceptable airspeed, Mach mumber, altitude, g, roll rate, wing sweep, speed brake operation, delivery angles, release modes, and minimum release intervals required for a specified aircraft/stores configuration.

Operational test and evaluation (OT&E): The field test, under realistic conditions, of any system (or key subsystem) such as weapons, equipment, or munitions for the key purpose of determining the effectiveness and suitability of the weapons, equipment, or munitions for use in combat by typical military users; and the evaluation of the results of such tests.

Range bias: Criteria determined by the user that defines whether the weapon delivery system is biased in range; that is, acceptable long/short weapon impacts about the target.

Safe escape/safe arming: Safe escape is the minimum release altitude providing the delivery aircraft acceptable protection from weapon fragmentation or detonation at the preplanned point. Safe arming separation is the selection of a minimum safe fuze arm time setting that provides the delivery aircraft acceptable protection from weapon fragmentation if early detonation should occur.

Separation: Termination of all physical contact between any part of a store and an aircraft; or between a store and suspensions equipment.

Safe separation: The parting of a store(s) from an aircraft without exceeding the design limits of the store or the aircraft or anything carried thereon, and without damage to, contact with, or unacceptable adverse effects on the aircraft, suspension equipment, or other store(s) both released and unreleased.

Acceptable separation: Acceptable store separations are those which meet not only the "safe" separation criteria, but also meet pertinent operational criteria. For instance, guided weapons as a minimum must remain within control limitations consistent with mission effectiveness. Conventional weapons, bombs, should not experience excessive angular excursion which induce ballistic dispersions adversely affecting weapons effectiveness or causing bomb-to-bomb collisions.

Pairs: The simultaneous separation of stores from two separate stations on an aircraft.

Ripple (or train): The separation of two or more stores or submunitions one after the other in a given sequence at a specified interval.

Salvo: The simultaneous separation of stores from multiple stations on an aircraft.

Multiple: The number of stores released simultaneously from aircraft store stations.

Quantity: The total number of stores released per pass by an aircraft.

Ejection: Separation of a store with the assistance of a force imparted from a device, either external or internal to the store.

Selective jettison: The intentional separation of stores or suspension equipment, or portions thereof (such as expended rocket pods), no longer required for the performance of the mission in which the aircraft is engaged.

Emergency jettison: The intentional simultaneous, or nearly simultaneous, separation of all stores or suspension equipment from the aircraft in a preset, programmed sequence and normally in the unarmed condition.

Submunition: Any munition that, to perform its task, separates from a parent munition.

Suspension equipment: All aircraft devices such as racks, adapters, missile launchers, pylons, used for carriage, employment and jettison of aircraft stores.

15.2.2 Importance of Analysis and Ground Tests

Stores compatibility testing falls within a category of testing often referred to as exploratory testing. In this class of tests (which also includes post-stall and spin testing, flutter testing, some forms of propulsion testing, and the like), the primary goal is to obtain data that reduces the uncertainty about a design feature or goal for the system. The fact that uncertainty exists suggests that there is risk in obtaining data to reduce or even eliminate the unknown facts. The risk is typically roughly proportional to the level of uncertainty; more uncertainty implies more risk to either or both the program and the test vehicle/crew. Stores certification testing always comes with risks that lie within a lower bound, if a store has demonstrated compatibility on another similar airplane. At the other end of the risk spectrum, the test team may be called upon to certify a brand new store on an aircraft design that is also under development.

To reduce risk to tolerable levels for exploratory flight demonstrations, the team has no more important tools than analysis like modeling and simulation (M&S), specialized wind tunnel tests, ground vibration tests, and other such analysis and ground tests. The test team must not only appreciate the value such tools bring to the table; they must learn how to utilize them effectively and place judicious confidence in such analysis and background testing. Much of this analysis and ground testing is done to verify design predictions, but

Chapter 15 Stores Certification Testing 177

if planned and utilized properly, it also provides considerable insight to guide the flight test planning. Figure 15.31 illustrates a specialized wind tunnel test of this type. The M&S tools often help predict behaviors that are not necessarily intuitive. These predictive tools are of considerable benefit, especially when corroborated by wind tunnel tests that help confirm the validity of model predictions. Though such analyses are essential to adequate preparation for flight tests, full details on the use of such tools is beyond the scope of this chapter. The interested reader can consult appropriate references[40,41] for further information. However, specific examples of how some of the tools have been used are given to help guide the flight test team.

Fig. 15.31 F-4E Separation Wind Tunnel Tests (US Navy Photo)[42]

15.3 CAPTIVE CARRY COMPATIBILITY

Captive carriage testing ensures the compatibility of the store with the airplane throughout the envelope to which the store is to be certified. It is an unusual store that does not degrade the airplane's behavior in some way; captive carriage testing ensures that the store does not compromise the airworthiness of the host airplane while ensuring that the store and its carriage hardware can survive flight stresses.

While surely simpler than the release and separation problem, carriage of an external store is not without its own challenges, and deserves careful attention. The purpose of a captive carriage effort is certification for carriage, recognizing that external carriage has engineering implications spanning performance, flying qualities, loads, fatigue, flutter, vibration, thermal properties and acoustics. To the extent that tests are required to clear a store configuration for flutter, performance or flying qualities, tests proceed according to the same methods addressed in other chapters covering those topics.

The carriage envelope for a store often exceeds the release envelope. Hence, the endpoint of testing may be much faster or at a higher load limit than any intended release. Additionally, some stores may be mounted so as to never be released, in which case captive carriage tests comprise the full certification trial.

15.3.1 Planning and Preparation

A captive carriage certification program ranges from a half-day of flying to several months, entailing several dozen flights. Complexity depends upon whether the test is a new store on a new airplane, an old store on a new airplane, a new store on an old airplane, or an old store on an old airplane. An "old store" means one successfully certified and operated on another airplane. Tests requiring the greatest scope are those where a store ("old" or "new") exceeds the mass properties of previously cleared stores, triggering the need for loads and flutter testing. Again, the captive carriage certification process is largely indifferent to whether a store will ever be released. Captive carriage thus pertains to military weapons carriage and release, as well as external stores certified for carriage on civil airplanes, such as optical or electronic sensor packages or external fuel tanks.

15.3.1.1 **Analysis**. The analysis required to certify a new store for carriage can range from mere hours of analysis to months of ground and flight testing. Stores certified on other airplanes, and fitting within the weight and geometric properties of other certified stores may occasionally be certified on the basis of analysis and ground fit tests alone. This approach is sometimes called "certification by similarity". A unique umbilical or arming wire arrangement might then require but a couple hours of flying. In contrast, massive or aerodynamically unique stores often require substantial person-hours devoted to preflight analysis of loads, flutter, aerodynamics, and flying qualities, followed by weeks or months of flying to clear the loading for flutter.

The flight test engineer requires ground analysis to identify portions of the flight envelope where analytical models can both be easily validated, and to choose those maneuvers and flight conditions where uncertainty is greatest. Few programs can afford exhaustive captive carriage flight tests; separation trials are expensive enough. The ultimate certification of the store for carriage is likely to be based on analysis which is then selectively validated with in-flight demonstration points.

The flight test team should expect and request the following information[40] from the tasking agency and supporting engineering specialists (perhaps vendors):
- store description
- mass properties (weight, center of gravity, and mass moments of inertia), including expected range of variation
- functional description to include carriage arrangement and the routing of any umbilicals or arming wires
- aerodynamic data
- electromagnetic compatibility data
- structural analysis
- environmental analysis
- systems safety
- transmitter and receiver details
- preparation, loading, arming, and download procedures
- proposed carriage clearance envelope, to include speed, load factor, configuration and maneuvering limits

15.3.1.2 **Ground Testing**. Armed with this information, the flight test team is ready to plan the ground and flight phases of captive carriage tests.

15.3.1.2.1 *Test Article Selection*. Preparatory to either captive or release testing, mass properties of the test stores must be verified to ensure they are representative of production articles. This verification includes weight, center-of-gravity, and mass moments of inertia. The behaviors for which stores are responsible are sensitive to mass properties and the test team must know that the test items match production specifications. Sometimes, it is even helpful to know whether an individual serial number store is on the high or low end of production tolerances. Where feasible and when sensitivities are known, stores that are worst case within a production population should be selected as test articles.

15.3.1.2.2 *Fit and Function*. Fit testing entails more than simply ensuring the fit of the store on the carriage equipment. The store, of course, must mate properly on the carriage equipment, including functional routing of any power or avionics umbilicals. Also, the store must not interfere with servicing panels on either the airplane or adjacent stores. A store that fits in a computer-aided design may be awkward to load, download, or service. The full range of plausible loading crew arrangements and loading hardware require must be exercised to identify anticipated operational challenges away from experts on the test team, loading in ideal conditions. Electrical and plumbing connections and confirmation of hook engagements should be confirmed visually with ease. Most importantly, loading, preparation and servicing must not expose ground personnel to possible injury. Electrical interlocks must ensure that release cannot be inadvertently triggered and firing circuits must be confirmed cold by stray voltage checks prior to making any electrical connection. The test engineer overseeing fit and function testing and his team must carefully think through and document appropriate procedures before carrying out the tests. A checklist is mandatory, even though it may change as the test progresses!

15.3.1.2.3 *Vibration and Acoustic*. If flutter testing for the stores is needed, a full ground vibration test (GVT) is required as discussed in Chapter 12. Even if flutter is not a concern, GVT is often still warranted to characterize the modes susceptible to resonance short of flutter. While stores flutter testing principally seeks to determine the influence of the store on the airplane, vibration and acoustic testing seeks to identify the influence in all directions: the store on adjacent stores and airframe, and well the airframe and adjacent stores on the store under test.

The under-wing environment can be brutal for externally mounted stores, particularly those with sensitive electronics. Engine noise, moving shock waves, airframe buffet and vibration, and hard landings all send tremendous high frequency energy into carried systems. The test team should insure that a store is subjected to an appropriate "shake-and-bake" environmental test prior to arriving on the flight line. The challenge, however, is that the airframe structure has vibrational characteristics of its own (dependent on both frequency and magnitude) which are not likely replicated at the component environmental test level. Moreover, a massive store may shift airplane structural modes in unpredictable ways, interacting with the airplane structure or with the flight controls. Hence, ground vibration testing, with the store mounted on-wing, is often necessary to characterize the integrated system response to either acoustic noise or to vibration. These data feed analytical models to better predict in-flight behavior. Noise and vibration are usually separate phenomena, due to differences in the source of exciting energy, but required instrumentation and test hardware are so similar that these tests are typically performed at the same time.

Some practicalities deserve emphasis. Results of these tests are very sensitive to torque applied to sway braces, so it is vital that loading crews apply the specified torques; otherwise ground tests are not representative of flight configurations. Additionally, though seemingly insignificant in weight, all connecting hardware (umbilicals and arming wires) must be in place for ground tests, in order to remove uncertainty.

Instrumentation for noise and vibration testing typically consists of transducers, accelerometers and strain gauges mounted to the store and to supporting carriage hardware. Some GVT instrumentation is *not* flight worthy, and must be used solely for the ground tests. Other instrumentation sensors are part of the flight instrumentation suite. Sometimes these sensors may be calibrated using the more sensitive equipment used solely for ground tests. It is both more efficient and safer if the same engineering personnel are involved in both ground and flight tests, using identical display and analysis software. An engineer who sees the #3 accelerometer on station 4 exhibit a nominal behavior during GVT, usually spots anomalous in-flight behavior more quickly if he or she also saw that accelerometer on the same displays in the same scaled units during ground tests.

15.3.1.2.4 **_Electromagnetic Compatibility and Hazards of Electromagnetic Radiation to Ordnance_**. Electromagnetic compatibility (EMC) and Hazards of Electromagnetic Radiation to Ordnance (HERO) testing ensures that an operational electromagnetic environment does not create signals within the store that either interferes with its function, or worse, inadvertently releases or fires the store. The host airplane, adjacent airplanes, adjacent stores, or ground/shipboard facilities are a source of signals whose strength might induce a current in an otherwise disabled circuit. Military test installations have dedicated teams to perform these tests for both military and civil customers. For the small civil flight test team, without access to these resources, EMC testing entails operating all airplane avionics systems through all their modes, emphasizing those with transmitters, to ensure any electronics within the store are not degraded by the host airplane avionics, or transmitters operating in the vicinity of the test site.

15.3.1.2.5 **_Environmental_**. Environmental testing is typically performed at the subsystem level. This testing should include operation throughout the flight temperature extremes. High temperature testing is particularly challenging due to the difficulty in reproducing flight aerodynamic effects, which may include both local heating and cooling, particularly at high Mach number. If significant temperature effects on store avionics or sensors (particularly electro-optics) are a concern, any instrumentation installed for ground tests should be flight worthy so that differences in instrumentation do not complicate comparisons between ground and flight results. Additionally, since concerns usually exist regarding cold-soaking or heat-soaking, the flight test team should be part of ground test planning so ground test results permit direct comparison with flight data. Ground tests are particularly challenging if the airframe is expected to provide the store with cooling or heating (either air-cooled or liquid-cooled).

15.3.1.3 **Captive Compatibility Flight Profiles**. Captive compatibility flight profiles (CFP) are qualitative assessments of the effects of a store configuration on aircraft flying qualities and on structural integrity of the store[40]. Aircraft and stores are usually not instrumented at all or, at best, carry minimal instrumentation for such qualitative evaluations. The three types of qualitative assessments most often recommended, along with applicability, summarized test requirements, and limitations of each assessment, are listed in the Table 15.1.

Table 15.1 Types of Captive Compatibility Flight Profiles (CFP)[40]

Type of CFP	Applicable to stores if:	Required testing:	Limitations
Handling Qualities (Test 251)	Shape, mass, inertia properties, or configuration not analogous to previously certified or tested item/aircraft	When higher limits (airspeed or load factors) than analogous configurations	Not to expand flying qualities or design limits of the airplane
Structural Integrity Test (Test 252)	No previous flight testing to analogous flight limits		No one-of-a-kind developmental stores not intended for certification
Endurance Test (Test 253)	No previous flight testing done, analogous to the required flight limits	Lanyard routing significantly different than previously flown	

More detailed techniques for CFP sorties of each type are discussed in the Guide to Aircraft Stores Compatibility[41]. Each of the following summaries closely mirrors advice in this guide. Four types of analyses and/or reviews used in such evaluations: weight and CG review, structural review, flutter analysis, and aerodynamic analysis.

15.3.1.3.1 *Weight and CG Review*. For this review, aircraft weight, most forward, and most aft center of gravity locations are necessary. If the aircraft aerodynamic center (AC or neutral point) can be predicted analytically using existing data on similar stores configurations, the aircraft AC and CG locations shall be used to guarantee an acceptable static margin. During this review, aircraft internal configurations (electronic countermeasures gear, and armor plate) ground rules establishing allowable store release sequence variations and whether or not empty rocket launchers, dispensers and spray tanks are to be jettisoned or retained are established. Subsequent aircraft modifications that add or remove equipment affect the weights and static margins computed during this review; such changes invalidate this analysis since the weights and stability margins are no longer representative of the modified aircraft. Therefore, this compatibility evaluation of a stores configuration does not replace disciplined use of operational aircraft flight manual weight and balance information to confirm that a specific aircraft/stores configuration is within allowable weight and CG limits. When the aircraft AC cannot be calculated reliably, wind tunnel testing is a requirement.

15.3.1.3.2 *Structural Review*. For each configuration the store loading at each carriage station is compared with allowable strengths of the bomb rack and the aircraft carriage station. Estimates of allowable strength must address fatigue and fracture failures, as well as static failures. Total station store loading is checked against the allowable structural limits. An aeroelastic analysis that defines side force, yawing moment, rolling moment, lift, pitching moment, and drag for the store installation is essential so allowable aircraft roll and g limits for each stores configuration can be set. External store loads determined analytically or from wind tunnel tests may exceed the static strength of bomb racks and/or airplane carriage stations. Allowable limits are established and an acceptable alternate stores configuration is defined, if necessary. The allowable structural limit load for a hung store may be a lower percentage of failing load if established by tests or analysis. Often a critical yaw condition exists when one of six stores carried on a multiple rack is released, creating a five store configuration which is not symmetrical fore and aft. If stores are attached to a multiple carriage rack, at least the following checks are made:

(a) Individual store loads against-individual ejector unit allowable structural limits.
(b) The forward and aft cluster loads against beam allowable structural limits.
(c) Any configuration of stores on a multiple carriage rack obtained through normal release or short loading (taking off with a partially filled rack).
(d) Any configuration of hung stores on a multiple carriage rack obtained from a full or partial loading.

15.3.1.3.3 *Flutter Analyses*. For these analyses, applicable stores are grouped according to weight, pitch radius of gyration, and longitudinal CG – approximating inertial similarity for each group. Then, if the aerodynamic configurations are also similar, flutter properties are likely to be similar. Grouping such similar aircraft/store combinations allows analysts to concentrate on ranges of store weight, pitch radius of gyration, and CG location for both symmetric and asymmetric flutter, ultimately identifying critical areas where flutter is most likely to occur. Comparison with analysis and test results for related or similar store configurations is also helpful. This grouping of inertially similar configurations is also recommended for evaluation of mechanical instability of either the store on the pylon/rack combination or other vibrational anomalies.

15.3.1.3.4 *Aerodynamic Analyses*. A preliminary operating flight envelope of the aircraft with external stores is usually predicted for modern designs using the most up to date modeling and simulation. As a first order approximation, limits for each store configuration are extrapolated from limits flown on other aircraft with similar capabilities and similar stores configurations. Such extrapolated limits should be used conservatively; a careful buildup from more benign configurations to limiting values in the test sequence is particularly important in such cases.

15.3.1.4 **Captive Flight Testing**. MIL-HDBK-244A[41] recommends that captive flight profiles exceed the typical mission duration by 50%. Furthermore, an additional 30 minutes sustained is recommended at low altitude and high dynamic pressures in order to ensure exposure to the most severe vibrational loading. Hence, the minimum scope captive carriage certification consists of two flights. The first is dedicated to sample testing the envelope for flying qualities and loads, the second evaluates high dynamic pressure dwell. This duration testing is required because of stores' elevated susceptibility to vibration and the consequent effects on fatigue, and to ensure the integrity of wiring and umbilicals. These two flights should be performed as a single test evolution. The stores and aircraft are inspected between flights, but the stores and its umbilicals should not be disturbed, and carriage hardware should not be adjusted (no tightening of sway braces). If the stores are disturbed or if weather should interfere, then the test should be repeated from the beginning.

The following points should be included on a captive carriage test card[40,41].

Chapter 15   Stores Certification Testing   183

- Taxi and ground handling
- Takeoff & Landing
- Maximum Mach number
- Maximum dynamic pressure
- Integrated handling qualities test block at select flight conditions. Each of these is warranted because of the destabilizing effect most stores have on both the pitch and directional stability.
  - ❖ Full stick roll performance
  - ❖ Steady heading sideslips
  - ❖ Pitch and rudder doublets (to evaluate oscillatory characteristics)
  - ❖ Wind-up turns (to evaluate maneuvering stability and buffet onset)

Low speed handling and stall

Maximum load factor at corner speed (likely inducing the highest buffet loads)

Maximum negative load factor.

Maximum rolling load factor

Transonic throttle chops. Fuselage mounted stores intended for carriage at transonic speeds are susceptible to hammer shock pressure fields when an engine either stalls or the throttle is abruptly chopped to idle as inlet spillage increases abruptly in both these cases.

Postflight inspection of the store and all carriage hardware ensures that no damage has occurred to either the store or airplane. Either air loads or vibration should not have dislodged all umbilicals or arming wires. Loose light arming wires or an umbilical can cause considerable damage at high airspeeds.

15.3.1.5 Test Instrumentation. Specialized instrumentation for captive carriage can range from none (for benign integrity checks) to elaborate systems capturing full-flying qualities parameters, avionics bus traffic, high frequency multi-axis accelerometers for vibrations, acoustic sensors, and strain gages for loads.

High-speed cameras (motion picture or video) installed for separation trials can also be used for capturing vibrational and buffet response. Film-based systems are prone to very short run-times due to the high frame rate needed, in which case the film or video tape can be conserved until the worst case point is identified from accelerometers (typically, buffet levels). The point is then repeated with film running to capture film/video of the behavior.

If captive carriage tests are to be followed by separation trials, the captive carriage tests should include exercising all instrumentation supporting separation trials. Few tests are as expensive as a separation test flight, and it's dismaying to discover post-flight that the critical camera missed the drop.

15.3.1.6 Telemetry. Even though the 1998 guidance on CFP[40] states that captive carriage flights are largely qualitative, telemetry packages have now been miniaturized and video cameras have become more capable so it is now possible to collect useful data on store interface details and handling qualities during CFP sorties. Jordan[43] describes the use of miniaturized, relatively low cost instrumentation designed specifically for captive carriage flights. The instrumentation suite recorded pull forces on the AV-8B/JDAM 1760 umbilical release lanyard hook and characterized some stability and control parameters. The AV-8B is one of the programs to use this second generation Captive Carry Miniature Telemetry unit (CCMT) during captive carriages to clear GBU-12 configurations. Two of these units, each hosting three tri-axial accelerometers and six strain gages, are configured to transmit 6,250 samples/second, quite an adequate rate to provide captive carry loads data. A strain gage sensor measuring a sway brace load to be telemetered through one of these channels is the sensor of choice. The CCMT is small enough and sensors are alos that the telemetered data is quite useful during CFP flights. Though such units are costly, they can be reused and, if wider usage develops, the cost is likely to go down, making them affordable for more purposes. Similar units can provide data to support handling qualities tests in the CFP series.

15.3.2 Risk Mitigation

Captive carriage flights are not usually considered high risk by stores compatibility test agencies; usually these flights do not incur the same risk as separation, employment, or jettison tests. Nonetheless, precautions are appropriate.

15.3.2.1 **Test Limits**. Since CFPs are typically flown as the first stage in a larger effort, the initial test envelope must be based on predictions and/or similarity with other cleared configurations. The test sequence must be chosen to build up from benign test events to events having increasing levels of risk (and more useful results) until the desired end point is reached.

15.3.2.2 **High Speed Risks**. More often than not, captive carriage sorties are meant to expand the dynamic pressure and Mach number envelope with incremental steps of added risk at each test event. These two variables often depend on fundamentally different physics and result in characteristics not anticipated. It is usually more productive, therefore, to lay out a test matrix based on systematic variation of these two variables. In the past, test planners often built a test matrix around a series of constant altitude points with little regard for parameters (or more likely, groups of parameters) that govern the physical phenomena being investigated. Use of techniques like "Design of Experiments" (DOE)[44] can bring focus to the test matrix, help optimize the data collection, and reduce the risks by minimizing exposure to anomalous behavior at high speeds.

15.3.2.3 **Safety Chase**. Safety chase is indispensable to captive carriage testing. Damage that occurs in flight can easily escape the notice of the test aircraft's pilot. For many of the airplanes and stores in question, the pilot may not be able to see any part of the store. While he/she might feel the sudden loss of a store, through the airplane, it is doubtful that the pilot would notice a lost fin, a flailing arming wire or umbilical, or an open access panel. The chase pilot is indispensable to the test event, both for the safety of the test airplane and its crew and as an observer/data collector for the test. If a panel or fin separates, only the chase pilot is likely to note the time and flight condition at which the loss occurred. During those flights in which close formation is impractical or unsafe, the chase pilot must frequently inspect the store and its ancillary hardware for integrity.

15.3.2.4 **Configuration Limits**.

15.3.2.4.1 *Asymmetric Risks*. Prior to 1990, strike aircraft typically flew with symmetrical loading and typically released entire loads of unguided weapons simultaneously. Deliberate asymmetric loads were uncommon, and typically small in magnitude. The move to large precision guided munitions, which could be effectively employed as a single weapon, resulted in airplanes frequently carrying and employing weapons while in an asymmetric loading configuration. It is now common to release a single, highly effective weapon, thus deliberately accepting an asymmetric load. The prevalence of airplanes commonly flying in asymmetric loads led to much more complete investigations of spin susceptibility for most fighter/trainer aircraft developed after 1972[45].

Asymmetric loads create a variety of interesting handling challenges. The principal hazards posed by asymmetric loadings are increased likelihood of departures and spins, along with prolonged recoveries from out-of-control events. If a 2000 pound store is suspended ten feet left of the centerline, a left rolling moment of 20,000 foot-pounds (ft-lb) is created at 1-g that must be balanced by aileron (left aileron down). This kind of loading and control input also produces a left (pro-spin) yawing moment due to both the added drag of the store and adverse yaw from the downward-deflected aileron. For example, F-15 departure and spin susceptibilities were found to be dependent on loading asymmetries greater than 5000 ft-lb and aerodynamic recovery from smooth spins was more difficult with such asymmetric loadings[46].

15.3.2.4.2 *Heavyweight Risks*. Carrying stores always adds weight and inertia; if stores are carried externally, drag can be increased significantly. Consequently, both performance and handling qualities are likely to be affected adversely by the added weight, inertia and drag. The added weight means that even in conventional maneuvers the airplane is operating at higher angles of attack and therefore is exposed more often to departure events. Operating at elevated angle of attack along with the higher inertias also usually reduces control effectiveness, as does having to utilize part of the available control surface deflection to offset any asymmetry to the loading. Air-to-air engagements, air-to-ground weapons deliveries, air-to-air refueling, and landings simply become more difficult and demanding with the added weight of stores, particularly asymmetric stores.

Christensen[44] emphasizes the importance of brake performance for sorties when heavy weight landings are necessary during captive carry flights. He suggests monitoring brake performance and keeping track of brake energies absorbed. It is also wise to have well-planned control room monitoring of such landings with emergency personnel notified in advance; high energy stops often lead to hot brakes and must be anticipated.

15.3.3 **Execution Lessons**

There are a multitude of lessons to be learned from past experience in conducting captive carry flight profiles. Christensen[47] again offers cogent general advice regarding three different areas before giving specific examples and lessons to be learned from them. First, recognize that a complete and comprehensive CFP test series involves flight tests at or near both structural and performance limits. Diving entries to obtain some of the desired test conditions are to be anticipated. Care must be taken not use a dive angle that is excessive and simulated practice runs should be considered for both the pilot and the control room team. Constant Mach numbers in steep descents can easily exceed structural limits if this kind of simulated profile is not done to prepare all participants for potential emergencies and/or desired recovery actions. Points in the transonic range can be particularly troublesome since the longitudinal sensitivity of the airplane/store configuration is typically higher than that of the basic airframe. A second area of concern is the structural load during accelerated maneuvering, like loaded rolls and rolling pullups with asymmetric stores. Adjustable g-limiters (if available) or trim can be helpful in remaining within the desired g-limits during deliberate step roll inputs. Practicing these maneuvers, both by the flight crew and with the control room monitors, is a necessary form of discipline for captive carry compatibility tests. Finally, never forget that flying qualities of the airplane carrying stores are often degraded; perhaps we should even call this phase of the testing "captive incompatibility" testing to underscore the fact that these flights are deliberately searching for anomalies in the airplane/store envelope. As one might expect, most of the specific examples to be discussed involve such "incompatibilities".

15.3.3.1 **F-16 Conformal Fuel Tank Degraded Flying Qualities**. With conformal tanks the F-16 behaved different in flight tests than wind tunnel data predicted. The predicted lateral-directional flying qualities with conformal tanks (Fig. 15.32) were quite similar to the clean airplane. However, in steady sideslip tests with the conformal tank installed maximum sideslip reached 21°, even though the sideslip typically peaks at about 15° in the basic airplane. Additionally, at low altitudes (where most air-to-ground deliveries are made), the airplane with conformal tanks reached maximum lateral control input before maximum rudder was attained, another indication of reduced directional stability with this configuration. Also, roll rate was unacceptably degraded by excessive adverse yaw in the conformal tank configuration. Adjustments were made in both lateral and directional axes in response to these findings from CFP evaluations of flying qualities.

Fig. 15.32 F-16 Conformal Tank Configuration (Copyright release by David Raykovitz)[48]

15.3.3.2 **F-16 Joint Standoff Weapon (JSOW) Degraded Flying Qualities**. When the F-16 carried the JSOW, the flying qualities of the airplane/store were degraded. Even with two of these stores mounted symmetrically, the pitch sensitivity of the configuration resulted in pitch oscillations during almost all phases of flight including level flight. Moreover, when the AGM-154 (Fig. 15.33) was carried asymmetrically, significant aileron and rudder trim were required throughout the transonic flight regime. These findings led to additional flight envelope limitations with this store loaded.

Fig. 15.33 AGM-154 JSOW (US Navy Photo – PM Bo Flannigan)[49]

15.3.3.3 **Store Structural and Fastener Failures**. It is also common for the captive carry environment on the airplane to affect a store or its components adversely. The AIM-9 Sidwinder repeatedly lost the cover of the seeker head due to the more intense vibrational environment when carried on the F-15. The fasteners were replaced and fixed in place with an adhesive to overcome the problem. F-16 CFP showed that the GBU-10 tail fins failed (Fig. 15.34) when the bomb was attached to a new pylon with an integrated countermeasures dispenser. Redesign cured the problem. Turbulent airflow caused the Joint Direct Attack Munition (JDAM) fin to move during captive carry flights. This anomaly (Fig. 15.35) could have caused the munition to strike the airplane or an adjacent store during release even though the shift in trim did not adversely affect the store during CFP. The fin locking mechanism was redesigned.

Fig. 15.34 GBU-10 Tail Section Failure (USAF Photo)[47]

Fig. 15.35 JDAM Fin Trailing Edge Movement During Captive Carriage (USAF Photos)[47]

15.3.4 Summary of Lessons Learned from Captive Carry Examples

These examples illustrate the value of a disciplined and meticulous captive carriage effort preceding any form of separation testing. They underscore the point that flaws found in captive carriage profiles can preclude problems during separation flight tests.

15.4 SEPARATION COMPATIBILITY

Store separation testing is the "capstone" to stores testing and is usually the most hazardous component of stores certification. The danger revolves around three primary questions: (1) Does the store positively move away from the aircraft and adjacent stores upon release? (2) Is there any tendency to parallel the flight path of the carrier aircraft ("formation flying" is not desirable in a stores separation!). (3) Is there any motion imparted to the store (like store-to-store collisions) at release that might undermine weapon effectiveness? The end result as suggested in Fig. 15.30 is to provide the ultimate user of the aircraft/store combination reliable limits and clear instructions on appropriate release techniques.

15.4.1 Planning and Preparation

The key to success with all hazardous tests is careful planning and preparation. Planning starts with understanding how the store is to be used. For military purposes that involves understanding the intended operational environment. For commercial stores, the envelope of intended use is smaller and more benign, but the same principles apply. Separation testing considers how a store is to be employed and demonstrates safe and effective releases throughout the envelope. Inevitably, separation demonstrations do uncover unsafe conditions but finding the envelope boundaries is done with risks mitigated as completely as possible. Surprises are still likely to occur; but careful planning and preparation minimizes unexpected anomalies. Jettison tests are designed to ensure that stores and/or their suspension equipment can be released with no damage to the aircraft, though these releases do not need undisturbed trajectories for reliable weapon deliveries. A jettison separation matrix anticipates that combat engagements often occur at high speeds and at high g loadings. Larger jettison envelopes give better chances for survival in combat.

Christensen[47] describes three approaches to planning separation testing based on the available analysis, the amount of time available, and the supporting analysis and ground testing done: (1) End Point/Worst Case demonstrations, (2) Build-up tests, and (3) Verification testing. End Point/Worst Case tests are typically based on analogy principles. If a

Chapter 15 Stores Certification Testing 189

similar store has been certified on a similar airplane (the store might be new, but have only minor differences from an earlier model – somewhat like growth versions of the AIM-9 Sidewinder missile that were tested on the F-15 before the advanced missile went into production). This approach has often led to surprises; not all incremental changes produce incremental changes in store release behavior. The build-up approach is commonly used for new airplane/store configurations when there is no time nor money available for full-blown modeling and analysis. It typically trades additional flight test events (higher cost) to minimize risk for analysis and wind tunnel time. Nowadays, the verification approach is favored by most program managers. It depends heavily on modeling and simulation, backed up with computational analyses and wind tunnel data, to reduce uncertainty and lower risk. The goal is to do just enough flight testing to verify that an analytical model accurately predicts separation behavior. Developing a model like this, with the computational tools now available and the wind tunnel data that can be obtained, has become rather common place. The task is demanding for the analyst, but surprises for the flight test team are fewer and less costly.

15.4.1.1 Analysis

15.4.1.1.1 *Flow Field Mapping and Carriage Load Predictions*. Computational fluid dynamics (CFD) has made great strides and virtually every agency engaged in stores certification makes use of such tools. But, the ability to completely capture separated and interfering flows at all Mach numbers in such CFD models is still not perfected. So, few (if any) programs involving new aircraft and/or new stores are based solely on CFD algorithms without both ground and flight test data to corroborate predictions. A detailed discussion of purely theoretical prediction methods is outside the scope of this chapter, but the interested reader can refer to Arnold and Epstein[50], where prediction techniques used throughout the NATO countries are summarized, for a broader theoretical treatment. Roberts[51] concluded in a later paper, citing several examples, that "reliance on a single tool, method, or approach should be avoided".

15.4.1.1.2 *Theoretical Trajectory Predictions*. The most common methods of predicting store separation trajectories with computational methods couple time-dependent solutions of aerodynamic equations and dynamic equations. However, flow physics are extremely complicated and simplified (linearized) equations do not adequately represent aerodynamic forces and moment, making the dynamical equations minimally effective for predicting flow fields with interacting bodies. Panel methods assume inviscid flow and do not allow credible modeling in the transonic regime where the nonlinear potential flow methods change from elliptic to hyperbolic differential equations. However, panel methods suffered from rather primitive gridding techniques in the early 1980's[50,52,53]. More recent progress in use of CFD modeling is illustrated in results from stores separation modeling for the F/A-22. It is worth noting that, Baker, et al.,[54] are careful to point out that there are three primary reasons that a high fidelity model from CFD analyses do not yet accurately predict stores trajectories: (1) factors not known and therefore not included in the model, (2) factors only partially understood and therefore approximated in the model, and (3) factors that changed after the last (wind tunnel, in this case) model verification.

Another example of modern analytical stores separation prediction uses a model that integrates the Newton-Euler equations for 6-DOF rigid body motion[55]. Separation of the

JDAM from the F/A-18C was modeled using two different CFD techniques and both show relatively good correlation during initial stages of store release. Figures 11 and 12 in AIAA Paper 2003-1246[55] show the time-dependent positions calculated using these techniques and Figures 7-10 and 13-16 from the same paper illustrate the comparisons of angular rates and miss distances. These results are for Mach numbers in the transonic range, typically the most difficult flight regime to model theoretically for separation characteristics.

15.4.1.2 **Ground Testing**. As Baker[54] suggests, any model must be verified and a useful stores separation model uses such data from whatever source is available. Stores separation literature lists four possible data sources other than flight test to consider in validating prediction models. Test organizations[50] use one or more of these approaches with, of course, variations in technique between each organization.

15.4.1.2.1 *Articulated Dual Sting Wind Tunnel Tests*. In the United States much of the stores separation wind tunnel testing done is carried out in the Arnold Engineering Development Center (AEDC) wind tunnels. In Europe similar facilities are operated by either industrial or government agencies with similar capability to that of the AEDC tunnels. One of the more popular techniques used by the USAF Seek Eagle office utilizes these two stings with a store model mounted on one sting (which is free to move in six degrees of freedom) and the aircraft model mounted on a second sting with limited movement. An internal balance in the store model on the first sting measures the aerodynamic forces and moments on the store under a specified set of tunnel conditions and relative orientation and position between the aircraft model and the store model. In the Captive Trajectory System (CTS) mode of operation the measured store forces and moments are used along with store mass properties, ejection forces, and rate damping forces to move the store sting to a computed position for a specified increment of time. These incremental movements emulate the separation trajectory and allow quasi-steady measurement of the dynamic trajectory. Proponents say this "...technique provides the most accurate experimentally determined aerodynamic data for a position in the trajectory..."[50]. One of the primary limitations of the CTS approach is the cost of obtaining a large enough data base to allow sensitivity analyses of separation trajectories for the entire matrix of variables (aircraft parameters like Mach number, angle of attack, dive angle, and load factor along with store variations likely to occur such as store mass properties, ejector performance, and fin deployment times). British facilities do use a technique similar to CTS in their wind tunnels operated by the Aircraft Research Association (ARA) Limited. The Netherlands relies on the National Research Laboratory (NLR) for store separation predictions and they apparently favor methods described in subsequent sections to verify predictions, along with measurements from a full scale flight test captive store load measuring scheme, which NLR believes is more accurate than theoretical or wind tunnel measurements. The French industry seems to prefer theoretical prediction techniques, but their industrial wind tunnels have the capability to produce data similar to the CTS information from the AEDC tunnels. In a preliminary report on Swedish separation trials for the export version of the Gripen describe only simulations and camera installation on the aircraft, with no mention of wind tunnel data collection for their multinational stores separation effort[56].

15.4.1.2.2 *Dynamic Drop Wind Tunnel Tests*. As suggested by the name, (also sometimes called "freedrop") this wind tunnel test technique records store position and orientation through the available instrumentation. High speed video cameras or miniaturized telemetry packages on the store can be used. Arnold and Epstein[50] point out that compressibility and viscous effects should be matched to produce model aerodynamic forces and moments that match full scale forces and moments. Froude scaling generally gives like aerodynamic results between model and full scale only for Mach numbers less than about 0.8. But assuring that the aerodynamics match between model and full scale for transonic and supersonic flows (the ones of critical interest in many store separation efforts for high performance airplanes) must maintain Mach number equality. This matching in the wind tunnel must be done at the expense of some other parameter. "Heavy" and "light" model scaling are two different ways to achieve Mach number equality. In "heavy" model scaling both the mass of the model and the velocity of the flow are increased over that for Froude scaling to achieve Mach number equality. This change affects induced angle of attack and aerodynamic damping effects for rotational motions. The amplitude of angular motions for the model are likely to be too large and oscillations are likely to be under damped. Light model scaling addresses the angular motion disparity by keeping the mass ratio of Froude scaling, along the velocity ratio, but assuming that the gravitational constant can be increased. This increase has been artificially accomplished by using magnetic fields or by accelerating the aircraft sting apparatus away from the store. Arnold and Epstein[50] indicate that that "heavy" model scaling is preferred in most of the stores separation literature. Since the wind tunnel must be stabilized for each set of data, information for only one flight condition can be collected at each stabilized point. So, dynamic drop tests to cover the entire test matrix can be very expensive for an aircraft that carries and employs a variety of stores throughout a subsonic and supersonic flight envelope. This expense and the difficulty of choosing and fabricating models for the "best" scaling for a specific aircraft store combination are two significant drawbacks for dynamic drop (or freedrop) wind tunnel tests.

15.4.1.2.3 *Grid Data Wind Tunnel Tests*. Typically, articulated dual sting facilities can be operated in either CTS mode or grid mode. Grid mode is used to collect interference aerodynamic force and moment data by subtracting the freestream store aerodynamic conditions from those measured with the store at a preselected relative position and orientation. Thus, this technique is a flow field technique; by gathering data throughout the separation flow field, store trajectory predictions can be calculated off-line. The analyst can then examine the sensitivity of these trajectories to specific variables. Sometimes a few CTS runs are used to select the grid patterns and a flow field volume large enough to cover the most extreme trajectories is then completed in grid mode. In comparison to CTS mode tests, a larger amount of data can be collected and the cost for wind tunnel support is considerably reduced, while sensitivity analyses are easier and less expensive to complete. Arnold and Epstein[50] describe this technique in more detail and offer a number of references for the interested reader.

15.4.1.2.4 *Flow Angularity Wind Tunnel Tests*. Another flow field mapping technique is the flow angularity approach where a velocity probe is usually mounted on the dual articulated sting on the sting that normally carries the store. This probe is moved to various locations throughout the aircraft flow field volume large enough to include all anticipated store trajectories. The interference coefficients are generated from this flow field data base rather than measuring them directly by considering only the local flow conditions at the nose and tail of the store. Instead of inserting a probe into the flow field, laser velocimeter measurements have also been used to map the flow field, thus eliminating the effect of the velocity probes itself. Flow angularity fields can also be extracted from grid mode data. This methodology is only good enough to give a first order approximation of the store aerodynamic characteristics for trajectory prediction46 so an extension of the flow angularity technique is often attractive.

15.4.1.2.5. *Influence Function Wind Tunnel Tests*. The influence function has been extended to include more store elements of the store than just the nose and the tail. Calculating the aerodynamic characteristics using a larger number of store elements requires that a store shape be "calibrated" in the wind tunnel by traversing the store through a known flow field and then measuring aerodynamic forces and moments at each point in the traverse. Once the store is thus "calibrated" it can be passed through an unknown flow field and the influence coefficients from the calibration can be used to solve for the aerodynamic coefficients by matrix manipulation. Getting and maintaining an accurate calibration for a store is both expensive and if the shape is ever damaged, the calibration must be repeated. Influence coefficients have also been calculated from CFD codes (notably results for the GBU-15 on the F-15) to produce acceptable trajectory predictions, though Arnold and Epstein warn that shock strength prediction in early CFD models could undermine this approach[50].

15.4.1.2.6. *Static Ejection Tests*. Wind tunnel tests are by no means the only kinds of ground tests to which stores should be subjected before stores separation flight tests begin. A total of approximately twenty such tests are outlined in Appendix A of MIL-HDBK-1763[37] and the interested reader (storer designers and test planners, specifically) should start with this basic set of qualification tests. Since most of these tests are utilized as subsystem proof tests, we will select only two or three as examples because of their close relationship with flight test safety and with setting up a flight test matrix that is sensible. Static ejection tests from this group of ground tests are designed to verify reactive loads, lanyard functions, and arming control system reliability for stores released from Ejector Release Units (ERUs). Forced ejection to guarantee safe separation of stores has become much more common as the speed and Mach number envelopes of aircraft-carried stores has increased. The characteristics of interest include accelerations, angular velocities, and attitude of the store as it reaches the end of the ejector stroke. The loads reacted to the aircraft structure by the ERU are of keen interest to the structural dynamicist and the store dynamics at the end of the ejector stroke dictate the initial conditions for the engineer responsible for predicting store trajectories at all flight conditions. This type of testing, often called "pit" testing, can expose problems with lanyard routing, umbilical connectors, ejector components, or store control system deployment and/or actuation. Often a lower cost Separation Test Vehicle (STV) with some of the expensive subsystems not installed is used in static ejection tests (and sometimes in early separation flight tests). The test article is often a one-time use device, even though cushioning material is often used when these stores are ejected, sometimes from the aircraft itself during these ground tests. Care must be taken to verify that the STV is representative in mass properties (weight, cg location, and moments of inertia) to the actual store even for purely static ejection tests on the ground.

15.4.1.2.7. *Aeroelastic Ground Vibration Tests (GVTs) and Aeroelastic Effects Tests.* Structural dynamics and aeroelastic phenomena are covered more generally in Chapter 12. However, these effects often drive stores separation flight tests and emphasis is necessary to underscore the importance of consulting with aeroelastic specialists in planning separation flights. These tests often demand specialized instrumentation that should be installed on the test aircraft as well as the store and the typically used sensors often need to be installed while the test aircraft is on the assembly line. So, aeroelastic preparations may be long-lead items for the test team. GVTs usually have a specific test plan for the GVT itself, but the stores separation engineer needs to be aware of what stores configurations are selected for the GVT series. The ancillary specifications are spelled out in the basic handbooks[40,41], but careful coordination is essential to get optimally useful stores separation information.

Aeroelastic Effects Tests are similarly specialized (and very expensive) wind tunnel tests, usually designed and conducted by structural dynamicists. Models for such wind tunnel tests are quite expensive and complicated, especially since the interaction between the unsteady aerodynamics and the vibrating structure depend on more similarity parameters than do rigid body wind tunnel results. Getting the scaling laws right for the stores models is crucial to understanding how the component parts of the aircraft/store configuration behave aeroelastically. Again, stores separation flight test engineers must carefully coordinate with structural dynamicists to guarantee that the most useful configurations are chosen; often the "worst case" scenarios for the structural dynamicist come from critical stores configurations. Close ties with these specialists is strongly suggested so that the results of these tests can be properly folded into the stores separation flight test plan. Finally, these tests, along with CFD that includes critical dynamic structural deformations, may indicate where specialized sensors need to be installed for stores separation tests.

15.4.1.2.8. *Gun/Rocket/Missile Firing Tests.* Integrating a gun onto an airplane and having it operationally usable has been a problem since Fokker, Boelcke, Garros, Saulnier, and others (Section 15.1.2) first attempted it. A total of eleven different effects of firing a gun are to evaluated in ground tests before flight tests according to DoD guidance[40]. Taken together these factors are meant to ensure safety, operational performance of the gun, and compatibility with the airplane and its environment. A good example of one of these factors (that led to loss of an aircraft and serious injury to the test pilot) is the gun gas ingestion that resulted in a dual engine flameout at low altitude and high airspeed during A-10 developmental testing.

Similar reasons dictate that ground firing of rockets and missiles be conducted to "establish baseline data prior to flight tests[40]." Once again, careful and complete coordination between the rocket or missile subsystem engineers conducting these initial ground tests and the flight test engineer planning the flight test demonstrations for the aircraft/stores combination is imperative.

15.4.1.3 **Instrumentation and Data Collection.** Stores separation data requirements, for both ground tests and flight tests, set high goals for instrumentation and data collection experts. We have already considered some of those special requirements (Section 15.3.1) but separation testing adds to these sensor constraints.

15.4.1.3.1 *Special Instrumentation*. The store and its suspension equipment get special attention during release dynamics. Measurements often needed on the bomb rack include load cells to measure reaction loads, along with hook opening movement and "weapon away'" sensors. For missile or rocket launches, ignition times and burn times, along with metric for thrust levels, may be necessary data. Recoil effects when firing guns engine effects, control system reactions, and aircrew comfort levels may all be important the separation engineer. Virtually every separation test event must take place with aircraft release conditions known and time-synchronized with the store release data. Head up display video is one means of acquiring this kind of information. Last, and perhaps most important, store motion at and immediately after release must be recorded. Telemetry packages that transmit store position and attitude relative the release aircraft are commonly used in flight test control rooms today. Another, often the preferred, way to capture this trajectory information is with an onboard camera. Video cameras are the preferred type, partly because of smaller size/volume and partly because the medium lends itself to data reduction better than the older high speed motion picture cameras. A minimum frame rate of 200 frames/second is suggested.

15.4.1.3.2 *Test Data*. Targets are quite commonly painted on stores during separation releases to help establish store position and orientation in video or film data traces and more than one camera view is better able to capture the complete store trajectory. Condensation around a store can obliterate camera view, which is why rules for test conduct usually stipulate clear, dry air as a high priority during critical separation tests. Relative humidities of about 70-80% at test altitude have been troublesome in the past so quality weather data at altitude in the release zone is important to success, especially for transonic releases. Condensation is particularly hard to assess when onboard video or film is used unless the information is telemetered to the ground.

15.4.1.4 Test Conduct Considerations.

15.4.1.4.1 *Test Articles*. If an STV is to be used in a separation flight test, it must be aerodynamically representative as well as having the same mass properties as the production article. If deployable fins are used on the operational store, they should be included in this STV, though guidance fins are often fixed in such tests. If the store to be released has onboard guidance, store test articles must be representative of the latest and best verified of the guidance algorithms.

15.4.1.4.2 *Photo Chase*. Photo chase coverages of separation flights "...provide the best 'big picture' of overall separation characteristics..."[44] even though onboard video or film are often the primary source of data. Photo chase is strongly recommended for separation tests, even though these chase flights are very demanding and occasionally quite dangerous. A chase aircraft can advise the release airplane on condensation issues if necessary. Positioning of the photo chase aircraft depends on release conditions; slightly aft and below the release aircraft on the side away from the store if the photographer can get a clear view of the entire release trajectory is usually best for wings level releases, even in a dive. Even though the photographer can usually zoom in to get detail (but zooming makes steady camera position more difficult, especially if the release is in maneuvering conditions), the chase aircraft usually positions close enough to identify store details, yet far enough away to keep most of the aircraft in the video or film frame. The chase pilot must be ready to break away if a store does not separate cleanly and damages the release airplane. Ordinarily, the photo chase also serves as safety chase, which is to be discussed further in Section 15.4.2.

15.4.1.4.3 *Control Room*. The ground team controlling separation tests, as is true with all hazardous tests, should be trained, proficient, and completely disciplined in their actions. Not all agencies utilize telemetry for separation flight tests, but for both safety and technical risk reasons, the practice is more common today than it was several years ago. Many stores released today have their own guidance package (as mentioned previously); monitoring the health of deployable fins, status of the store propulsion system, and the actual control inputs during the entire trajectory are all technical issues needed in near real-time by many programs today. For those occasions where multiple releases are attempted during a separation test flight, reasons for proceeding from one point to the next event can be evaluated in near real time on the ground with expertise drawn from specialists in making that go-no-go decision. A carefully trained, tightly disciplined control room team can often prevent wasting precious test assets by monitoring store health from the ground.

15.4.2 Risk Mitigation

Risk management is build upon the foundation that careful planning and anticipation of potential failures can help prevent unacceptable consequences. According to the current Risk Management Guide for the U.S. DoD[57]:

"Risks have three components:
- *A future root cause (yet to happen), which, if eliminated or corrected, would prevent a potential consequence from occurring,*
- *A probability (or likelihood) assessed at the present time of that future root cause occurring, and*
- *The consequence (or effect) of that future occurrence.*

A future root cause is the most basic reason for the presence of a risk. Accordingly, risks should be tied to future root causes and their effects." The basic principles of risk management require thoughtful assessment and consideration of ways to reduce both the likelihood of "a future root cause" occurring and the consequences when such an event does occur. Risk mitigation is the usual phrase for describing the whole of this process, though in risk management terms it is more tightly defined as lessening the adverse impacts of risk. For stores separation flight test activities, an effective risk management plan is concerned with all aspects of the project – budget, schedule, technical, and safety. For our purposes, we largely stick to technical and safety aspects of the flight test activities. Of course, failing to anticipate and mitigate a safety risk or a technical risk and planning adequately to reduce the effects such "root cause" events might have on a program, can and often does lead to budget and schedule consequences as well.

15.4.2.1 *Separation Collisions*. Three types of collisions make up the most common safety "root causes" in separation flight tests. Aircraft performance (larger Mach number envelopes, higher thrust-to-weight ratios, and better maneuvering capabilities) has driven the stores certification community to examine each of these unwanted interactions to for potential risks to the separation program. Gravity-drop or ballistic releases are much less common; ejector mechanisms are used on almost all high performance airplanes today to release or jettison stores and the racks that carry stores externally. As propulsive power went up, more stores could be carried and it became routine to carry multiple stores on the same pylon or internal carriage rack. Releasing multiple stores safely led to the need for evaluating safe intervals for multiple releases. Collision potential hinges on at least three factors: high airspeed, load factor, and release interval. However, as Arnold and Epstein emphasize, store separation problems including collisions of all types, can occur throughout the operating envelope[50].

15.4.2.1.1 *Store-to-Aircraft Collisions*. In this context store-to-aircraft collisions imply the released store impacts some part of the airplane itself (as opposed to its carriage equipment) like the wing, fuselage, or empennage. When a store is separated, especially with pneumatic or mechanical ejection forces, the objective is for the store to exit the flow field of the airplane as quickly as possible. Several cause factors might be responsible for failure to move away from the aircraft; the ejector rack forces might be two low, the rack or rail and the supporting structure itself may be flexible enough to cause binding. Low load factors (less than 1g) may reduce the ejection force since g loads usually add to the ejection force. Remaining in the aircraft flow field can cause drastic changes in the behavior of a store's trajectory. The BLU-1 in one separation test was ejected from an F-105 and initially started down, pitched up when it did not immediately clear the flow field, and was then swept upward smashing into the horizontal tail. Empty fuel tanks (large, light stores) frequently exhibit this kind of behavior; such collisions of empty or near empty fuel tanks have collided with an A-7 and an A-37. These examples[47] illustrate the store-to-aircraft collision and that it can occur to low speed and high speed airplanes in very different release conditions.

15.4.2.1.2 *Store-to-Pylon/Rack Collisions*. Store-to-pylon/rack collisions are the type most likely to occur; they could be lumped with store-to-aircraft collisions, but the frequency of this type of collision sets it apart. These kinds of collisions occur before the store clears the adjacent supporting structure or the launching rail, often within 200 milliseconds after release. But Store-to-pylon/rack collisions are much less likely to be catastrophic than when a store impacts some other part of the airplane. Such collisions are likely to destroy the ballistic repeatability that is desired for ordnance; this kind of collision often does nothing more to the aircraft itself.

15.4.2.1.3 *Store-to-Store Collisions*. Store-to-store collisions produce effects that may be as catastrophic as store-to-aircraft collisions are and the affected stores are likely to behave erratically during the remainder of their trajectory. Explosions do occasionally occur from such collisions, but explosions are much less likely to occur than simply loss of ballistic accuracy. The B-52 can release up to 84 MK-82 bombs and it is common for side-to-side store-to-store collisions to occur. These collisions typically do not cause explosions nor to they significantly alter the ground impact pattern of such a multiple bomb release. Most store-to-store collisions do not pose a hazard to the carrier but one type does. If one store "drafts" on another one (that is, one store is directly behind another), an explosion becomes more likely because the aft store speeds up and the bomb fuzing may arm the weapon. This sort "drafting" has been observed on ripple releases from a Multiple Ejector Rack (MER-10) at 450 KCAS.

15.4.2.2 Build-Up Approach.

A cardinal principle of safety in hazardous flight testing is the concept called the build-up approach. Given that trajectory prediction for store separation testing is never completely accurate (ripple releases at short intervals, jettison of pylons and racks with or without stores on them in an emergency, or partially full fuel tanks with little or no baffling), it is always necessary to resort to "brute force" testing for some purposes. In this context the term "brute force" indicates flight testing done with incomplete analysis and ground testing done to give reliable predictions for the proposed releases. Tests based on analogy with similar stores on the same aircraft and in similar flight regimes are "brute force" under this use of the term. We neither suggest nor advocate flight tests with no previous analysis or estimation of likely results by the use of this term. That testing is not the "brute force" testing under discussion; it is simply flight test folly. So, given an incomplete set of predictive analysis and a firm need date or a cost constraint that is untouchable, a build-up approach involves starting as near to a known benign release condition as possible. Analogy with a similar store on the same airplane might supply this test condition, for example. This condition is likely to be release of a single store from stabilized, straight and level flight. The next step is to consider the envelope extremes that are the end goals for the certification. A careful examination of what variables are likely to be of importance (perhaps using analogy with the similar store again) in moving incrementally out to each of the end points. Dynamic pressure, Mach number, angle of attack, normal load factor, aircraft cg position, and the like are aircraft conditions that are likely paths to the envelope extremes of interest. For the store, mass distribution, cg migration, fin deployment (if any), multiple releases, ripple release interval, and such like are parameters that be explored one event at a time in this "brute force" approach. As this cursory list of parameters (and it is by no means comprehensive for your project) shows, this tack can produce a very large, expensive test matrix. The risk mitigation offered by such an incremental build-up is substantial, but do not expect a regular increment in any one or two parameters to be logical. When store separations lead to catastrophes they usually occur at "cliffs" in the parameter space. So if you suspect a test event approaches a boundary, take smaller incremental steps.

15.4.2.3 Probability of Impact.

Christensenr[47] suggests the use of a matrix (Fig. 15.36, next page) used by the Air Force Seek Eagle Office to subjectively, if not quantitatively, evaluate risk levels for each build-up increment in the test sequence. This risk assessment tool is parallel in concept with Air Force guidance on risk assessment[57].

15.4.2.4 Use of Inert Stores.

Using inert stores can lower risk by eliminating danger of early detonation in case of arming failures and/or store-to-store collisions. However, some liquid-filled stores pose a special problem. Inert (simulant-filled) firebombs proved to have very different trajectories than those filled with napalm during A-7 testing. Consequently, the USAF now uses only napalm-filled firebomb canisters for separation testing, though inert fuses are used. The general rule of thumb is to use inert stores for separation testing whenever it is practical.

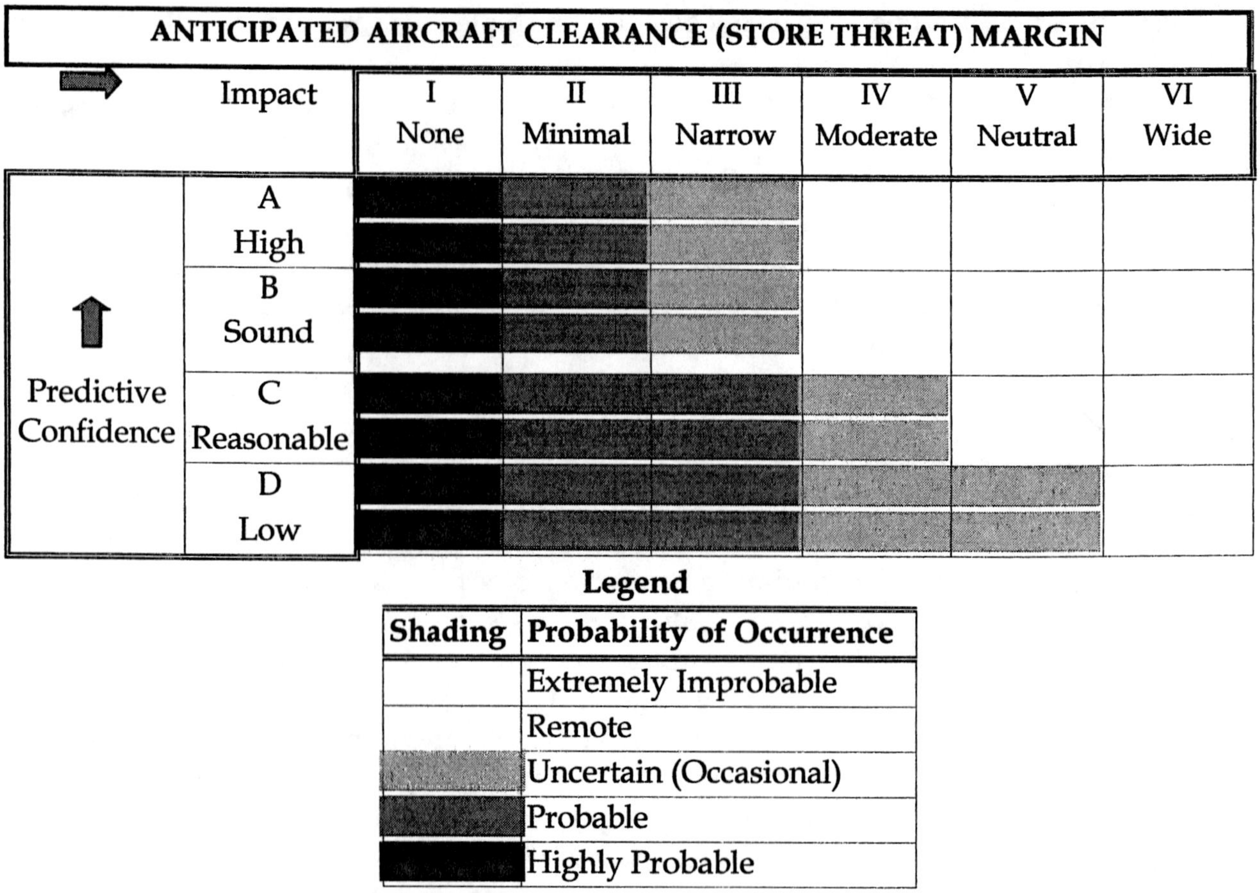

ANTICIPATED AIRCRAFT CLEARANCE (STORE THREAT) MARGIN							
	Impact	I None	II Minimal	III Narrow	IV Moderate	V Neutral	VI Wide
Predictive Confidence	A High						
	B Sound						
	C Reasonable						
	D Low						

Legend

Shading	Probability of Occurrence
	Extremely Improbable
	Remote
	Uncertain (Occasional)
	Probable
	Highly Probable

Fig. 15.36 Probability of Occurrence Matrix (Adapted from Christensen[47,54])

15.4.2.5 **Store Subsystem Failure Modes.** Stores have become very sophisticated vehicles with their own subsystems that must operate reliably and repeatably. Stores with folding fins that do not open quickly and simultaneously have posed problems in the past. The MK-20 Rockeye had fins that are designed to open within 50 milliseconds after release. USAF data for the A-7D/Rockeye certification effort showed that because of variation in the springs that opened the four Rockeye fins, the fins almost never opened simultaneously, which led to store-to-aircraft collision with the aft fuselage of the A-7. One solution is to guarantee that any folding fins open simultaneously with an interconnect mechanism. The GBU-10 series has interconnected fins, though some of these stabilizing devices open so slowly that certification programs on several different aircraft planned separation test events on the assumption that the fins would not be open at all to be conservative in separation trajectory prediction. More modern guided stores with full-up internal guidance packages like the AGM-154 JSOW have some capability to recover from separation dynamics that are less than optimum, risk mitigation efforts must concentrate on a careful and complete analysis and prediction effort to provide reasonable risk reduction. These principles were reinforced in the F-16/BRU-57 Smart Rack/AGM-154 certification effort[51] Initial separation tests for guided weapons are often carried out using Separation Test Vehicles with the control surfaces locked in a fixed position, but such a practice does not always guarantee lower risk.

15.4.2.6 **Preflight and Postflight Inspection.** Complete and meticulous inspections by both the aircrew and the ground crew are a vital part of test team discipline for stores certification tests. Even on captive carry flights and certainly on separation test sorties, these inspections ascertain that arming wires, umbilicals, and lanyards are properly routed and connected and are part of the attention to detail expected in a professional test organization. Preflight inspections are usually emphasized, but postflight inspections often provide invaluable information that cannot be observed either from the cockpit or on the telemetry. Store fins move in flight, lanyards move to positions where they may bind, and vibrations cause loose screws in store components and suspension equipment. Both aircrew and ground crew are responsible for executing these inspections with care and completeness – just one aspect of test discipline that mitigates risk.

15.4.2.7 **Forward-Firing Weapons.** Missiles, powered stores, rockets, and guns with trajectories that carry the store in front of the airplane deserve special attention. Anticipation of potential emergencies includes more than just reviewing procedures to follow in case of a store-to-aircraft collision (like controllability checks and safety chase inspections); forward-firing weapons are more likely to cause engine flameouts (due to gun gas ingestion or a guided weapon that does not guide properly, striking the engine inlet). One of your authors "enjoyed the privilege" of serving as safety chase on a GAM-83A Bullpup firing in which the missile left the rails on an F-100 and immediately pitched up and looped back over the flight. The bad news is neither pilot had thought of such an eventuality; the good news is the Bullpup did not successfully complete its "attack loop" without hitting either aircraft after the safety chase called for a break maneuver to evade the "attacker".

15.4.2.8 **Safety Chase.** There probably is no set of hazardous flight tests where a safety chase is more important than stores separation missions. No one else can quickly and fully confirm that the release aircraft has completed a safe release or jettison event. Usually the chase pilot can also offer uniquely useful and welcome advice regarding hung stores. He can also give almost immediate help after a store-to-aircraft collision damages the release aircraft. Of course, experience, discipline, and good judgment are key attributes for the chase pilot and considerable formation experience is needed. The short description of a safety chase pilot is "indispensable to risk mitigation".

15.4.2.9 **Hung Stores.** Procedures and safe orbit areas in case a store does not release or worse, partially releases, perhaps dangling by one attach point, should be part of the test planning for stores separation tests. Safety chase can offer advice based on his observations, but assuming that a hung store is unsafe is always prudent. Preplanned routes that avoid populated areas and isolation areas where ground crews can safely take care of the unexpended, perhaps even armed, ordnance are essential to any planning that addresses this risk. Assuming worst case scenarios for all hung stores and having considered appropriate actions to take are the only prudent ways to mitigate the risk they pose.

15.4.3 **Execution with Discipline (Professionalism)**

In keeping with our persistent references to test discipline, it is appropriate to summarize some of the operating rules that have proven essential to safe conduct of stores certification missions; this advice comes from past experience (and mistakes) that in many cases should not be repeated.

15.4.3.1 **Clearing the Range**. The procedures for guaranteeing that a given drop area is clear differ for each set of stores separation sorties because the release parameters, the store characteristics, the failure modes, and the guidance scheme (or lack of one) are all part of what kind of footprint must be cleared. All these considerations are folded into the planning necessary for safe conduct of separation tests and should be part of the test plan. Often it is prudent to have chase aircraft conduct visual sweeps of the "firing box" just before starting planned test events. Water ranges often require a "boat sweep" to be sure boats have not strayed into the drop area and its buffer zone.

15.4.3.2 **Radio Communications**. The radio communication that takes place on a well-planned and professionally executed stores separation sortie reflects the discipline of the team. Christensen suggests some calls[47] that make up a good script and recommends that both aircrew and control room personnel (if applicable) rehearse use of the script. Examples of his suggestions include "cleared to arm" , "cleared to release" (both from range control or the test conductor), "on conditions" (test aircraft), "chase ready" (chase aircraft), "30 seconds" (test aircraft or controller), "10 seconds" (test aircraft or controller), "ready..., ready, ... pickle/fox" (test aircraft), "weapon away" (chase aircraft), and then "clean and dry" (chase aircraft after inspecting the test airplane. During this entire sequence the transmissions must be short and any participant must clear the frequency as soon as a call is made to leave room for any calls of "skip it" or "abort" issued usually by the chase pilot, the range controller, or the test conductor during the run-in to the release.

15.4.3.3 **Dynamic Separation Conditions**. Stores carried by modern aircraft are often designed for use in maneuvering flight, where the dynamic loads produce release conditions that may seriously affect the characteristics of separation. Usually, as part of the build-up, such releases are preceded by straight and level test events. Countermeasures pods and similar devices often are delivered as the aircraft maneuvers to defeat an attacker, either from the ground or from the air. Separation test matrices then must be built up around variables like dive angle, roll rate, or load factor. If a high fidelity simulator is available, especially where multiple aircraft can be emulated, practice sessions in the simulator may be helpful as well. Even after simulation both the test aircraft pilot and the chase aircraft pilot often need to rehearse the profile with one or more dry runs to work out positioning and to clear up any rough spots in the techniques required prior to making the actual release or jettison.

15.4.3.4 **Aircraft and/or Performance Limitations**. Typically, the desired end points for stores certification are at or near aircraft operating limits. It is the entire test team's focus to approach these limits having considered all possible consequences and having prepared for them as completely as possible. If dive angles are necessary to reach some of the flight conditions with stores loaded, as is often the case, the test pilot must have clear pull-back procedures if a limit (speed, load factor, or any other limit) is breached for any reason. Always expect a release failure for such conditions and practice recovery techniques ahead of the actual test, again preferably on a high fidelity simulator. The test pilot, chase pilot, and test conductor should work out the best possible course of action to withdraw from a failed separation event in the safest and quickest manner. Such procedures are included in the preflight briefing so every team member understands what to expect, both in the air and in the control room. In setting test limits for separation testing take into account such failure states and leave some margin for error if at all possible. For dynamic separation flight test events, carefully evaluate any combined structural loads that are produced during a rolling pullout; allowable load factors for such maneuvers are considerably lower, typically 60-65% less, than for a wings-level pullup.

15.4.3.5 Aircraft Response to Separation. Anticipate abrupt changes in load factor and often significant dynamic reaction for the test aircraft when a store, especially a heavy store, separates. Often these test aircraft responses will be opposite to those when a release proceeds normally; hung stores can behave erratically and such anomalies need to addressed before actual separation events are conducted.

15.4.3.6 Mission Impact. One of the primary responsibilities of the test pilot to the ultimate user, the operational pilot, is to document with clear, concise language in the Pilot's Handbook how the airplane behaves when a given store, rack, or pylon is released or jettisoned. What transients occur and how they are best handled is part of the descriptive language challenge for test pilots. Objectionable behavior at a specific desired end point in the certification plan may be grounds for restricting the operational envelope.

15.4.4 Stores Separation Lessons to be Learned

We now turn attention to some of the stores certification flight tests that offer key reminders of points just stressed under each of the previous two sections. These examples are only a small sample of events that teach how to conduct certification of stores more safely and more efficiently. There is no intent to offer negative criticism; remember, the only test teams with a mistake-free record are those who have accomplished nothing. This kind of work is almost always at the edge of the aircraft envelope for end points that are useful to the ultimate user. It is not a fundamental flaw to err, but it is fundamentally wrong to not learn from past experience, preferably before mistakes are repeated.

15.4.4.1 A-10 Secondary Gun Gas Ingestion. The primary lesson from this incident is that prior planning and a disciplined safety review process are imperative to sound risk management during stores certification tests. An A-10 test aircraft was lost and the test pilot severely injured during a gun-firing mission during developmental testing at Edwards AFB in 1978. Firing conditions were a 15° dive, 350 KIAS, and 5,000 feet. On the fourth pass, unoxidized gun gases formed a fireball in front of the airplane's 30 mm cannon and the secondary gun gases from this fireball were ingested into both turbofan engines, essentially starving the combustion process of oxygen. Both engines compressor stalled exceeding the temperature limits in both engines. The emergency procedure called for both engines to be cleared by shutting them down and restarting them. But at that altitude there was no time to cool down the engines and attempt a restart. The pilot was unable to restart either engine before reaching bailout altitude. Ejection was the only alternative left and the pilot experienced very high rotational acceleration at chute opening, breaking or cracking a bone in his neck. Though the possibility of engine failure due to gun gas ingestion was known by the test team, planning and practicing to mitigate against that risk was incomplete. The hazard was not thought to be significant enough to warrant such special planning and preparation. Later gun testing was flown at a higher altitude to give move time in case engine restarts were necessary.

15.4.4.2 **F-15E/GBU-12D/B** **Impact**. Roberts[51] describes a store-to-aircraft collision during certification of the GBU-12D/B laser-guided bomb on the F-15E that emphasizes the importance of simulation augmented by thorough wind tunnel testing and updated by comparison to results from flight tests. The incident release conditions were at $M = 0.97$ to clear to the maximum employment dynamic pressure from the left CFT, aft outboard station using a 78-inch lanyard. The fact that the conformal tank was installed means the separations were significantly different from previous releases and little was known about how store trajectories were affected by the CFT. Moreover, the racks had been modified (a MAU-12 for the inboard CFT stations and a MAU-190 for the outboard CFT stations) on the F-15C (emulating the F-15E). These modified racks had different stroke lengths and ejector piston spacing as part of the development effort on the BRU-46/47 rack. The relative performance of these racks had not yet been confirmed, but static testing had been done. There was uncertainty with respect to the end-of-stroke performance of these modified racks.

At release the store pitched up approximately 30° almost immediately, as shown in Fig. 15.37 when the stabilizing fins failed to deploy. In an unstable aerodynamic condition with the fins folded, the bomb climbed and struck the left horizontal tail of the release aircraft as shown on the right. A store pitch-up had been expected but the CTS wind tunnel trajectories were truncated after just a few feet of travel because of limitations in the wind tunnel hardware. It was assumed that motion beyond this limited travel would not be a hazard to the test aircraft. That assumption was clearly wrong for these release conditions, especially when the fins did not deploy as planned for the lanyard length used. As the photos show, the lanyard to deploy the fins was not fully extended in time to stabilize the GBU-12D/B before it flew back up into the horizontal tail.

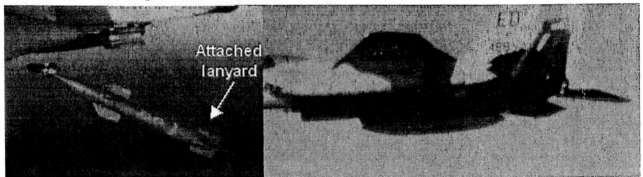

Fig. 15.37 GBU-12D/B Pitch-Up and Impact with Horizontal Tail (USAF Photos)[51]

This incident resulted in changes to the Seek Eagle Office separation prediction process to obtain more complete trajectory data, less reliance on CTS data while increasing both the CTS conditions tested and substantially adding to the interference grid data obtained off line, and an emphasis on sensitivity analysis for the preflight predictions. Since these tests, analysis with 6-DOF models has been considerably improved and less reliance on limited CTS trajectory data is common. The lanyard length was also shortened to deploy the stabilizing fins at first motion from the improved BRU-46 ejector rack. These changes allowed this laser-guided weapon to be certified to usable limits for release from the F-15E aft outboard CFT station.

Chapter 15 Stores Certification Testing 203

Fig. 15.38 CBU-89 Ripple Release from the B-1B[51]

15.4.4.3 **CBU-89 Tactical Munitions Dispenser (TMD) Releases from the B-1B**. Internal carriage of stores on racks often pose less severe separation constraints on weapon release but they have their own problems. The ripple release shown in Fig. 15.38 is from the B-1B's forward, mid, and aft bays at $M = 0.90$ (maximum release dynamic pressure) in 1995. A total of 30 weapons were released from this rack at a release interval of 40 milliseconds from the 10-carry module of this internal multiple carriage rack. This large number of dispenser stores released within a very short period of time often leads to store-to-store collisions with the attendant possibility of debris being ingested into the engines or premature store dispensing/arming of submunitions in the dispensers. Predictions for such dispensers must take into account relative positions of the multiple stores to deduce the bay combinations, rack download sequencing, and/or release intervals to be certified. Such a complex prediction problem demands investment in different wind tunnel approaches, including both specialized multiple sting tests for CTS and grid mode data. Up to three store model support mechanisms were used in this case. Ultimately, the 6-DOF simulation must integrate data sets to produce a credible prediction of store trajectories. Store-to-store contact analysis is added to the usual considerations of aircraft impact and store motion. Despite the complexity and the costs, a credible analytical tool was developed and is essential for such complicated store releases. In fact the analytical and wind tunnel work on this TMD led to similar prediction methodology by simply altering the mass distribution for the same external shape successor to this TMD, the Wind-Corrected Munitions Dispenser (WCMD). These dispensers were later certified with the same approach at somewhat lower investment. Ejector rack settings had to be altered for the more aft store cg location on the WCMD. The primary lesson is that an investment in simulation prediction support often pays dividends over and over again.

15.4.4.4 **F-16 Fuel Tank Jettison Store-to-Aircraft Collison**. The F-16, like many fighters that have been in operational service for some time, must have improved stores and modified carriage hardware certified to extend its service life. Often the modifications are small enough to clear the new configuration by analogy, but what is taken to be a minor alteration is sometime carried too far. When improvements were made in the pylon that hosts the 370-gallon external fuel tank by adding a Pylon Internal Dispenser System (PIDS) internal to the pylon, it was hoped that the increased thickness at the aft part of the

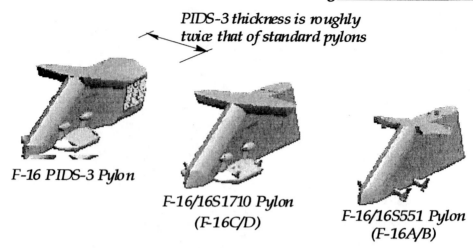

Fig. 15.39 PIDS Pylon Compared to Earlier F-16 Pylons (USAF Images)[51]

pylon (Fig. 15.39) would not adversely affect the jettison trajectory of the tank pylon combination. Unfortunately, the aerodynamic blockage attributed to the thicker PIDS-3 pylon caused the tank to translate further aft and strike the F-16 fuselage ventral fin with release conditions that were benign for the earlier pylon/tank jettison trajectory. Moreover, when the carriage of the AGM-158 Joint Air-to-Surface Stand-off Missile (JASSM) on an adjacent station, the jettison trajectory was further complicated. The JASSM rudder is folded against the left side of the store. This asymmetry in the adjacent store resulted in a store-to-aircraft collision (Fig. 15.40) when the tank was jettisoned from the right wing when release conditions were moved up only 25 KIAS, 0.05 Mach number from a successful release from the same station on the left wing. The cause for this difference was the stowed position of the JASSM rudder and the blockage effect of the thicker PIDS-3 pylon. The message is that relatively small changes may cause significant differences in separation trajectories and that they cannot be glibly rationalized. Figure 15.41 shows how much these slight asymmetries affected miss distance.

Time= 2.15 ms *Time= 2.35 ms* *Time= 2.85 ms*
Impact + 1 frame *Impact + 4 frames* *Impact + 15 frames*

Fig. 15.40 Impact of the 370-gallon Fuel Tank with the F-16 Ventral Fin (USAF Photos)[51]

15.4.4.5 **F-15E/AGM-130 Near Miss.** Another modification to update an older airplane's capabilities highlights the dangers of false assumptions and missed opportunities. Adding a guidance package to a store with known characteristics would seem to be a recipe for improved separation clearance. However, the failure possibilities that are introduced with active store flight control systems has proven this assumption false. As Roberts puts it[51]: "A flight control system can add exponentially to the many hazards that must be evaluated for the risk they pose during store separation." The F-15E certification of the AGM-130A (a modification that essentially strapped a rocket motor on the GBU-10) was proceeding well in March 1992 when the desired release envelope was cleared after three missions. Jettison tests had also been completed, but the predictive simulation did not correlate well with the actual trajectory. The next event was an incremental build-up step in dynamic pressure (to maximum employment dynamic pressure at a Mach number of 0.95) with the release from the left wing pylon. The result of this release is shown in Fig. 15.42. Too much confidence was placed in the capability of the guidance system to arrest motion of the store, especially when released in a demanding transonic flight condition. The fix for this store on this airplane was to adopt a preset fin deflection for each wing station. These settings precluded buildup of high angular rates when the store was too close to the airplane. Even after changes were made to the autopilot, the store was certified with different employment and jettison limits. Finally, this sort of corrective action cannot be readily applied to other weapons certified on several different platforms or distributed on multiple carriage stations. The Seek Eagle Office also incorporated and autopilot simulation capability in their predictive tools and generally became more wary of control system idiosyncrasies, failure modes, and operating limits.

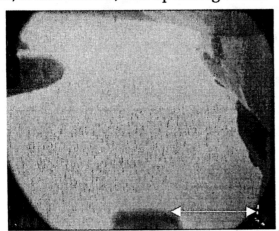 

Fig. 15.41 Comparison of the Fuel Tank Jettison from Opposing Wing Stations with JASSM on Adjacent Station (USAF Photo)[51]

Fig. 15.42 Pitch-up of AGM-130 Released from F-15E Left Wing Pylon (USAF Photo)[51]

15.4.4.6 **F-15C/AGM-120A Jettison Miss**. Another F-15 incident calls attention to both test team discipline and attitude toward what are perceived to be benign release conditions. The AIM-120A CTS data from the wind tunnel indicated there could be an unsafe motion, but this information was set aside in favor of the 6-DoF grid mode results which did not demonstrate this tendency. Evidently the simulation trajectories were not compared to all the wind tunnel predictions nor were there apparently enough sensitivity analyses done. Consequently, when the jettison vehicle had a longitudinal cg considerably aft of the nominal value, no red flags were raised. The jettison sortie was planned with two jettison events, one from the forward missile station at an intermediate build-up dynamic pressure and a second from the aft missile station at the planned end-point in dynamic pressure. The first missile was jettisoned with a fin that was unlocked in captive flight. That fin broke off immediately after release and the missile remained hazardously close to the release aircraft as it climbed and translated laterally. It cleared the aircraft wing tip by only a few feet (Fig. 15.43). The chase pilot did not recognize that the trajectory was anomalous and did not abort the mission at that point. The second jettison attempt led to the missile pitching nose-up and damaging the aircraft's horizontal stabilizer.

Fig. 15.43 Near Miss of the F-15 Wing Tip by a Jettisoned AGM-120A (USAF Photo)[51]

At several points in the process there were opportunities for both engineers and test crew to prevent this incident. First, the CTS data at Mach 0.95 was ignored for the off-line simulation. Second, the grid data upon which the simulation relied did not have data points at Mach 0.95 in it, even though CTS results suggested it was important to do so. Thirdly, the chase aircrew were not fully aware of the wind tunnel results and did not recognize the aberrant trajectory during the first of the two jettison events. Finally, careful monitoring of telemetered video data (not available for this test) would have also precluded attempting the second jettison when the fin separated from the first store. Now the Seek Eagle Office recommends use of video telemetry to the control room, a denser wind tunnel test matrix, and sampling of CTS trajectory data throughout the intended store separation envelope to ensure that off-line simulations are verified against all available data. The AIM-120A was ultimately certified with demonstrations to lower jettison limits than this test event attempted.

15.4.4.7 **TA-4J** **Safety** **Chase** **Impact.** The original F/A-18 Full Scale Development program was in full swing in 1981, clearing diverse weapons loads for fleet carriage and release. A jettison test constituted a build-up test point preparatory to release of a Mk-82 from the rack. This test was required in order to ensure that the weapons and racks could be safely jettisoned from benign 1-g flight conditions in the event that emergency jettison of stores and racks was later required either during the test program or in fleet service. The jettison entailed separating the carriage rack (with attached bomb) rather than the bomb alone. The rack/bomb configuration was highly unstable due to its irregular shape and was expected to tumble unpredictably. A safety/photo chase airplane, a TA-4J, accompanied the test airplane and the pilot had been briefed as to the unpredictable behavior of this rack/bomb combination. Unfortunately, the chase pilot didn't appreciate the extent of the hazard zone. The rack-plus-bomb diverged immediately outboard upon release, shearing the right wing from the chase airplane. The tumbling chase plane was engulfed in flames, followed by the ejection of pilot and photographer, which makes for dramatic video. Fortunately, neither crewmember was seriously injured. The video has since been obligatory training for those pilots at Patuxent River who conduct or chase separation testing.

15.4.5 Summary of Lessons Learned from Store Separation Examples

These few examples are by no means comprehensive; there are plenty of other incidents to learn from in the literature[47,50,51]. But, several recurring themes arise:

- Store certification planning must be treated as a risk management issue. This principle applies to every person involved in and every facet of the preparation for stores testing.
- Risk aversion demands that M&S be valued for its ability to broaden and deepen sensitivity analysis, but the limitations of the predictions must be recognized, reviewed, and verified at every step in the process. Plan for failures, especially if the probability of occurrence is high or if the consequences are catastrophic.
- Planning and preparation must leave an indelible mind-set on every individual directly involved in the conduct of stores tests – an attitude that does not try to generalize when not warranted and that recognizes the potential impacts of even "small" modifications to the store, to the aircraft, or to suspension equipment.
- Make maximum use of available instrumentation and data so risk areas can be quantified and prudent actions taken to mitigate the risk.

15.5 CONCLUSION

This chapter is merely a starting point for those who intend to become proficient in flight testing stores for compatibility with their carrier aircraft. We have attempted to summarize decades of history, hundreds of incidents (some funny, some tragic) that illustrate the principles to be observed, and some of the most difficult interfering flow problems that exist in aerodynamics. The goal has been to encourage those interested in this unique element of flight test to be professional in every sense of the word. Because this chapter is distinctly introductory, understanding what is here is only the beginning of mastering the science and art of stores separation.

REFERENCES

1. Woodman, H., **Early Aircraft Armament: the Aeroplane and the Gun up to 1918**, Arms and Armor Press, London, 1989. [ISBN 0 85368 990 3] (author's Amendment and Updating list, July 28, 1993: http://www.quarry.nildram.co.uk/Woodman.htm)

2. "Vickers F.B.5 Gunbus": http://en.wikipedia.org/wiki/Image:Vickers_F.B.5._Gunbus.jpg#filehistory

3. "Rosebud's WWI and Early Aviation Image Archive" - Aeroplanes and Airships before 1920": http://www.earlyaviator.com/archive1

4. Gustin, E., "The WWII Fighter Gun Debate: Precedents": http://www.geocities.com/CapeCanaveral/Hangar/8217/fgun/fgun-fr.html

5. "Oswald Boelcke": http://en.wikipedia.org/wiki/Oswald_Boelcke

6. "The Dicta Boelcke": http://www.jastaboelcke.de/index_02.htm

7. Genth, T., "Fotoalbum WWI": http://www.angelfire.com/hi5/tgetth/Fotole.html#fw1

8. Runarsson, K., "German Aircraft Armament Unit Structure during the First World War: The Bomb Armament and Unit Structure of Ground Attack Forces": http://www.brushfirewars.org/weapons/weapons/aircraft_armament/wwi_german/ww1_german_aircraft_armament_1.htm

9. "Aerodromes, Archie, Recce and Bombs": http://www.earlyaviator.com/archive3 htm#aero

10. "Handley Page Type O": http://en.wikipedia.org/wiki/Handley-Page_O/400

11. "Gotha G": http://www.answers.com/topic/gotha-g

12. Savine, A., "Ilya Mouromtez, I. I. Sikorsky": http://www.ctrl-c.liu.se/misc/ram/ilyamour.html

13. "Messerschmitt Bf 109G": http://www.nationalmuseum.af.mil/shared/media/photodb/photos/050218-F-1234P-084.jpg

14. "P-38 Lightning": http://www.af.mil/shared/media/photodb/photos/021001-O-9999G-005.jpg

15. "Kawanishi N1K2-J George": http://www.nationalmuseum.af.mil/factsheets/factsheet_media.asp?fsID=525

16. "Messerschmitt Me 262": http://www.nationalmuseum.af.mil/shared/media/photodb/photos/040820-F-1234P-081.jpg

17. Boyne, W. J., **Messerschmitt Me 262: Arrow to the Future**, Schiffer Military/Aviation History, Atglen, Pennsylvania, 1994.

18. Dorr, R. F., and Donald, D., **Fighters of the United States Air Force: from World War I Pursuits to the F-117**, Temple Press/Aerospace Publishing Limited, London, 1990.

19. "Republic P-47 Thunderbolt": http://www.af.mil/shared/media/photodb/photos/020930-O-9999G-013.jpg

20. "Junkers Ju-87 Stuka": http://www.nationalmuseum.af.mil/shared/media/photodb/photos/050406-F-1234P-041.jpg
21. "Junkers Ju 87G "Gustav"": http://www.warbirdsresourcegroup.org/LRG/ju87g.html
22. "The SBD - Dauntless": www.aviaistorija.puslapiai.lt
23. "Piaggio P.108": http://en.wikipedia.org/wiki/Piaggio_P.108
24. "Piaggio P.108 Heavy Bomber": http://www.comandosupremo.com/P108.html
25. "Avro Lancaster": http://en.wikipedia.org/wiki/Avro_Lancaster
26. "Henschel Hs 293": http://en.wikipedia.org/wiki/Henschel_Hs_293
27. "Heinkel 177 'Greif' (Griffin)": http://www.simviation.com/fsdcbainhe177.htm
28. "Henschel Hs 293 Air-to-Ground Guided Bomb": http://www.walter-rockets.i12.com/missiles/hs293b.htm
29. "Air Force Link/Photos/History/1940s": http://www.af.mil./photos/index.asp?galleryID=161&page=3
30. "Cold War": http://en.wikipedia.org/wiki/Cold_War#End_of_the_Cold_War
31. " Full Bomb Load": http://www.f-100.info/images/f-100c_bank.jpg
32. " Firing Bullpup": http://www.f-100.info/images/f-100d_bullpup.jpg
33. "F-4 Phantom II": http://www.af.mil/shared/media/document/AFD-060726-020.pdf
34. "Sukhoi Su-25": http://www.airliners.net
35. "Sukhoi Su-25": http://en.wikipedia.org/wiki/Sukhoi_Su-25
36. "Grumman A-6 Intruder": http://www.dodmedia.osd.mil/Assets/2000/DoD/DD-SD-00-00382.JPEG
37. "B-52 Stratofortress": http://www.af.mil/photos/index.asp?galleryID=12
38. "AGM 129 ACM": http://www.af.mil/news/airman/0104/missi3b.html
39. "F-117A Nighthawk"(U.S. Air Force photo by Staff Sgt. Derrick C. Goode): http://www.af.mil/factsheets/factsheet_media.asp?fsID=104
40. "Aircraft Stores Compatibility: Systems Engineering Data Requirements and Test Procedures", MIL-HDBK-1763, Department of Defense, Washington, DC, June 1998.
41. "Guide to Aircraft/stores Compatibility", MIL-HDBK-244A, Department of Defense, Washington, DC, , April 1990.
42. "Aircraft Stores Separation", http://www.nswccd.navy.mil/div/about/galleries/gallery3/067.html.
43. Jordan, M., "Adding New Instrumentation to Aircraft Platforms", SFTE NATO RTO-SCI-162 Symposium *"Flight Test – Sharing Knowledge Experience"*, Warsaw, Poland, May 2005, pp. 12-1-12-12. http://www.summitinstruments.com/common/pdf/NavAir1.pdf.
44. "NIST/SEMATECH e-Handbook of Statistical Methods, http://www.itl.nist.gov/div898/pri/section1/pri1.htm, updated July 2006.

45. Dungan, G. D. and Clark, C. M., "Flight Testing Large Lateral Asymmetries on Highly Augmented Fighter/Attack Aircraft", AIAA Paper 1994-2138, Seventh Biennial Flight Test Conference, Washington, DC, June 1994.

46. Wilson, D. B. and Winter, C. P., "F-15A Approach-to-Stall/Stall/Post-Stall Evaulation", AFFTC-TR-75-32, Air Force Flight Test Center, January 1976.

47. Christensen, K.T. "Chapter 9, Stores Certification", *Pilot's Handbook for Critical and Exploratory Flight Testing,* Society of Experimental Test Pilots, Lancaster, CA, 2003.

48. http://www.maj.com/cgi-bin/gallery.cgi?i=1010162.

49. "AGM-154 Joint Standoff Weapon": http://www.navy.mil/management/photodb/photos/030327-N-5781F-004.jpg

50. Arnold, R. J. and Epstein, C. S., *Store Separation Flight Testing,* AGARD Flight Test Techniques Series, Volume 5, Neuilly-sur-Seine (France), April, 1986.

51. Roberts, W. E., "USAF Store Separation Lessons Learned: Limitations of Modern Tools and Methods," Society of Flight Test Engineers 36th Annual Symposium, Fort Worth, TX, Oct. 2005.

52. Tinoco, E. N., "PANAIR Analysis of Supersonic/Subsonic Flows About Complex Configurations", paper presented at Computational Fluid Dynamics Workshop, University of Tennessee Space Institute, Tullahoma, TN, Mar. 12-16, 1984.

53. Dix, R. E., Morgret, C. H., and Shadow, T. O., "Analytical and Experimental Techniques for Determining the Distribution of Static Aerodynamic Loads Acting on Stores", AEDS-TR-8312, Calspan Field Services, Inc., Arnold Engineering Development Center, TN, Oct. 1983.

54. Baker, W. B., Jr., Keen, K. S., and Morgret, C. H., "A Case Study of a Modeling and Simulation Approach to Store Separation on the F/A-22", paper presented at Military Operations Research Society Workshop, Quantifying the Relationship Between Testing and Simulation, Kirtland AFB, NM, Oct 16, 2002. http://www.mors.org/meetngs/test_eval/presentations/C_Baker.Keen.Morgret.p

55. Murman, S. M., Aftosmis, M. J., and Berger, M. J., "Simulations of 6-DOF Motion with a Cartesian Method", AIAA Paper 2003-1246, AIAA Aerospace Sciences Meeting, Reno, NV, Jan. 2003.

56. Erlandsson, A., Bergsten, R., and Joubert, J. W. J., "An Effective and Multinational Gripen Store Separation Team", SETP Paper Number 1965, Society of Experimental Test Pilots, Lancaster, CA, 2004.

57. "Risk Management Guide for DoD Acquisition", 6th Edition (Version 1), Department of Defense, Washington, DC, Aug. 2006.

Chapter 16
SPECIAL HANDLING FLIGHT TESTS

Chapter 9 in Volume 1 treated test techniques for basic open-loop dynamic stability, and closed-loop handling qualities. Those methods, in concert with evaluating representative mission tasks, typically provide a robust evaluation of the handling qualities of most airplanes over the majority of their flight envelope. There are several particularly challenging, yet vital, regimes in which handling qualities testing deserves extended discussion. In the first two cases, the U.S. Federal Air Regulations levy specific demonstration requirements, due to the criticality of the results on flight operations. In each of these cases, flight demonstrations introduce unique hazards requiring deliberate mitigation by the test team.

16.1 MINIMUM CONTROL AIRSPEED

Minimum control airspeed (V_{mc}) often profoundly influences daily operation of a multiengine airplane, since V_{mc} determines rotation speed (V_r), refusal speed ($V_{refusal}$), and balanced field length (*BFL*). The hundreds of takeoffs and landings performed without an engine failure are thereby constrained by the remote possibility of an engine failure during takeoff and landing, perhaps restricting the airports and runways from which the airplane can prudently operate. Virtually all agencies that certify or develop multiengine aircraft explicitly spell out that minimum control speeds both in the air an on the ground (V_{mc_a} and V_{mc_g}) are to be verified and demonstrated by testing[1,2,3,4,5]. Consequently, V_{mc} may drive routine operational utility and must be determined accurately by flight test.

16.1.1 Theoretical Foundations

V_{mc} reflects the full complexity of the airplane's lateral-directional stability, controllability, and flying qualities. Indeed, it may drive the lateral-directional design for many multiengine airplanes. Surprisingly, the topic is completely omitted from the most popular undergraduate and graduate-level flight dynamics texts[6]. Consequently, it is helpful to review static lateral-directional stability and the steady heading sideslip maneuver to lead up to asymmetric power effects, V_{mc}, and its determination by flight test.

16.1.1.1 Review of Steady Heading Sideslips (SHSS).
Static lateral-directional stability is best understood by considering a steady-heading sideslip, a flight condition of considerable utility, and a maneuver used in flight test to numerically evaluate static lateral-directional stability. The SHSS provides both insight into the ever-present coupling between lateral and directional stability, and is a stepping stone to understanding the engine-out control problem and the more complicated phenomena responsible for lateral-directional dynamic modes. Unlike the static longitudinal problem, where equilibrium largely balances gravity forces against aerodynamic forces and moments, coupling between aerodynamic forces and moments dominates the lateral-directional trim problem, while the center of gravity location is a second order effect.

Sideslip frequently manifests itself in either slips or skids. A skid is a dynamic maneuver in which the airplane is banked away from the sideslip [for example, left wing down bank ($\phi < 0$), with positive sideslip ($\beta > 0$)]. A skid is useful principally in improving the roll performance of sluggish airplanes. Some radio-controlled models, for example, have no ailerons and roll the airplane by deflecting a rudder to generate sideslip. This sideslip generates roll rate through dihedral effect. The same mechanism is used in many tactical jet fighters which lose roll control power from their ailerons at high angles of attack, yet roll smartly with rudder.

The slip, or more formally, the steady heading sideslip (SHSS), is a static maneuver in which aerodynamic moments are all balanced to generate an equilibrium condition, using asymmetric deflection of the roll and yaw control surfaces. The SHSS, as suggested by the name, is explicitly constant altitude and airspeed, with no pitch, roll, or yaw rates. Operationally, steady-heading sideslips are often used for cross-wind landings. Less commonly, they are also used in very tight formation flight to move the airplane laterally relative to the flight leader while keeping the wings parallel with the lead aircraft (examples include aerial refueling, a sailplane towed behind a tow-plane, or precision acrobatic demonstrations). Lastly, there are some rare missions in which slips are required to turn the airplane without banking (for example, where a sensor must be kept level with the horizon). As a flight test technique, the SHSS is used to measure lateral-directional stability derivatives, as well as aileron and rudder control powers (see Chapter 8, Vol. 1).

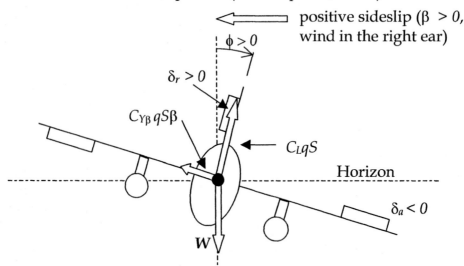

Fig. 16.1 Steady Heading Sideslip (with Symmetric Power)

The balance of the moments and forces is described in the following manner. There are four variables which appear in the force and moment balance: sideslip, aileron deflection, rudder deflection, and bank angle. Sideslip generates a yawing moment through yaw stiffness (C_{n_β}), a roll moment through dihedral effect ($C_{\mathcal{L}_\beta}$), and a side force through the sail area of the fuselage and vertical tail (C_{Y_β}). The rudder generates an opposing yawing moment through the rudder control power ($C_{n_{\delta_r}}$), a rolling moment due to rudder cross-coupling ($C_{\mathcal{L}_{\delta_r}}$), and a direct side force ($C_{Y_{\delta_r}}$). The aileron generates a yawing moment through adverse/proverse yaw ($C_{n_{\delta_a}}$), a rolling moment through roll control power ($C_{\mathcal{L}_{\delta_a}}$), and a direct side force ($C_{Y_{\delta_a}}$), which is usually negligible). Bank angle contributes only a side force due to inclination of the lift vector ($C_L \tan\phi \approx C_L \phi$). Ignoring cross-coupling terms, we can think of the rudder as balancing the yaw stiffness, the aileron as countering the dihedral effect, and the bank angle as countering side force. This cross-coupling guarantees a coupled system of equations. (The development of this cross-coupled set of equations follows the development of Phillips and Niewoehner[7])

With two moment equations and one force equation, the following underdetermined system results for the range in which linear aerodynamic may be assumed:

$$\sum C_Y = C_{Y_\beta} \beta + C_{Y_{\delta_a}} \delta_a + C_{Y_{\delta_r}} \delta_r + C_L \phi = 0$$
$$\sum C_\mathcal{L} = C_{\mathcal{L}_\beta} \beta + C_{\mathcal{L}_{\delta_a}} \delta_a + C_{\mathcal{L}_{\delta_r}} \delta_r = 0 \qquad (16.1)$$
$$\sum C_n = C_{n_\beta} \beta + C_{n_{\delta_a}} \delta_a + C_{n_{\delta_r}} \delta_r = 0$$

Fixing any one of the independent variables (typically sideslip as below), permits solution of the linear system for the remaining three.

$$\begin{bmatrix} C_{Y_{\delta_a}} & C_{Y_{\delta_r}} & C_L \\ C_{\mathcal{L}_{\delta_a}} & C_{\mathcal{L}_{\delta_r}} & 0 \\ C_{n_{\delta_a}} & C_{n_{\delta_r}} & 0 \end{bmatrix} \begin{bmatrix} \delta_a \\ \delta_r \\ \phi \end{bmatrix} = - \begin{bmatrix} C_{Y_\beta} \\ C_{\mathcal{L}_\beta} \\ C_{n_\beta} \end{bmatrix} \beta \qquad (16.2)$$

Solving for the required deflections and bank, per unit sideslip:

$$\begin{bmatrix} \delta_a / \beta \\ \delta_r / \beta \\ \phi / \beta \end{bmatrix} = - \begin{bmatrix} C_{Y_{\delta_a}} & C_{Y_{\delta_r}} & C_L \\ C_{\mathcal{L}_{\delta_a}} & C_{\mathcal{L}_{\delta_r}} & 0 \\ C_{n_{\delta_a}} & C_{n_{\delta_r}} & 0 \end{bmatrix}^{-1} \begin{bmatrix} C_{Y_\beta} \\ C_{\mathcal{L}_\beta} \\ C_{n_\beta} \end{bmatrix} \beta \qquad (16.3)$$

For the range of small sideslip angles (nominally < 10 degrees), the solutions to this linear system are legitimate for static, 1-g flight conditions, though both crew and passengers may find the sideslip uncomfortable. Additionally, these solutions are detrimental to the airplane's range and endurance, as the minimum drag condition is typically found at zero sideslip (aligning both the fuselage with the flow, as well as minimizing the drag caused by deflected control surfaces).

The student should be able to qualitatively describe the physical origin of each coefficient in the above system of equations and its common sign. The student should also be able to qualitatively describe the expected signs of the controls required to stabilize in a sideslip. The student pilot's mantra "wing-down, top rudder" is beneficial to the engineer as well, meaning that stabilizing in a slip, such as for a cross-wind landing, entails holding the upwind wing down slightly with aileron, and balancing the yawing moment with opposite rudder. So, for positive sideslip, the right wing is down (positive bank angle), the right aileron is raised (negative aileron deflection) with right stick and left rudder (rudder trailing edge left is a positive deflection with a left pedal input). From the pilot's perspective, a SHSS requires crossed controls, in this case right stick/yoke and left pedal. The reverse is required for negative sideslip.

As a flight test technique, the SHSS reveals several features of the lateral-directional stability important for good handling qualities. The U.S. Military Standard for the flying qualities of piloted airplanes[7] requires that rudder pedal forces increase linearly with sideslip angles less than 10°, beyond which the rudder force may lighten (though not reverse). Federal Air Regulations[4] require that the rudder pedal force increase steadily up to the maximum appropriate value of sideslip.

Both civil and military requirements prohibit a decrease in the rudder pedal force to the extent that the rudder is blown by the side slip to its limit deflection, and opposite pedal force is required to pull the rudder back to neutral. This behavior is described as "rudder lock" and is caused by a stalled vertical tail. Compare the YB-17 and B-17E models depicted in Figure 16.2. The highly raked strake forward of the vertical tail on the E-model was expressly incorporated to eliminate rudder lock, by creating a vortex in the presence of sideslip that would then blow to the downwind side of vertical tail, keeping the flow attached at large sideslips and maintaining rudder power.

Fig. 16.2 Evolution of the B-17 Vertical Tail (USAF Photos[8,9])

The most common occurrence of rudder lock today is in sailplanes but it can occur on any airplane with a large horn balance and a reversible control system. The English Electric Canberra, later operational in the USAF as the Martin B-57, was such a design. In fact, V_{mc_a} for the original Canberra (with no hydraulic boost for the rudder) was on the order of 160 knots and the maximum rudder force necessary to achieve this V_{mc_a} was over 400 pounds – well over double the rudder force allowed under current military and civil certification limits. Like the B-17, the prototype Canberra vertical tail and rudder[10] were modified to reduce the size of the horn balance (to get away from rudder lock) and to remove the ventral strake on the fin. The production EB-57 (derived from the Canberra) vertical tail and rudder are shown in Fig. 16.3, with annotations to show the changes made.

Fig. 16.3 Modifications to B-57 Vertical Tail and Rudder (adapted from USAF Photo[11])

Example 16.1 The F-14 Tomcat is a twin-engine supersonic strike fighter designed for carrier-based operations with the US Navy (Fig. 16.4). Its aerodynamic performance across its flight envelope depends on its variable geometry design, permitting both very high supersonic operation at aft wing sweep, and excellent slow speed handling and performance with the wings forward. The flight controls include all-moving stabilators (*stabilizer* + *elevator* = *stabilator*) for pitch and high-speed roll control, and a conventional rudder on each of the two vertical tails. Slow speed roll control is provided principally by half-span spoilers, supplemented by differential stabilator. Spoilers distinctively produce proverse yaw, rather than the adverse yaw exhibited by conventional ailerons. The stability derivatives for the power approach configuration are approximated below (gear and flaps down). For the range of sideslip angles from -5 to +5°, find the bank angle and control deflections required to stabilize in a SHSS at an approach speed of 150 mph.

$S_w = 565$ ft² $\qquad c_w = 9.8$ ft $\qquad b_w = 64.12$ ft

$y_{eng} = 4.5$ ft $\qquad W = 54000$ lbs.

$C_{Y_\beta} = -0.0218$ deg⁻¹ $\qquad C_{\ell_\beta} = -0.0024$ deg⁻¹ $\qquad C_{n_\beta} = 0.0021$ deg⁻¹

$C_{Y_{\delta_r}} = 0.0047$ deg⁻¹ $\qquad C_{\ell_{\delta_r}} = 0.00024$ deg⁻¹ $\qquad C_{n_{\delta_r}} = -0.00172$ deg⁻¹

$C_{Y_{\delta_a}} = 0.00251$ deg⁻¹ $\qquad C_{\ell_{\delta_a}} = -0.00179$ deg⁻¹ $\qquad C_{n_{\delta_a}} = -0.00038$ deg⁻¹

Solution:

First solve for the required lift coefficient:

$$C_L = \frac{W/S}{0.5 * \rho * V^2} = \frac{54000 \text{ lbs}/565 \text{ ft}^2}{0.5 * 0.00238 \text{ slugs/ft}^3 * (150 \text{ mph} \frac{5280}{3600})^2} = 1.66$$

Solving using the system of equations above (the π/180 term converts radians to degrees):

$$\begin{bmatrix} \delta_a/\beta \\ \delta_r/\beta \\ \phi/\beta \end{bmatrix} = -\begin{bmatrix} C_{Y_{\delta_a}} & C_{Y_{\delta_r}} & C_L \\ C_{\ell_{\delta_a}} & C_{\ell_{\delta_r}} & 0 \\ C_{n_{\delta_a}} & C_{n_{\delta_r}} & 0 \end{bmatrix}^{-1} \begin{bmatrix} C_{Y_\beta} \\ C_{\ell_\beta} \\ C_{n_\beta} \end{bmatrix}$$

$$= -\begin{bmatrix} 0.00251 & 0.0047 & 1.66*\pi/180 \\ -0.00179 & 0.00024 & 0 \\ -0.00038 & -0.00179 & 0 \end{bmatrix}^{-1} \begin{bmatrix} -0.018 \\ -0.0024 \\ 0.0021 \end{bmatrix} = \begin{bmatrix} -1.14 \\ 1.47 \\ 0.62 \end{bmatrix} \text{ (°/° or rad/rad)}$$

Therefore, over the range of sideslips for which linear behavior can be reasonably expected, each degree of positive sideslip requires -1.14° of aileron (right aileron up), 1.47°s of trailing-edge-left rudder, and 0.62° of right bank. Negative sideslips require the opposite sign input. Had this airplane exhibited adverse yaw, as is more common, the required rudder would be slightly smaller, but the signs would remain the same. (The student should think through this statement.)

Fig. 16.4 F-14 Tomcat (U.S. Navy Photo – Nathan Laird, Photographer[12])

16.1.1.2. <u>Engine-Out Operations and Static Minimum Control Airspeed</u>. Minimum control airspeeds apply principally to the case of multiengine airplanes with a critical engine off, and the remaining engine(s) at maximum power. Typically, this condition is a concern because of the challenges entailed with takeoff or landing with a failed engine. However, some airplanes shut down an engine or engines in cruise flight to improve either range or endurance (a common practice for the US Navy's maritime patrol airplane, the turboprop P-3 Orion, when loitering on station).

Minimum control airspeeds also are a concern with high thrust-to-weight propeller airplanes, particularly small Unmanned Air Vehicles (UAV). Since UAV configurations may omit either rudders or ailerons, the lateral-directional control can be overwhelmed by engine torque at low speeds. This problem is treated in detail by Phillips[6].

Understanding the V_{mc} problem is vital to the designer as it typically sizes the rudder on a multiengine airplane. If the rudder and vertical tail are inadequately sized, V_{mc} will set the lower bound for both takeoff and landing speeds, and hence the required field length. Even if the wings and high lift surfaces can achieve much lower speeds, pilots avoid deliberately flying a multiengine airplane below its V_{mc}. Operationally, the difficulties of maintaining control at slow speeds have killed pilots in airplanes that otherwise had excellent reputations as premier designs (the P-38 Lightning and F-14 Tomcat, notably). Consequently, V_{mc} is the principle reason regulatory agencies require specific training before a multiengine rating is issued.

For jet or rocket powered aircraft which generate small inherent yawing moments, the critical engine is either of the two outermost engines. For airplanes with propellers, yawing and rolling moments are likely both from the propeller hub and from the asymmetric flow over fuselage and surfaces. In such cases, careful thought or tests are required to discern the critical engine (though commonly it will be the right-most engine for clock-wise rotating propellers, as described below). Civil regulations offer specific guidance on determining the critical engine[3,4,13,14].

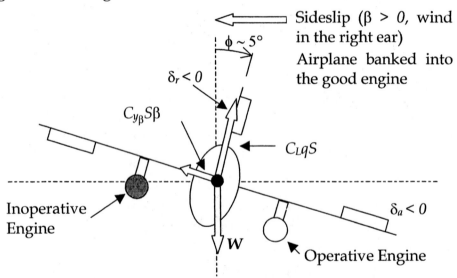

Fig. 16.5 Steady Heading Sideslip (Asymmetric Power)

The simplest asymmetric power problem is cruise flight with a jet engine contributing only a yawing moment. The non-dimensional yawing moment contribution due to thrust may be calculated several equivalent ways:

$$\Delta C_{n_{eng}} = -\frac{T}{\frac{1}{2}\rho V^2 S_w}\frac{y_{eng}}{b_w} = -C_T \frac{y_{eng}}{b_w} = -\left(\frac{T}{W}\right)\frac{y_{eng}}{b_w}C_L$$

where the thrust coefficient C_T is defined:

$$C_T \equiv \frac{T}{\frac{1}{2}\rho V^2 S_w}$$

Note that with y described in the stability axis system, an operative engine on the left wing would have $y < 0$, with a positive contribution to the moment. Also, remember that the asymmetric thrust challenge posed by loss of the left engine results from the operative *right* engine. Finally, for thrust systems essentially parallel to the flight path $C_T \approx C_D$.

Using the thrust coefficient, the following lateral-directional static system results:

$$\sum C_Y = C_{Y_\beta} \beta + C_{Y_{\delta_a}} \delta_a + C_{Y_{\delta_r}} \delta_r + C_L \phi = 0$$

$$\sum C_l = C_{l_\beta} \beta + C_{l_{\delta_a}} \delta_a + C_{l_{\delta_r}} \delta_r = 0 \qquad (16.4)$$

$$\sum C_n = C_{n,\beta} \beta + C_{n,\delta_a} \delta_a + C_{n,\delta_r} \delta_r - \frac{y_{eng}}{b_w} C_T = 0$$

Again, even with the flight condition (and hence thrust and lift coefficient) fixed, four variables persist, meaning that the airplane can be statically trimmed over a range of sideslips. Typically the minimum drag condition is still at zero sideslip. Though zero rudder deflection, aileron deflection, and bank angle were solutions for zero sideslip with symmetric power, non-zero control deflections are required to balance asymmetric thrust at zero sideslip.

$$\begin{bmatrix} C_{Y_{\delta_a}} & C_{Y_{\delta_r}} & C_L \\ C_{l_{\delta_a}} & C_{l_{\delta_r}} & 0 \\ C_{n_{\delta_a}} & C_{n_{\delta_r}} & 0 \end{bmatrix} \begin{bmatrix} \delta_a \\ \delta_r \\ \phi \end{bmatrix} = \begin{bmatrix} 0 \\ 0 \\ 1 \end{bmatrix} \frac{y_{eng}}{b_w} C_T$$

$$\begin{bmatrix} \delta_a \\ \delta_r \\ \phi \end{bmatrix} = \begin{bmatrix} C_{Y_{\delta_a}} & C_{Y_{\delta_r}} & C_L \\ C_{l_{\delta_a}} & C_{l_{\delta_r}} & 0 \\ C_{n_{\delta_a}} & C_{n_{\delta_r}} & 0 \end{bmatrix}^{-1} \begin{bmatrix} 0 \\ 0 \\ 1 \end{bmatrix} \frac{y_{eng}}{b_w} C_T \qquad (16.5)$$

For equations 16.5, the thrust required for level flight occurs at a specified speed. Furthermore, since the airspeed is an input to the problem, the thrust coefficient provides the most direct formulation.

The static minimum control airspeed in the air (V_{mc_a}) is the minimum airspeed at which all airplane moments can be balanced in a steady heading sideslip with the operative engine at its maximum power. (For afterburning powerplants, a minimum control airspeed is defined for both maximum unaugmented thrust and maximum afterburner. The F-14 example that follows will illustrate the significance of this issue.) A minimum control speed exists because for a jet, the yawing moment due to thrust is nearly constant with variation with airspeed, yet the aerodynamic moments available from controls is proportional to dynamic pressure. The propeller is often a more severe problem since the thrust likely *increases* at slower speeds. As dynamic pressure decreases, aerodynamic controls lose moment-generating capacity to balance the thrust asymmetry. Static V_{mca} is typically defined by the speed at which either the rudder or the aileron reaches its physical limit (saturation). Because bank angle and sideslip angle also contribute forces and moments to the equilibrium state, certifying agencies typically stipulate no more than 5° of bank angle. It may not be evident why a small bank angle permits slower minimum control speeds, but the example that follows demonstrates the truth of this assertion.

The minimum control speed can be evaluated by rearranging the equation above with ϕ_{max} fixed and the lift coefficient treated as a dependent variable. T_{max} is the maximum asymmetric thrust with the critical engine out.

$$\begin{bmatrix} C_{Y,\beta} & C_{Y,\delta_a} & C_{Y,\delta_r} & \phi_{max} \\ C_{l,\beta} & C_{l,\delta_a} & C_{l,\delta_r} & 0 \\ C_{n,\beta} & C_{n,\delta_a} & C_{n,\delta_r} & -\left(\dfrac{T_{max}}{W}\right)\dfrac{y_{eng}}{b_w} \end{bmatrix} \begin{bmatrix} \beta \\ \delta_a \\ \delta_r \\ C_L \end{bmatrix} = \begin{bmatrix} 0 \\ 0 \\ 0 \end{bmatrix} \quad (16.6)$$

Since minimum control airspeed is defined by the saturation of either the rudder or the aileron, the analysis must consider the deflection of both control surfaces. One approach is to decrease the airspeed until either of the control deflections reaches its limit and then apply equations 6.17 to analyze the equilibrium condition.

$$\begin{bmatrix} C_{Y,\beta} & C_{Y,\delta_a} & C_{Y,\delta_r} \\ C_{l,\beta} & C_{l,\delta_a} & C_{l,\delta_r} \\ C_{n,\beta} & C_{n,\delta_a} & C_{n,\delta_r} \end{bmatrix} \begin{bmatrix} \beta \\ \delta_a \\ \delta_r \end{bmatrix} = \begin{bmatrix} -\phi_{max}\left(\dfrac{W}{S}\right) \\ 0 \\ T_{max}\dfrac{y_{eng}}{b_w S} \end{bmatrix} \dfrac{2}{\rho V^2} \quad (16.7)$$

Caution is necessary since eqn. 16.7 assumes the control powers are constant across their full range of motion. While this assumption is legitimate in the normal operating range within ±10° of neutral, linearity commonly breaks down prior to the saturation limits of most airplane control surfaces (typically 25-30°). Since control power usually decreases approaching this saturation limit, the above method has the undesirable effect of underestimating minimum control airspeed.

EXAMPLE 16.2 The F-14B Tomcat V_{mc} further offers an example of a linear analysis of the type discussed above. Using the stability derivative data from example 16.1 for the F-14 in the approach configuration, estimate the static V_{mc_a} for the F-14B with the increased thrust of each F-110 engine: 16,100 lbs of thrust in maximum dry power, and 27,000 lbs in maximum afterburner. Each engine is offset 4.5 ft from the airplane centerline and contributes a pure yawing moment.
Solution
The V_{mc_a} formulation above is used to write the control deflections and sideslip angle as a function of airspeed, thrust, and bank angle.

$$\begin{bmatrix} \beta \\ \delta_a \\ \delta_r \end{bmatrix} = \begin{bmatrix} C_{Y\beta} & C_{Y\delta_a} & C_{Y\delta_r} \\ C_{l\beta} & C_{l\delta_a} & C_{l\delta_r} \\ C_{n\beta} & C_{n\delta_a} & C_{n\delta_r} \end{bmatrix}^{-1} \begin{bmatrix} -\phi_{max}\left(\dfrac{W}{S}\right) \\ 0 \\ T_{max}\dfrac{y_{eng}}{Sb_w} \end{bmatrix} \dfrac{2}{\rho V^2}$$

$$= \begin{bmatrix} -0.0218 & 0.00251 & 0.00470 \\ -0.0024 & -0.00179 & 0.00024 \\ 0.0021 & -0.00038 & -0.00172 \end{bmatrix}^{-1} \begin{bmatrix} -5\left(\dfrac{\pi}{180}\right)\left(\dfrac{54000 \text{ lbs}}{565 \text{ ft}^2}\right) \\ 0 \\ \dfrac{(27,000 \text{ lbs})(4.5 \text{ ft})}{(565 \text{ ft}^2)(64.12 \text{ ft})} \end{bmatrix} \dfrac{2}{(0.00238 \text{ slugs/ft}^3)V^2}$$

$$= \begin{bmatrix} -1.49e5 \\ -1.67e6 \\ -5.66e4 \end{bmatrix} \dfrac{1}{V^2} \quad (V \text{ in fps})$$

Using sea-level standard density, the required deflections and sideslip angles were determined at both zero and five degrees of bank (into the operating engine), for both the dry power and afterburning cases. Results have been plotted with the airspeed converted to KEAS in Figs. 16.6 and 16.7.

Assuming a limit maximum rudder deflection of 25°, the V_{mc_a} for dry power is 103 KEAS, well below the airplane's typical approach speed of 125-135 KEAS (allowing for 5 degrees of bank into the operating engine). The V_{mc_a} for maximum afterburner is 153 KEAS, well above the airplane's typical approach speed and approaching the range for takeoff. For this reason, the F-14B and F-14D models, equipped with the F-110 engine, were prohibited from use of afterburner with the landing gear down.

Next, observe the substantial reduction in required control deflection with a small bank angle. The result is very sensitive to bank angle (~6 knots/°). For this reason both civil and military specifications allow for a small bank angle, but prohibit trying to balance the airplane with a more uncomfortable attitude. This bank angle restriction applies to design, specification compliance, and demonstration for certification, but not emergency operation, in which the pilot is free to fly the airplane by whatever means deemed necessary to retain control.

Finally, V_{mc_a} occurs with some sideslip on the airplane, the yaw stiffness helping balance the engine's yawing moment. The constraint is balancing the forces and moments with the least amount of required control deflection, even if that means flying with sideslip; the goal is control, not drag or passenger comfort. So, this problem is strikingly different from the engine-out cruise problem in which minimum drag constrained the sideslip to be zero, and the controls were deflected as required.

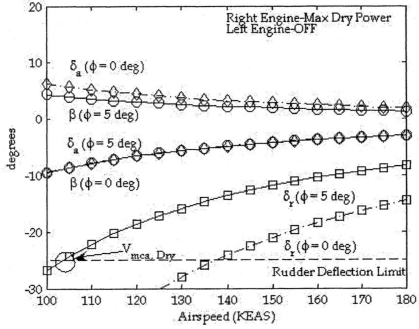

Fig. 16.6 F-14 Lateral-Directional Trim with Asymmetric Power (Right Engine - Dry Maximum Power)

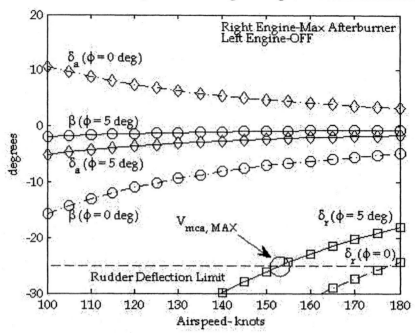

Fig. 16.7 F-14 Lateral-Directional Trim with Asymmetric Power (Right Engine - Maximum Afterburner Power)

Notice one important and non-intuitive feature of the V_{mc_a} problem. High gross weights adversely affect most performance metrics, such as takeoff, landing, single-engine climb, and the like. For the V_{mc} problem, lighter gross weights pose the most danger, as can be seen with weight, W, in the denominator below the maximum thrust. Most multiengine airplanes adjust their approach speed to operate at a specified angle of attack (and lift coefficient), thus lower gross weights permit the airplane to operate at slower speeds, and hence lower dynamic pressures. If the approach lift coefficient is fixed, then the control deflections required to balance the airplane increase with decreasing gross weight! The result is that an airplane is at most risk for loss of control at lighter weights. The most taxing maneuver is therefore executing a single-engine waveoff or missed approach due to the lower weights, and hence slower speeds, at the end of a mission than are observed on takeoff. The operating handbooks for many multiengine airplanes therefore recommend landing 10-15 knots faster than the typical approach speed if one engine is out. For an airplane normally landing at 100 knots, an extra 10 knots increases the moment from the rudder power by 21%!

Minimum control airspeed testing is among the most hazardous of flight test operations, as *de facto*, the test team must probe the boundaries of controllability. Because of the strong sensitivity of thrust to density altitude, V_{mc} testing must be done at low altitude with little margin for surprises. Note that the deadliest flight test accident in the United States since 1990 was the Lockheed C-130 High Technology Testbed performing V_{mc_g} trials. Seven crew members perished when a rudder actuator failure apparently led to loss of control when the aircraft became airborne during the final high speed taxi run[15]. A significant finding of the NTSB investigation was that a Systems Safety Review that included failure states in the actuator was not accomplished. This accident reminds every test team that both V_{mc_a} and V_{mc_g} testing are hazardous operations.

16.1.1.3. **Determining the Critical Engine.** For most airplanes, minimum control airspeed depends on which engine is failed. It is almost always the case for propeller airplanes, where the operating propeller generates rolling, yawing, and pitching moments which vary with angle of attack. Some jets also have such thrust-dependent moments (though this trait is unusual). Regulatory and certification requirements for V_{mc} dictate demonstrations using the worst case ("critical") engine failure. The Flight Test Guides supporting Part 23 and Part 25 of both FAA and EASA certifications give rather complete explanations of how to determine the critical engine for any multiengine airplane[3,4,13,14].

In the case of the right-hand turning propeller (the propeller rotates clockwise when viewed from behind), the down-going blades on the right see a local increase in angle-of-attack, while the up-going blades see a decrease. The blades consequently vary in their aerodynamic loading, during their rotation, increasing in both lift and induced drag on the descending (right) side, and decreasing on the left. A right-hand propeller therefore exerts left yawing moment, positive lift, and left rolling moment on the airframe. The critical engine is, therefore, the left-most engine, because an operating propeller on the right wing produces a left yawing moment due to its lateral location, an aggravating left yawing moment from the propeller loading, a left rolling moment due to engine torque, and a left rolling moment due to the vertical lift on the hub displaced from the airplane's roll axis. The rolling moments are significant to the V_{mc_a} problem because additional compensatory aileron typically contributes an adverse yawing moment.

16.1.1.4. **Other Factors.** The roll-control mechanization may exert a surprising influence on the dynamic V_{mc} problem. During an engine failure at high power setting, the resultant roll rate may actually mask the underlying yaw rate, causing the pilot to respond with aileron, rather than rudder. If the ailerons exhibit significant adverse yaw, countering the roll rate with aileron can actually erode control since adverse yaw adds to the yawing moment from asymmetric power. For example, dynamic V_{mc_a} for late-model F/A-18 Hornets was lowered during development by *decreasing* aileron authority available to the pilot at certain slow speed conditions, mitigating severe adverse yaw.[5]

An aft center of gravity can increase V_{mc_a} due to the shortened lever arm from c.g. to the tail. Hence certification authorities typically require V_{mc_a} demonstrations to be performed with the c.g. at or near the aft limit.

Flap settings can likewise influence the V_{mc} necessitating testing at each viable takeoff and landing configuration.

Some airplanes (notably those with reversible rudder systems) have a V_{mc_a} which occurs because the maximum allowable rudder force (150 pounds) is encountered prior to full rudder deflection.

Finally, the state of the failed engine must be established. For jet airplanes, a locked rotor, rather than a windmilling engine, can increase V_{mc_a} by several knots. However, the locked rotor case is very difficult to replicate in flight tests. Consequently, a calculated correction factor is usually added to flight test results obtained with a wind-milling engine. For propeller airplanes, V_{mc_a} varies dramatically depending upon whether the propeller on the dead engine is feathered, wind-milling, or locked. Certifying agencies may require determination of V_{mc_a} in all viable states, and operating limits then depend upon the functionality and dependability of auto-feathering systems.

16.1.1.5. Dynamic V_{mc_a}. Static V_{mc_a} is but one of several minimum control airspeeds. The U.S. Federal Air Regulations and the European Certification Standards[1,2,3,4] stipulate that V_{mc_a} may not exceed 1.2 times the stall speed at the maximum takeoff weight. The definition used in these civil regulations describes a dynamic maneuver started at maximum power on all engines with a throttle chop to idle on the critical engine. The pilot must then be capable of neutralizing any resultant rates and returning the airplane to a steady heading sideslip with no more than 5° of bank angle (though 5° may be exceeded during the dynamic portion of the recovery), no more than 150 pounds of rudder force, and no more than 20° of heading change. Flight test literature describes this as "dynamic V_{mc_a}" as opposed to the static V_{mc_a} described above[16]. Dynamic V_{mc_a} is more demanding than static V_{mc_a} since the rudder must overcome not only the thrust asymmetry, but also whatever yaw rates may have built up as a result of pilot's reaction time (2 seconds is assumed in flight test). Consequently, dynamic V_{mc_a} is expected to be a higher airspeed than static V_{mc_a}. Dynamic V_{mc_a} prediction requires high fidelity man-in-the-loop simulation, since the pilot interaction with both the displays and controls is an integral element in the airplane's behavior. The reader of a specification must clearly understand which V_{mc_a} is intended. The FARs, for example, clearly intend dynamic V_{mc_a} as indicated by the included definition, though the adjective "dynamic" is not used.

16.1.2 Flight Testing for V_{mc_a}

Several factors complicate the experimental determination of V_{mc_a}. First, thrust produced by both propellers and jet engines is highly dependent on density altitude. In both cases, lower altitudes mean greater thrust and test results at altitude are not directly representative of the values near sea level. Yet, since it is the boundary of controllability that is sought, substantial risk exists that control might be lost at low altitude. Consequently, an extrapolation is necessary to correct the results down to those altitudes at which testing would be foolhardy. Additionally, because of the risk entailed, minimum control speed demonstrations are flown with minimum essential crew aboard. For many airplanes, an abrupt upset near stall may not be recoverable from altitudes as high 2,000 feet. Secondly, V_{mc} may not be dependent on airspeed alone, but may be very sensitive to variations in angle of attack due to either propulsive or aerodynamic effects. In this case, V_{mc_a} also varies with weight.

16.1.2.1. Determining Static V_{mc_a}. Static V_{mc_a} testing is fairly straightforward[1,3,8,13]. The goal is determination of the lowest speed at which the airplane can be held to zero pitch, roll, or yaw rates with the desired thrust asymmetry. The airplane should be established at the test altitude in the appropriate configuration at a speed 20-40 knots above the predicted static V_{mc_a}. Note that the stall speed for maximum takeoff weight may exceed V_{mc_a}, in which case V_{mc_a} does not exist for those weights. In this case, the tests must be repeated at bands of weights in order to determine the heaviest weight at which V_{mc_a} exists. Once the power is set to the desired asymmetry, the nose should be raised to establish a slow (1 knot/sec) deceleration. Throughout the deceleration, the pilot should attempt to hold the airplane to zero body rates, with no more than 5° of bank away from the "failed" engine. The speed at which the airplane can no longer be held to zero body rates is the static V_{mc_a}.

Once a steady heading sideslip can no longer be maintained, the throttles should be immediately matched, and the nose lowered to build airspeed. Airplane and crew are in a dangerous state, and positive control should be promptly restored. Trying to arrest rates with aileron should be particularly avoided until the thrust asymmetry is removed. At this speed and configuration, aileron inputs may contribute more adverse yawing moment than rolling moments, and an abrupt departure can ensue.

Because thrust from both jet engines and propeller-driven powerplants decreases with increasing density altitude, V_{mc_a} likewise varies considerably with altitude, and lower test altitudes provide the most accurate measure of sea level behavior. So, static V_{mc_a} should be determined by stepping down in altitude (for example 10,000, 5,000 and 2,000 ft MSL), and then extrapolated to the sea level using the method described below. Each test point can be completed in several minutes, and results can be quickly obtained. Results should be ratified by repeating the point several times, preferably with different engines. This technique ensures that no unforeseen factors introduce asymmetries. For example, the mechanization or rigging of the rudder, might give a limit deflection that varies slightly from right to left.

16.1.2.2. Determining Dynamic V_{mc_a}. Dynamic V_{mc_a} accounts for a finite response time of the human pilot [1,3,8,13]. When operating at airspeeds near the static V_{mc_a}, the first indication of an engine failure is likely to be uncommanded roll and yaw rates in response to the asymmetric thrust. Control, in this case, entails not only neutralization of the asymmetric moments, but also driving to zero whatever body rates developed during the delay in pilot response. Countering established body rates further stresses the flight control system and dynamic V_{mc_a} is typically higher than static V_{mc_a}. Dynamic V_{mc_a} is the lowest speed at which a pilot is expected to respond to an abrupt engine failure and restore the airplane to steady, level flight without exceeding reasonable limits of bank or heading change. The civil certification regulations define V_{mc_a} as the lowest speed at which the airplane can be recovered to steady flight (< 5° of bank) from a total loss of power on the critical engine, without exceeding 20° of heading change or 150 pounds of rudder force. The standards also limit the bank angle reached to no more than 45° during the dynamic portion of the recovery. Again, these regulations do not use the term *dynamic V_{mc_a}*, though the context clearly indicates this type of V_{mc_a} is intended.

Dynamic V_{mc_a} testing proceeds slowly and requires considerably more test time than static V_{mc_a}. Static V_{mc_a} is determined first. Determining dynamic V_{mc_a} requires repeatedly testing for a satisfactory recovery at progressively slower airspeeds, until the airspeed is found at which recovery exceeds any of the established bounds. The lowest airspeed for which the airplane can be satisfactorily recovered sets the dynamic V_{mc_a} for that altitude. As with the static V_{mc_a}, altitude must then be stepped down to some comfortable minimum (2,000 ft is common), and the same extrapolation process used to identify the sea level value.

The airplane is stabilized in a climb through the test altitude at the speed to be tested, preferably with all engines at full power. The critical engine is then abruptly brought to idle (or shut-down, if required by the certifying authority). After a two-second delay to simulate pilot's reaction time, the controls are manipulated to regain level flight, without adjusting the throttle. If pilot is able to maintain/regain control, using no more than 45°

degrees of bank, then speed is reduced and the procedure repeated until the controllability boundary is found. In a multi-piloted airplane, the throttle chop is done by the co-pilot, the intent being to faithfully simulate the time required for the pilot at the controls to successfully diagnose and respond to such a failure. In a single-piloted airplane, where the pilot in control clearly knows what is coming, this effect is simulated by deliberately inserting a delay in the response (2 seconds is common practice).

High thrust-to-weight airplanes are particularly challenging (specifically those equipped with afterburners). With both engines stabilized at maximum power, either the climb angle can be unrealistically high, or the airplane rapidly accelerates out of the speed range under test. An alternate method is to start with matched throttles at some moderately high setting (for instance, maximum nonafterburning power), and then advance one, while chopping the other. Repeatable results within a reasonable time-to-test demand that the technique be carefully refined and practiced in an appropriate simulator.

The dynamic character of this testing can also produce unexpected results if the airplane has strong adverse yaw (a *yawing* moment due to deflected *roll* control surfaces). Because of roll-yaw coupling, the pilot may first perceive and respond to the airplane's roll with lateral stick rather than rudder in response to the yaw; that is, roll rate masks the pilot's perception of yaw rate. Then, with strong dihedral effect, adverse yaw can dramatically aggravate the asymmetric yawing moment due to thrust, potentially propelling the airplane into a departure (at low altitude). If an airplane has significant adverse yaw, then, once the V_{mca} is determined, the test should be repeated several times at the chosen V_{mca} speed with various combinations of aileron and rudder inputs to ensure the test team has not missed a sensitivity to pilot technique during the dynamic V_{mca} recovery. The regulatory or certifying agency defines which of these values is to be used in the development of takeoff and landing operating guidelines.

16.1.2.3. **Extrapolating V_{mca} to Sea Level.** The risk posed by low altitude loss of control necessitates a careful step-down in altitude, as described above. The lower testing is performed, the more accurate the sea level result. The risk, however, is severe and an abrupt departure at even 2,000 feet may not be recoverable prior to ground impact. A range of altitudes is usually flown, stepping down incrementally. As altitude decreases, expect static V_{mca} to increase. Interestingly, the U.S. Federal Air Regulation describing test methodologies for V_{mca} contradicts itself, indicating that "no corrections to V_{mc} are normally necessary" if demonstrated below 3,000 ft density altitude[1]. Then, on a later page, the same document enjoins, "...extrapolation is nearly always necessary."

Several methods are accepted for extrapolating V_{mca} data to sea level. Kimberlin suggests a straight-line extrapolation from a series of test altitudes down to the surface[17]. It is important to hold weight within a tight tolerance, as results are sensitive to both weight and center of gravity location. Most fixed wing test teams commonly plot test data against pressure altitude. For *Vmca*, it is important to use density altitude for the ordinate, and then to extrapolate to sea level (still in density altitude). Negative density altitudes are common in winter conditions, and, while other airplane performance attributes improve at negative density altitudes, V_{mca} degrades (is higher) for the same reason.

The certifying regulations[2,13] (Appendix 3 in both AC 23-8B and CS-23, Book 2) offer a second extrapolation method, which accounts for considerable variation in propeller efficiency (expected at slow speed). This method entails collection of V_{mc_a} data at several altitudes. The minimum control airspeed is assumed to vary as a function of thrust horsepower and density ratio as expressed in eqn. 16.8:

$$V_{mc_a} = \left[C(THP)\sqrt{\rho}\right]^{1/3} \tag{16.8}$$

Assuming thrust horsepower is available from the collected data; the constant C is calculated and then averaged across the test points taken over the altitude range. Then, given C and a functional description of the propeller efficiency over the airspeed range, $\eta(V)$, V_{mc_a} can be determined by numerically solving equation 16.8 for V_{mc_a}:

$$V_{mc_a} = \left[C(SHP)\left(\sqrt{\rho}\right)\eta\left(V_{mc_a}\right)\right]^{1/3} \tag{16.9}$$

16.1.3 Minimum Control Airspeed Ground (V_{mc_g})

Many multiengine airplanes also have a ground minimum control airspeed (V_{mc_g}), the minimum speed during a takeoff roll at which the rudder has adequate control power to track the runway centerline when an engine fails. This requirement is most common with airplanes lacking nose-wheel steering, due to either castering nosewheels or having a "conventional" landing gear ("taildragger"). Admittedly few multiengine "tail-draggers" remain in active service, and those still flying do not usually need test flights. While thrust asymmetry may be eliminated by aborting the takeoff and pulling all engines to idle, there are some situations (dependent on weight, configuration, weather, runway length, field elevation, and the like) in which it is safer to takeoff, followed by an engine-out landing, rather than attempt a high-speed abort. Analytical prediction of V_{mc_g} is beyond the scope of this discussion due to complexities of nose-wheel steering and gear dynamics, all of which dramatically affect ground handling. As with dynamic V_{mc_a}, prediction of V_{mc_g} depends upon a high fidelity simulation, followed by flight test validation.

Flight tests for V_{mcg} are performed in all appropriate takeoff flap configurations at maximum certified takeoff thrust and at the aft center of gravity limit. The aft c.g. limit is important due to both the shorter rudder moment arm and the destabilizing effect of reducing the weight on the nose gear. The tests are conducted by repeating throttle chops (or fuel cut-off, if necessary) at progressively lower speeds, until the airplane deviates more than 30 feet laterally for its original track.[4,5,13,14] The test technique is therefore similar to the testing performed for dynamic V_{mc_a}. For this test, the engine controls or propeller feathering can not be manually adjusted to keep the airplane within this 30 feet lateral limit. Rudder pedal alone should be used for directional control, with no braking performed until control is confirmed with rudder pedal alone and without manipulation of the engine controls from their take-off setting (auto-feather is permissible). These tests should only be performed with negligible crosswinds.

Two hazards prevail during V_{mc_g} testing. The first is similar to testing for V_{mc_a}, in that the point of the testing is to explore the bounds of controllability. Accepting only negligible cross-winds is important for both data quality and safe conduct. The second hazard is blowing a tire during a test event. Considerable heat can build up in the brake assembly and tires during repeated high speed braking. The temperatures of the wheel assemblies should be monitored during and after every braking event to ensure that tire temperatures are within limits and to minimize the likelihood of a tire or brake assembly blowing up during a subsequent event. An occasional lap around the pattern can dissipate a considerable amount of brake energy. However, do not raise the landing gear; a blown tire inside the wheel well can be catastrophic.

Fig. 16.8 Mini Guppy Turbo (Guppy 101 – Photo courtesy of Daren Savage)[18]

Finally, as with any such test with high risk levels, be sure the system safety review is complete and comprehensive. An accident occurred during the certification testing of the Mini Guppy Turbine (Guppy 101), a modification to a Boeing 377 by Aero Spacelines Incorporated, to carry outsize cargo. On the sixth takeoff of the day at Edwards AFB in May 1970, this turboprop version crashed on takeoff, killing all four crewmembers. The accident investigation left the cause as undetermined, though at least two parts of the rudder assembly were found broken in the wreckage[18]. This tragedy drives home the point that a thorough systems safety review is always in order for V_{mc_a} and V_{mc_g} testing.

16.2 CROSS-WIND LANDING

Nature seldom cooperates to the extent that 'winds-down-the-runway' can be expected for a majority of an airplane's landings. Cross-wind landing limits are typically in the range of 20-25 knots for most airplanes, to provide the flexibility to land under all but the harshest cross-wind conditions. Civil regulations[1,2,3,4,13,14] require demonstration of controllability up to $0.2V_{S0}$. Notice that whenever cross-wind limits are discussed, the limits express the maximum component of the wind perpendicular to the runway. The absolute value (magnitude) of the wind is allowed to exceed the crosswind limit, provided the perpendicular component remains within the bounds.

The test team has three prominent questions it must answer for the certifying or procuring agency, and the customer.
- Does the airplane demonstrate adequate controllability throughout the approach, round-out, flare, touch-down and roll-out?
- Do static and dynamic landing gear stresses remain within design limits?
- Do the flying qualities or structural loads data commend a particular pilot technique? (If so, it must be detailed in the pilot's operating handbook.)

The first and last questions are answered by building up incrementally in crosswind velocities, using diverse methods to identify the technique most suited to a particular model's gear and control attributes. Some airplanes can handle high side loads experienced from landing in a crab; others land in a slip to minimize landing gear side loads. Some airplanes use aerodynamic braking safely in strong crosswinds; others require promptly lowering the nose to retain directional control. Some airplanes permit large weight asymmetries, mandating a series of build-up tests, with the heavy wing both upwind and downwind. The test matrix, in addition to providing the requisite build-up, adds complexity need to answer such questions, giving the ultimate user substantiated recommendations on the safest way to employ and operate the airplane.

Mechanization of the nosewheel steering can be very important. Some airplanes have nose-wheel steering systems that are not selectable; the rudder pedals move the nose wheel whether weight is on the wheels or not. Other systems activate when the main gear touches down. Still other mechanizations must be deliberately selected to 'On' by the pilot. The pilot must be fully aware of how the system operates. Touchdown with a deflected nose wheel can result in alarming unintended yaw rates. If nosewheel steering must be selected 'On' by the pilot, then testing dictates a speed at which selection of nosewheel steering is appropriate.

The biggest challenge of crosswind testing is scheduling the desired winds. Large military and commercial programs often deliberately schedule remote site testing where high winds can be reliably obtained. Rogers Dry Lake bed at Edwards AFB in California and Reykjavik, Iceland are popular with good facilities. Both sites provide the additional benefit of uncluttered runway environments, where minor excursions from the landing runway are not likely be catastrophic. As with other types of testing, building up incrementally in crosswind values provides both confidence in the airplane and increased likelihood of consistency from participating pilot(s). Unfortunately, test teams frequently are forced to accept whatever winds arise naturally at the chosen site, gathering crosswind landing data episodically.

Several distinct hazards deserve attention during build up to peak crosswind landings. The foremost concern is controllability at each increment. The pilot and test conductor monitors control deflections especially if saturation limits are approached while still short of the demonstration objectives. Scrutiny of telemetered real-time data can significantly improve safety by alerting the pilot and test team to remaining control authority margins. Even at the demonstration end point, both the test pilot and the operational pilot seek some additional available control authority in order to respond to the gusts that frequently accompany high crosswinds.

As with any testing involving multiple full-stop landings, the risk of a blown tire increases due to accumulation of heat in the tire and brake assemblies, as well as any structure in contact with the brake assembly. This danger reinforces the suggestion that actively monitoring pressures and temperatures in real time is mandatory. Additionally, crosswind landings often create uneven tire wear; frequently, ground inspections also monitor tire wear during crosswind landing tests.

16.3 HANDLING QUALITIES DURING TRACKING (HQDT)

Modern flight control systems introduce complex dynamic effects in the control path that were not present when airplanes were controlled by cables and pushrods. Hydraulic actuators, structural filtering, nonlinear features such as spoilers, display latency, and high-order feedback control all contribute to system dynamics that can interact in unpredictable ways. Degradations in controllability are not necessarily graceful, and 'handling qualities "cliffs exist" where slight increases in pilot gain can transform a docile airplane into a "bucking bronco". At a 'cliff', flying qualities do not gradually degrade; they collapse, usually destructively.

A number of attempts have been made by researchers to identify engineering criterion that provide *a priori* assessment of the behavior of such complex configurations: the Bandwidth Criterion, the Neal-Smith criterion, and Equivalent Systems modeling are just three such approaches. While classical flying qualities analysis was superbly described in MIL-F-8785C[19], these higher-order methods were better captured by this document's successor, MIL-STD-1797B[5]. Hodgkinson gives insightful treatment to these methods in a more accessible document[20]. However, the focus of these methods is prediction and analysis, not flight test. They might alert a test team to likely problems, but predictions do not correlate perfectly with airplanes that have good handling qualities. The false alarm rate of such techniques is not zero, so a good airplane is occasionally forecast to have problems; nor do they perfectly predict poor handling episodes. For example, subtle but important inaccuracies or simplifications in the analysis model can lead to such conclusions. In environments where high pilot gain combines with little margin for error, the test team must plan and probe for such 'cliffs' while minimizing risk and ensuring that dangerous attributes do not escape detection. It is too late when such characteristics are encountered in operational service.

Handling Qualities During Tracking (HQDT) elevates pilot gain by creating a demanding closed-loop tracking task[21]. The purpose is not direct replication of a realistic task, but rather increasing pilot gain above that expected with realistic tasks, in hopes of revealing handling issues. The distinction from traditional tracking handling qualities is subtle but important. In traditional flying qualities testing, a mission task is flown, with the pilot attempting to achieve some pre-determined task tolerance. The Cooper Harper rating is determined by evaluating the perceived effort (compensation) required by the pilot to achieve either the desired or acceptable performance bounds. The question is not "…can I drive the tracking error to zero?", but rather, "…how hard must I work to achieve desired (or acceptable) performance?" The premise is that the pilot does not want to work any harder than necessary to achieve desired performance, because there are other activities that deserve his/her attention (navigating, communicating, situational awareness, and operating the weapons systems).

For HQDT, the pilot concentrates on driving the tracking error to zero in order to force the pilot gain as high as possible. The goal is not providing a Cooper-Harper rating to quantify his/her opinion, but the goal is to reveal undesirable dynamics of the high-gain pilot and airplane combination. Consequently, the task need not be directly mission representative. Some examples illustrate the utility of this approach.

An airplane's first several landings are cause for apprehension; low and slow on approach is a poor time to discover poor flying qualities. The concern and risk can be mitigated by a brief formation flight exercise at moderate altitude prior to descent for landing. The point is to try to fly a precise formation position, in the landing configuration and speed, while the lead airplane maneuvers gently (20° bank angle turns, with smooth reversals). Longitudinal control is best evaluated from a cruise position (45° bearing line), and lateral control is best evaluated from a trail position, stepped down to remain directly aft of the leader. In either case, no overlap of the two airplanes is allowed, so that momentary loss of control does not result in contact. Striving to chase the lead airplane through gentle turn reversals, and small shifts in flight path angle, stresses the flight control system with the pilot actively in the loop above that which is expected on a straight-in approach to landing. Benign behavior on the HQDT drill permits the pilot to proceed to landing with elevated confidence in the airplane's capability to allow the corrections involved with tracking centerline and glideslope. If problems are encountered during the HQDT drill, the pilot can sample other configurations and speeds to find a combination that results in the best predictability.

The HQDT methodology has been applied by the German flight research institute (DLR) for both fixed wing and helicopter air-to-ground tracking, using a ground based lighting system in which the test aircraft is required to shift abruptly between illuminated ground targets[22]. The purpose is again to stress the flight controls at levels of pilot gain exceeding that expected during routine tasks. McRuer includes a list of candidate tasks for evaluation using the HQDT methodology as an important mitigation for exploring and uncovering adverse Airplane Pilot Coupling[23].

16.4 AERIAL REFUELING

Aerial refueling poses its own unique challenges to handling qualities testing due to the heightened risk of loss of control while close (or connected) to another airplane. For many air vehicles, aerial refueling demands more precise control of the aircraft than anything else done in flight, and exercises the flight controls strenuously (in terms of required actuator bandwidth). While landing and formation flight allow tolerances on the order of feet, in flight refueling typically demands positioning the receiver within inches.

First, there are two prevalent refueling methods used by the military. Even those familiar with one approach are unlikely to be familiar with subtle differences in the other method. Both methods have been and are likely to remain in use for many years, because operational constraints forced the adoption of one or the other for different services at different times. The following discussion is intended principally for ground engineering personnel who have only a vague idea of how aerial refueling is carried out.

Fig. 16.9 KC-10 Refueling F-117 with a Flying Boom (USAF Photo)[24]

Fig. 16.10 KC-130 Refueling F/A-18s Using Probe and Drogue (USMC Photo)[25]

The U.S. Air Force (plus Israel, Turkey and the Netherlands) often employs boom refueling for fixed-wing airplanes (Fig. 16.9), though on occasion have used probe and drogue refueling (Fig. 16.10). Essentially all U.S. rotorcraft, US Navy/USMC fixed-wing, and most European air forces also utilize probe and drogue refueling. The two systems are distinguished by answering the question: "Who is responsible for precision control at after contact?" In boom refueling, the boom operator is responsible for precisely flying the boom probe into the refueling receptacle wherever is located on the receiver airplane. Refueling heavy bombers and transport airplanes strongly favored this solution, given the maneuverability of the boom and the lack of maneuverability for such large airplanes. The receiver aircraft/pilot has the easier task of maintaining position within a comparatively large allowable box (a foot, or so, in any direction). USAF fighters often use this system because it is the system used by their tankers, though it still has the added advantage of lowering workload for the receiver pilot. Flow rates in this scheme also exceed those available from probe and drogue systems, but this capability is not fully utilized or needed by much of the USAF fighter fleet[26].

Boom refueling does have several constraints. First, it is not a feasible system for refueling helicopters, so even the USAF employs some drogue tankers in support of those helicopters whose missions require refueling. Secondly, it requires tankers large enough to accommodate a boom and operator.

Because of the U.S. Navy's requirement for small tankers capable of landing at sea, that service long ago settled on probe and drogue refueling. While some dedicated tanker airplanes were designed and fielded, the US Navy depends, in part, on 'buddy-store' refueling, where a small tactical fighter carries a refueling pod aloft capable of refueling like-sized airplanes (its 'buddies'). The fuel off-loaded from a tactical sized tanker is tiny compared to that available from heavy aircraft, but embarked operations frequently necessitate short-notice emergency refuelings within five miles of the ship. Though the USAF tanker fleet has probe and drogue capabilities, it is simply not always feasible to depend on land-based tankers for this operational capability.

Probe and drogue refueling demands very precise receiver control during the initial contact, which together with the low flow rates, pushed the USAF bomber force toward the boom method. In the approach to contact with the drogue basket, the receiver pilot must control the probe position to within approximately ±2 inches horizontally and vertically. The slightest amount of turbulence or prop wash, coupled with hose dynamics, means that even the most experienced and proficient pilots routinely miss attempts at basket engagement. Basket oscillations of up to a foot or more at frequencies up to 1 Hz are common. Moderate turbulence necessitates patiently waiting a minute or more for a quiet couple of seconds, prior to initiating contact. Once engaged in the drogue basket, the tolerances relax considerably, and basket motion is largely damped by corrections by the engaged receiver.

Aerial refueling testing entails two principle goals. First, the functionality of the fuel systems must be certified (either tanker or receiver, depending upon the system under test). Secondly, satisfactory flying qualities must be established through out the refueling sequence and at every likely receiver position, including those merely transient to/from the actual contact position. The contact itself is affected by the complex interaction of receiver handling, receiver sight-picture (day and night), tanker flow field, and drogue or boom dynamics. Flying qualities vary significantly across the altitude range. For large body tankers, the flow field may be much more severe at heavy tanker weights due to the higher required lift coefficient. Receiver dynamics and controllability are also a major factor as considerable weight change (and larger angles of attack) occurs when taking on maximum fuel loads in many aircraft. The F-100 refueling from a KB-50 hose and drogue tanker could not take on a full load of fuel from the fuselage hose; as weight increased, the receiver pilot had to use minimum afterburner to maintain contact. The required test matrix is also further swollen by the need to test diverse permutations of tanker and receiver configurations (for example, landing gear or landing flaps extended).

Test conditions and instrumentation required to establish fuel system functionality are established by engineers well-versed in both the fuel system architecture and the pressure ratings of its components. Certification requires operation in all possible modes, to include some likely failure states. Fuel system instrumentation must have suitable sample rates and dynamic range to capture any transitory pressure spikes that could compromise system integrity. This latter concern arises because of the significant flow rates involved; the momentum of 500 gpm being arrested by the abrupt closure of a tank shut-off valve can set off significant hammer shocks.

Aircraft flying qualities are a direct concern in this chapter. Airplanes whose handling is otherwise regarded as superb have struggled meeting the challenges of aerial refueling. As examples, the C-17 development uncovered a lateral and longitudinal Pilot Induced Oscillations (PIOs) during the approach to contact phase for aerial refueling[27], and the F-14 was noted for PIO when refueled from the center station of KC-135 tankers. Niewoehner documented severe PIO susceptibility during failure mode testing of F-14s, while engaged with several drogue tankers[28].

The HQDT procedure discussed in the introduction to this section is easily adapted to safely probe receiver aircraft handling prior to actual contact between the tanker and receiver. Furthermore, it costs only minutes per test point. For probe and drogue aircraft, the receiver should be established in formation with the extended probe several feet aft of the center of the basket. Boom-refueled aircraft start the exercise with the boom in sight of the pilot and with the receiver sufficiently in trail that contact with the boom is impossible. With a 'start maneuvering' call from the receiver pilot, the tanker pilot initiates 30° bank-to-bank S-turns, using roll rates of 5-10 degrees per second. The point is *not* that such maneuvers are representative of the refueling task, rather, it is a *more difficult* all-axis tracking task, exercising the lateral and longitudinal controls, as well as thrust. If no PIO susceptibility is uncovered in this tracking drill, then the receiver can call for the tanker to stabilize and then proceed to contact with high confidence of suitable handling.

If any PIO occurs, during either the tracking drill or an actual refueling, the receiver pilot must first ensure clearing away from the fueling apparatus (boom or drogue). Fighting the PIO is a recipe for disaster. The PIO usually clears itself once pilot gains are relaxed, and there is usual ample altitude below the tanker for aggressively descending away from the tanker.

One additional note for development airplanes carrying flight test instrumentation: every pilot with probe and drogue refueling experience has seen or experienced incidents where an off-center attempt at contact resulted in the drogue being 'lipped', after which the basket swings away from the receiver, and then swings back with significant speed. Cracked radomes, cracked windscreens, lost Pitot tubes or flow direction vanes are common results. The X-32 was unable to utilize aerial refueling during its flyoff competition with the X-35 due to the risk of damage to such equipment and perhaps swallowing the broken parts into the engine intake. Pilots quickly learn how to avoid such episodes on fleet aircraft. For the test pilot, the stakes are raised because of the installation of test nose booms with sensitive angle of attack and sideslip vanes mounted on them. This fragile equipment is often at risk while refueling. The test pilot must simply be more patient and less aggressive than when refueling an operational airplane.

16.5 SUMMARY

For the authors, this chapter bears the indelible imprint of personal experience, experiences in which they or colleagues narrowly missed catastrophe. The loss of control of multiengine airplanes to asymmetric power has killed far too many colleagues. Hence, demonstrating the boundary of control inevitably places test aircrews near the survivable limits in such flight tests. A mentor once related, "Spin testing wins awards and documentaries; it is flutter and V_{mc} testing that will kill you."

This chapter is intended to acquaint the new flight test practitioner with this risk-laden corner of the profession, its special vocabulary, the regulatory requirements, and the methods used. The best way to learn how to do this kind of work well is within a team of "gray-haired" practitioners who saw it done, did it themselves, and observed first-hand where the hazards lie.

REFERENCES

1. "Minimum Control Speed," Federal Aviation Regulation 23.149, Federal Aviation Administration, Dept. of Transportation, Washington, DC, Revised Jan. 2007.
   http://www.access.gpo.gov/nara/cfr/waisidx_07/14cfr23_07.html

2. "Minimum Control Speed," Federal Aviation Regulation 25.149, Federal Aviation Administration, Dept. of Transportation, Washington, DC, Revised Jan. 2007.
   http://www.access.gpo.gov/nara/cfr/waisidx_07/14cfr25_07.html

3. "Certification Specifications for Normal, Utility, Aerobatic, and Commuter Category Aeroplanes - CS-23", CS 23.149, "Minimum Control Speed," original issue, European Aviation Safety Agency, Nov. 2003.
   http://www.easa.eu.int/home/certspecs_en.html

4. "Certification Specifications for Large Aeroplanes - CS-25", CS 25.149, "Minimum Control Speed," Updated through Amendment 2, European Aviation Safety Agency, Feb. 2006.
   http://www.easa.eu.int/home/certspecs_en.html

5. "Military Standard, Flying Qualities of Piloted Airplanes", MIL-STD-1797B, Department of Defense, Washington, D.C., Feb. 2006.
   http://assist.daps.dla.mil/quicksearch/basic_profile.cfm?ident_number=70037 Note: This document is a controlled document and is only available to registered DoD users.

6. Phillips, W., **Mechanics of Flight**, John Wiley & Sons, Inc., Hoboken, NJ, 2004.

7. Phillips, W. and Niewoehner, R. J., "Effect of Propeller Torque on Minimum-Control Airspeed", **AIAA Journal of Aircraft**, vol. 43, No. 5, pp. 1393-1398, Sept-Oct 2006.

8. http://www.nationalmuseum.af.mil/shared/media/photodb/photos/060512-F-1234S-004.jpg.

9. http://www.nationalmuseum.af.mil/shared/media/photodb/photos/050520-F-1234P-017.jpg.

10. http://www.bywat.co.uk/canframes.html.

11. http://www.dodmedia.osd.mil/Assets/Still/1983/Air_Force/DF-SN-83-07782.JPEG.

12. http://www.dodmedia.osd.mil/Assets/Still/2006/Navy/DN-SD-06-15436.JPEG.

13. "Flight Test Guide for Certification of Part 23 Airplanes," Advisory Circular AC 23-8B, Federal Aviation Administration, Washington, DC, Aug. 2003.

14. "Flight Test Guide for Certification of Transport Category Airplanes," Advisory Circular AC 25-7A, Federal Aviation Administration, Washington, DC, Mar. 1998.

15. "Probable Cause Report," ATL93mA055, File No. 736, National Transportation Safety Board, Washington, DC 20594, Mar. 1994.

16. "Fixed Wing Stability and Control: Theory and Flight Test Techniques," United States Naval Test Pilot School, Rept. USNTPSFTM-103, Patuxent River, MD, Jan. 1997, p. 6.32 ff.

17. Kimberlin, R. D., **Flight Testing of Fixed Wing Aircraft**, American Institute of Aeronautics and Astronautics, Reston, VA, 2003, pg. 348.

18. Savage, D., "377 MGT Accident," http://www.allaboutguppys.com/, 2006.

[19] "Military Specification, Flying Qualities of Piloted Airplanes," MIL-F-8785C, Nov., 1980.

[20] Hodgkinson, J., **Aircraft Handling Qualities**, American Institute of Aeronautics and Astronautics, Alexandria, VA, Feb. 1999.

[21] Twisdale, T. R., and Franklin, D. L., "Tracking Test Techniques for Handling Qualities Evaluation," AFFTC-TD-75-1, Air Force Flight Test Center, Edwards, California, May 1975 (ADB006191).

[22] Hamel, P. "Recent and Future Aircraft-Pilot Coupling Research at DLR," Report No. IB 111-96/15, DLR Institut für Flugmechanik, Braunschweig, Germany, 1996.

[23] McRuer, D., *et. al.*, *Aviation Safety and Pilot Control: Understanding and Preventing Unfavorable Pilot-Vehicle Interactions*, National Academies Press, Washington, DC, 1997.

[24] http://www.nationalmuseum.af.mil/shared/media/photodb/photos/061006-F-1234S-005.jpg.

[25] http://theaviationzone.com/images/hercules/kc130/bin/kc130017.jpg. (USMC Photo).

[26] Bolkcom, C., "Air Force Aerial Refueling Methods: Flying Boom versus Hose-and-Drogue," Congressional Research Service, June 5, 2006 www.fas.org/sgp/crs/ weapons/ RL32910.pdf, cited 5 Jan 2007

[27] Iloputaife, Obi , "Minimizing Pilot-Induced-Oscillation Susceptibility during C-17 Development" AIAA-1997-3497, AIAA Atmospheric Flight Mechanics Conference, New Orleans, LA, Aug. 11-13, 1997.

[28] Niewoehner, R. J. and Minnich, S., "F-14 Dual Hydraulic Failure Flying Qualities Evaluation," *1991 Report to the Aerospace Profession*, Proceeding of the 35th Symposium, Society of Experimental Test Pilots, Burbank, CA, 1991, pp 4-15.

Chapter 17
PROPULSION SYSTEMS TESTING

Progress in propulsion technologies has accompanied and spurred progress across the domain of aeronautics technologies. The success of the Wright Flyer depended not only upon advances in aerodynamics and control, but also on a light weight engine of sufficient power. World War I airplanes were primitive aerodynamically and structurally, but were strongly constrained by the challenges of reliability, cooling and limited power. The huge performance leap between the two world wars was attributable to advances in all the aeronautics disciplines: structures, aerodynamics and propulsion. The 375 HP engine that powered the De Havilland D.H. 4 was supplanted 20 years later with a 3700 HP engine powering the B-29 Superfortress. We refer to the post-war years as the jet age, but again, the advances were across the spectrum of disciplines; the introduction of the jet engine was simply the most visible aspect. Airplane performance is really characterized by its excess power; weight reductions (materials and structures) and improved aerodynamics has pushed the required power curve down, while propulsion technologies have raised the power available curve.

By the centennial of flight, successful fixed wing airplanes had used human power, reciprocating engines, gas turbine jet engines, turbo-fans and ducted fans, propellers driven by gas turbine (turboprops), liquid and solid rockets, pulse jets, and electric drive propellers, in addition to several other exotic configurations. Boeing announced plans in the spring of 2007 to flight test of a fuel-cell powered light civil aircraft that same summer. A new generation of small commercial jet aircraft, called "microjets" or "very light jets (VLJs)" and carrying four to nine passengers, are powered by small turbofans that grew out of the NASA General Aviation Propulsion (GAP) program (1996-2002). These new engines have some of the highest thrust-to-weight ratios ever achieved in commercial turbofans (over 8 for the first prototype)[1]. The first of these VLJs, the Eclipse 500, received its provisional type certificate in July 2006.

Consequently, we surely have not seen the end of new propulsion development, but several preferred systems have shaken out, each with their respective economic or performance benefit within particular applications:

- Reciprocating engines with opposing cylinders and fixed or constant speed propellers dominate the general aviation market due to their comparatively low purchase cost and simplicity. Engine power per pound runs approximately 1.2 HP per pound, and stock engines are available up to 400 HP[2]. Higher horsepower reciprocating engines are found chiefly in the vintage airplanes.
- Turbine-powered airplanes with constant-speed propellers dominate the high end of general aviation aircraft through short-haul commercial transports, giving light weight and low operating costs for speeds up to approximately 300 KTAS. Engine power per pound ranges up to approximately 3 HP per pound, for engines up to 4700 HP[3]. Turbines perform well at altitudes where reciprocating engines need boosted manifold pressures, and they burn the same jet fuel as turbofan engines.
- High bypass ratio turbofans provide the best performance and economy for higher speed commercial and business jet applications.
- Low-bypass ratio after-burning turbofans have replaced pure turbojet engines in

most tactical military airplanes. (Some turbojets are still in service in older airplanes such as A-4 Skyhawks.)

This chapter discusses only these four propulsive schemes and their testing. While each represents very mature technologies, considerable flight test continues to be performed to develop and certify new and derivative propulsion systems, and their adaptation to new airplanes. All engine and propeller vendors have aggressive research and development programs seeking to reduce noise and emissions, while increasing efficiency, reliability, and maintainability. The propulsion marketplace abounds with small innovations where slight improvements in either cost-to-operate or environmental impact translate into market advantage and large sums of money in the marketplace. Consequently, propulsion testing abounds as bread-and-butter work for flight test teams.

17.1 THEORETICAL FOUNDATIONS

17.1.1 Thrust as Momentum Transfer

Whether a jet, propeller, or turbofan, airplane propulsion systems generate thrust through the acceleration of some mass of air. Thrust is the direct product of the mass flow rate ($\dot{m}$) and the increase in velocity (ΔV) of that mass of air: $T = \dot{m}\Delta V$. The same amount of thrust can therefore be achieved by either accelerating a large mass of air a small incremental velocity, or accelerating a smaller mass of air a larger incremental velocity. A propeller exemplifies the former (the V-22 Osprey being an extreme example); a large mass flow rate passes through the propeller disc, experiencing a comparatively small increase in velocity. It is the product that matters. The pure turbojet exemplifies the small mass flow rate, accelerated at a considerable incremental velocity.

The propulsive efficiency of a propulsive system is the thrust power ($P = TV_{true}$) divided by the total power, which is the sum of the thrust and wasted power.

$$\eta = \frac{TV_t}{TV_t + \frac{\dot{m}(V_e - V_t)^2}{2}} = \frac{2\dot{m}(\Delta V)V_t}{2\dot{m}(\Delta V)V_t + \dot{m}(\Delta V)^2}$$

$$= \frac{2V_t}{2V_t + \Delta V} = \frac{2V_t}{V_t + V_e} = \frac{2}{1 + V_e/V_t}$$

(17.1)

So, propulsive efficiency decreases as the velocity of the exhaust increases; the smaller the velocity increases across the propulsion system, the greater the efficiency. For systems with equivalent thrust, the system with the larger mass flow rate exhibits the higher propulsive efficiency. This observation explains why propellers continue to dominate the low speed range, and why modern commercial aircraft are powered by split spool high-bypass ratio turbofan engines, rather than pure turbojets. Both these systems maximize mass flow rate (turbojets typically exhibit thrust specific fuel consumption of ~ 1 lb thrust per lb/hr of fuel flow (1 lb-hr/lb), while turbofans typically exhibit ~0.6 lb-hr/lb). Equation 17.1 above ignores a number of significant factors affecting propulsive efficiency; its value is illustrative of the gross factors influencing propulsion system design.

17.1.2 The Propeller

The Wright brothers were first to recognize that propellers are best described as a rotating, twisted wing[4]. Prior to that point, the wood screw provided the relevant, but misleading physical analog. Once we realized that the propeller is a lifting surface, then many other insights from 2-D and 3-D aerodynamics apply. The forces at any given cross-section of the blade include a lift force normal to the local flow, and a drag force parallel with the local flow. While a 3-D wing in a steady flow sees the freestream and induced velocities (from the tip vortex), the propeller sees an additional velocity component due to the rotational velocity, which increases proportionally with the distance from the propeller hub. The angle ϕ is the arc tangent of the velocity triangle whose base is the local radius times the rotation rate (in radians per second). If the local incidence is β, then the local angle of attack is the difference: $\alpha = \beta - \phi$ (Fig 17.1). The goal in propeller design is building twist into each blade section such that the local angle of attack is near constant from root to tip. Additionally, the rotational rate imparts not only a local variation in incidence, but a significant variation in dynamic pressure as well. The result is an increase in loading from root to tip, until the load drops to zero near the tip.

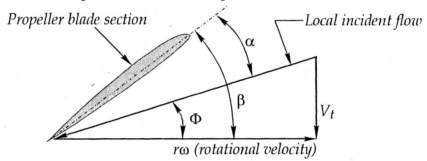

Fig. 17.1 Local Flow at Blade Section

Due to the uneven load distribution over a propeller disk, the efficiency is more typically characterized as the ratio of the thrust horsepower (THP) to the brake (BHP):

$$\eta_p = \frac{TV_t}{BHP} = \frac{THP}{BHP} \qquad (17.2)$$

Insight into two-dimensional aerodynamics also suggests limits of propeller performance. When the true airspeed decreases, then ϕ decreases and α increases, blade stall is approached, and both the parasite and induced drag on the propeller increase. Hence, we expect that at low speed, brake load on the propeller increases. At zero airspeed, such as at brake release on takeoff, the propeller load is maximal, and the THP and efficiency equal zero since THP is the product of thrust and airspeed. At the other end of the speed range, where velocity increases, ϕ increases with airspeed, decreasing α and decreasing the aerodynamic load on each propeller blade. The propeller will provide no thrust when α reaches α_{0L}. So, at high speed, we likewise expect propeller thrust to drop to zero. In between, the efficiency of a well-designed propeller peaks above 80%, as high as 88%.

The above discussion ignores the influence of compressibility. As tip speed approaches Mach 1, shock wave formation significantly increases the brake horsepower required, ultimately limiting the rotational speed and diameters of all propellers, regardless of power source.

Propeller efficiency normalizes with *advance ratio*, typically signified by *J*:

$$J = \frac{V_t}{nD} \qquad (17.3)$$

where n is the rotational velocity, D is the propeller diameter, and the units are consistent such that J is non-dimensional. Physically, the advance ratio is the number of propeller diameters the propeller travels downrange with each revolution. Figure 17.2 depicts the typical variation of propeller efficiency with advance ratio, as a function of β, the blade pitch angle (75% hub-to-tip is the common reference radius).

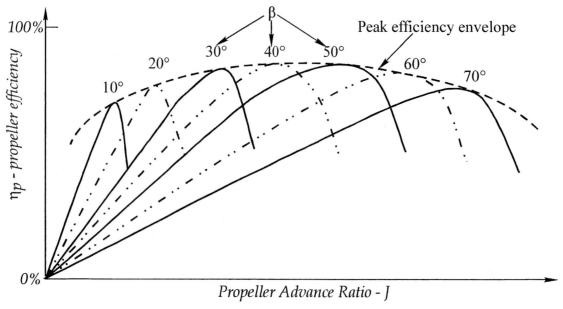

Fig. 17.2 Propeller Efficiency Map (adapted from[10])

Figure 17.2 implies that a fixed pitch propeller is designed for the speed at which maximum THP is desired. *Cruise propellers* have their peak efficiencies near the desired cruise airspeeds, in order to achieve maximum specific range at cruise. *Climb propellers* have their peak efficiencies at lower airspeeds to achieve maximum thrust at low speeds, finding most frequent application with amphibians, floatplanes, and banner/glider towplanes. This same figure portrays the value of variable blade pitch, enabling dramatically improved efficiencies over a wide range of advance ratios, and hence a large range of airspeeds. The variable pitch propeller was introduced shortly after World War I, followed shortly by variable pitch propellers with a governor permitting the pilot to set propeller speed, in lieu of propeller pitch. The governor, employing engine oil as the hydraulic medium, adjusts pitch automatically, increasing or decreasing the load on the propeller to maintain the commanded speed. While some fixed pitch propellers are found on historic or legacy airplanes, constant speed propellers are the overwhelming choice for both reciprocating engines and turbo-props.

17.1.3 The Internal Combustion Aircraft Engine

The first forty years of flight brought startling advances in internal combustion engine technology. Dramatic innovation characterized the period; the history is a very interesting study, and the reader is urged to consult several excellent books treating the subject for additional detail[5,6,7]. World War II was the climax of piston engine performance, and the debate centered around two legacy solutions to the challenge of cooling: inline, water-cooled engines, and air-cooled, radial engines. The best fighters in the European theater were water-cooled; the best fighters in the Pacific theater were air cooled. Neither configuration is prevalent today, though two radials are in limited production as of spring 2007. The Russian Vedeneyev M-14P is derived from a 1930s design and is common among the airshow circuit, producing 360-400 HP. A recent entry, Australian Rotec Engineering, produces 7- and 9-cylinder radial engines with electronic ignition rated at 110 and 150 HP.

Fig. 17.3 Lycoming IO-540 Opposed Six-Cylinder Aircraft Engine
(photo courtesy of Lycoming)

The performance and reliability of the turboprop engine has almost completely displaced the market for higher horsepower reciprocating engines. The low power end of the market remains robust due to the simplicity, cost and excellent maintainability of the air-cooled opposed cylinder engine (Fig. 17.3). This engine geometry dominates this market because of its flat orientation; an even number of cylinders lie horizontally in a plane, on opposing sides of a central crank. The result is a compact, low vibration configuration conducive to air cooling and providing an ideal geometry to fit within the mold-lines of the class of airplanes in which they're found. Other configurations, such as radials, suffer from awkward installations, obscuring the pilot's view over the nose.

Our discussion presumes the reader is familiar with the basics of the 4-stroke Otto cycle, and its distinctions from both 2-stroke gasoline engines and Diesel engines. All three of these engines surround us on the highway and in our lawn sheds. Four-stroke engines have long been the dominant for airplanes, but UAV applications are increasingly introducing both 2-stroke and Diesel engines. The former are available down to very small sizes, and the latter are desirable for shipboard and military applications where Diesel fuel provides superior availability and safety. NASA's General Aviation Propulsion program concentrated on the development of a 2-stroke, Diesel engine with 4 opposed cylinders and burning Jet A[1].

17.1.3.1 **The Basic Engine.** Airplane reciprocating engines introduce a number of features not commonly found in automotive 4-stroke (Otto cycle) engines, which warrant discussion. The reader desiring more detail than that provided below is urged to read Borden and Cake[8], or Hurt[9], from which most of the following summary was drawn.

The pilot directly encounters several differences between automotive and aviation reciprocating engines in the cockpit controls and displays. Most reciprocating engines have a selector switch for the ignition system allowing selection of either or both magnetos. The engine throttle quadrants typically includes not just a throttle (as in a car), but also a mixture control, a propeller speed control (for other than fixed pitch propellers), and cowl flaps. The display console includes common automotive gauges such as alternator ammeter, oil pressure and oil temperature, but also several unique gauges such as manifold pressure, propeller speed, and cylinder temperatures.

The more air burned, the more engine power generated. Hence, brake horsepower from a reciprocating engine is proportional to the product of the crank shaft speed, the displacement, and the mean effective pressure of the engine. A typical automotive engine achieves pressurization in the cylinder solely from the piston's compression ratio; the total compression in an airplane engine comes from ram compression, as free stream airflow is slowed in the induction system, from mechanical pre-pressurization (such as superchargers), and then direct compression in the cylinder.

The temperature at which knocking occurs limits the maximum total compression in any internal combustion engine. Knocking is caused by the detonation of the fuel-air mixture in advance of the combustion flame front, and may quickly damage or destroy the engine, due to the severity of the over-pressure spikes. Pre-ignition is caused by hot spots in the cylinder caused by carbon deposits or material imperfections in the cylinder head, valves or piston. Figure 17.4 depicts the pressure profiles for a normal combustion, as well as both detonation and pre-ignition.

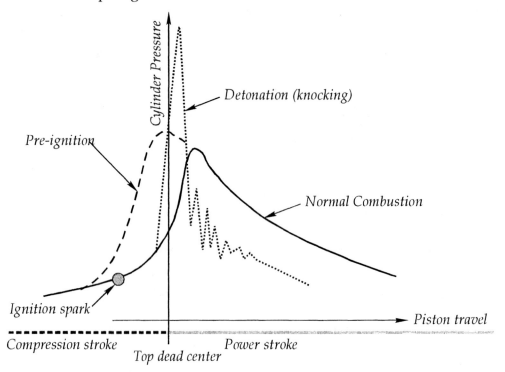

Fig. 17.4 Piston Pressure Profiles (Adapted from [10])

As in a car, the throttle for an aviation engine directly opens and closes a butterfly valve in the carburetor, controlling air flow into the intake manifold. With the throttle full open, then the intake manifold pressure is the ambient pressure at the air intake, minus flow losses through the induction system. The cockpit manifold pressure gauge (MAP) senses the absolute pressure in the intake manifold, which determines the mass of air drawn into the cylinder during the suction stroke. Because energy released in a piston cycle is proportional to the mass of air in the cylinder, the manifold pressure is hence a direct measure of the fuel-air density and the engine brake horsepower produced by the engine.

For a naturally aspirated engine *at fixed throttle*, in which the engine has no mechanism for boosting the pressure, the MAP (reflecting brake horsepower available), decays with the lower ambient pressure at altitude. Maximum takeoff power occurs at full throttle, and maximum continuous power (MCP) occurs at some manifold pressure less than 29" Hg (25"Hg is common). As the airplane climbs, the power is initially MAP limited, and the throttle must be advanced as the ambient pressure decreases to maintain MAP. The result will be an *increase* in power as temperature and manifold density increase. Eventually, the ambient pressure drops to the point that the maximum continuous MAP is no longer achievable at full throttle. Above that altitude (the full throttle height or FTH), the engine power is limited by the ambient pressure; the MAP will be ambient pressure minus the induction system losses. The effect is severe; a normally aspirated engine produces 70% of sea-level BHP at 10,000 ft, and 47% at 20,000 ft. Figure 17.5 depicts a typical power profile for a normally aspirated engine with a sea-level rating of 285 HP.

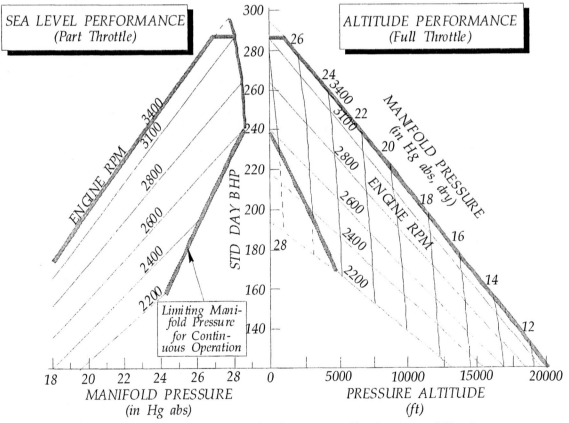

Fig. 17.5 Engine Power Chart for Normally Aspirated Engines

17.1.3.2 **Superchargers and Turbochargers.** Superchargers and turbochargers compensate for the power degradation that accompanies increasing altitude and/or provide for higher total engine compression at sea-level. The engine mechanically drives superchargers, while a turbine in the exhaust gas stream drives turbochargers. The former extracts engine horsepower to pre-compress the intake manifold, while the latter draws its energy from residual heat that would otherwise be dumped into the exhaust gases.

Superchargers appeared shortly after World War I, and were highly developed in airplane engines during World War II. The simplest supercharger is a single-stage radial compressor, mechanically driven at 6 to 10 times the crankshaft speed, and mounted on the engine block. The impeller is downstream of the carburetor and upstream of the intake manifold. The higher intake manifold pressure permits significantly increased sea level performance, and reduced degradation at altitude. Impeller speed is based upon desired operating conditions at which boost is desired. Lower impeller speeds provide sea level boosted power that tapers at lower altitude, while higher impeller speeds provide high altitude boosted power (Fig. 17.6, annotated as "high blow" or "low blow"). Pressurization available from a supercharger or a turbocharger enables MAP settings above the thresholds at which knocking (detonation) can be expected. Consequently, the pilot must restrict throttle setting to hold no more than the maximum normal power setting. Once above the critical altitude, full throttle produces manifold pressures less than the MAP limit. Figure 17.7 depicts the altitude performance of a supercharged engine; note both the achievable manifold pressures well in excess of standard sea level pressure and the limiting takeoff and normal MAP settings.

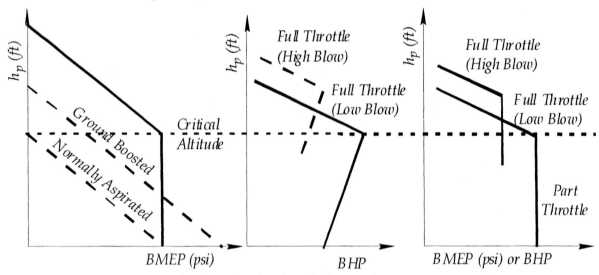

Fig. 17.6 Critical Altitudes for Supercharged Reciprocating Engines

More sophisticated pressurization systems incorporate multiple stages or multiple speed superchargers, or a turbocharger for pre-boost, followed by a supercharger for final boost. "High Blow" can sustain sea-level power to nearly 20,000 ft. Such compound systems almost always require an intermediate heat exchanger (an "intercooler") to reduce the intake manifold temperature to values that will not provoke pre-ignition or detonation.

Turbo-charging is not restricted to vintage "warbirds". The Continental TSIO-550C powering the Columbia 400 single has twin turbo-chargers, each with their own intercooler[10]. The turbochargers provide only an additional 10 HP at sea level, but permit this light airplane to operate up to 25,000 ft. Again, the interested reader is referred to Borden and Cake for an excellent and accessible discussion[9].

Chapter 17   Propulsion Systems Testing   243

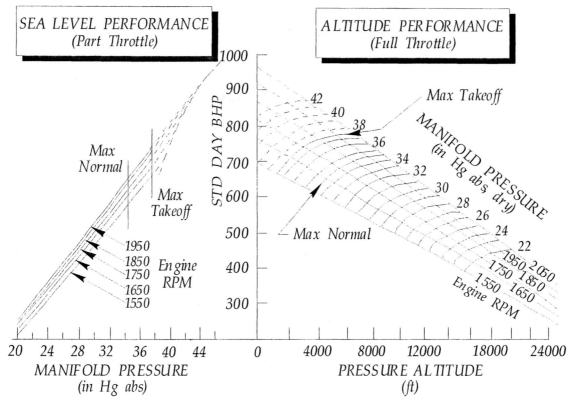

Fig. 17.7 Example Engine Power Chart for a Supercharged Engine

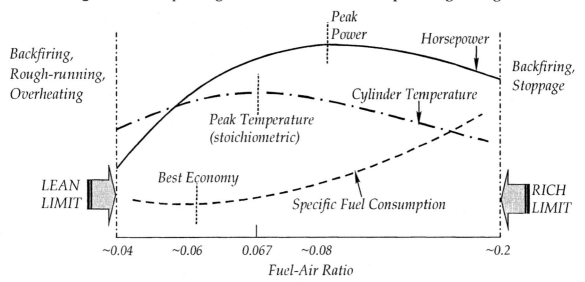

Fig. 17.8 Effect of Mixture Setting

17.1.3.3 **Fuel-Air Mixture.** Aviation carburetors or fuel injection systems operate like automotive carburetors and fuel injection systems. However, the broader operating conditions for an aviation engine necessitate direct control over the mixture setting in order to more precisely tune the engine to the desired thrust and efficiency. Cylinder head temperatures, or exhaust manifold gas temperature, peak at the point of stoichiometric burning, where both all of the oxygen and fuel are consumed; neither oxygen nor fuel are found in the exhaust. This peak occurs at a fuel-to-air ratio of 0.067 for gasoline. The stoichiometric peak does not however coincide with either of the operating points of most direct interest to pilots. Best power, for a given throttle setting, occurs slightly rich of peak, at fuel-to-air ratios between 0.074-0.080. This is because the flame propagates more rapidly through the richer mixture, resulting in a higher peak pressure, coinciding with top-dead-center. The best economy, the lowest power specific fuel consumption, occurs with the mixture slightly lean of peak, 0.060-0.065 for normal timing, and 0.055-0.061 with advanced spark setting. A cylinder head temperature gauge, or an exhaust manifold gas temperature gauge allows identification of the stoichiometric fuel/air ratio and setting the engine for either peak power or peak efficiency (Fig. 17.8).

The fuel economy of reciprocating engines is expressed in specific fuel consumption, c, whose English units are pounds mass of fuel per unit brake horsepower. Common cruise values range from 0.4 to 0.6 $\frac{lbs}{hour-HP}$, where the power is set to 30-60% of maximum[10]. Turbo-charged engines can run as low as 0.38 to 0.42 due to their recapture of exhaust gas energy. The best specific fuel consumption occurs at low RPM settings so as to minimize the power loss to friction.

17.1.3.4 **Ignition Systems**. For reliability and safety, magnetos are used for spark energy, isolating the ignition system from the aircraft's electrical system. These systems are similar to the ignition systems found on 2-stroke lawn equipment such as mowers, and enjoy tremendous reliability. Most airplane installations have two independent magnetos, for redundancy, each of which power a separate spark plug. Consequently, airplane reciprocating engines typically have two spark plugs per cylinder, rather than one, as is common in most automobiles. Two spark plugs provide redundancy and safety, and also give more rapid combustion due to simultaneous ignition from two locations within each cylinder, resulting in greater power. A pre-takeoff magneto check is a standard part of the high power run-up. The slight observed drop in RPM is due to the slower burn rate, which reduces the peak cylinder pressure, and the resultant power.

17.1.3.5 **Cooling**. Cooling reciprocating piston engines has challenged designers since the Wright Brothers, governing the lay-out of the cylinders. The two dominant arrangements, radials and in-lines, were each associated with their cooling strategy. Radial cylinders provided more even cooling of the cylinders, but imposed the cost of large cross-sectional frontal area. Many fighters of World War II opted for in-line engines to reduce cross-sectional area, necessitating water cooling with its weight, reliability, and survivability implications. Other configurations were produced, but these two dominated the hey-day of piston airplanes.

Most modern reciprocating engines are opposed engines, using air cooling. Cooling air is typically drawn into the cowling and forced down through the external cooling fins of each cylinder. Baffling in the engine compartment distributes cooling air to all cylinders and blocks alternative air paths through the engine compartment. All air passing through the engine compartment exits with a loss of total pressure, adding to total airplane drag. In the interest of drag reduction, no more air is desired than required to keep the engine within its design temperature limits. Cooling drag justifies careful attention to the baffling integrity, and explains the significance of cowl flaps on those airplanes so equipped. Closing cowl flaps reduces drag not so much because of profile drag, but rather because of limiting the momentum defect of cooling air passing through the engine compartment.

Engine oil also requires cooling, typically with a small air-oil heat exchanger, resembling an automotive radiator. Air induction systems may also force air over the oil system components, further cooling the oil, and heating the fuel-air mixture, both desirable effects[9]. Oil needs cooling, and heating the fuel-air mixture reduces the likelihood of carburetor or induction system icing. Engine and oil cooling, and induction system heating, are the subject of deliberate tests, described later in this chapter.

17.1.3.6 **Water Injection.** Low altitude performance of supercharged engines is limited by the intake manifold temperatures at which knocking (auto-detonation) occurs during compression. Water injection systems were common on supercharged and turboalcohol mixture drops the temperature of the intake manifold, allowing the engine to be run for short periods at manifold pressures at which auto-detonation and engine damage otherwise occur. Hence, water injection does not directly boost power, but simply allows higher power manifold pressures. If the water pump is turned on, most applications automatically inject the water with fuel at the appropriate throttle settings.

17.1.3.7 **Atmospheric Water Vapor.** Atmospheric water vapor (humidity) profoundly affects reciprocating engine performance[10]. High humidity can reduce engine power by as much as 12% and increase specific fuel consumption. Piston engine performance degrades because the water vapor displaces atmospheric oxygen. The carburetor float or fuel injection mechanism does not distinguish between air and water vapor, running the engine at a rich fuel/air ratio. Flight test data reduction carefully accounts for atmospheric conditions and adjusts results according to engine manufacturer direction. AC 23-8B requires normalizing engine performance to 80% humidity conditions[11].

17.1.3.8 **Propeller Integration.** The easy integration of engine and propeller is a strong point for piston engines. Reciprocating engines typically produce optimal power at crank speeds of 2000-4000 RPM, a range quite suitable for propeller operation. This speed allows direct connection to the engine crank shaft, avoiding the cost, complexity, weight, and maintenance challenges of a gear box. Both helicopters (with lower speed rotors) and turboprop airplanes (with much higher engine rotation speeds) use gear boxes to step down rotor or propeller tip speeds and avoid inefficiencies associated with shock waves. These aircraft pay a consequent penalty in propulsion system weight and complexity. With gear boxes, integration is more complicated than simply bolting the propeller onto a crankshaft flange. Constant speed propellers commonly use engine oil as both the actuating hydraulic medium for setting propeller pitch, and as lubrication for internal mechanisms, but these mechanisms are not as complex as the gear trains needed to convert to lower propeller or rotor speeds.

17.1.4 The Gas Turbine Engine

17.1.4.1 **The Brayton Cycle.** The Brayton cycle is elegant in its operating simplicity, but the implementation is much more complex. Conciseness demands that we restrict our discussion to reviewing principles, emphasizing those features that most directly affect the flight test engineer and pilot. All gas turbines include a compressor, a combustion chamber, and a power turbine. Turboprops, turbofans, and turbojets simply vary how energy remaining in the exhaust products after the high pressure turbine stage(s) is handled.

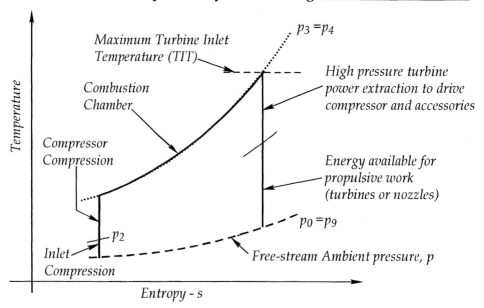

Fig. 17.9 T-s Diagram for Brayton Cycle Engines

The *h-s* or *T-s* diagrams best depict the physics of the Brayton cycle. Figure 17.9 depicts the ideal Brayton cycle without component losses. The rising curves are lines of constant pressure (isobars). A well designed inlet will decelerate the flow, providing the initial boost in pressure. The compressor provides the remaining pressure rise. Fuel is mixed with the flow immediately downstream of the last compressor stage, and then ignited in the combustion chamber, boosting the total enthalpy and temperature. A high pressure turbine then extracts energy from the flow to power both the compressor and the accessories (oil, fuel, and hydraulic pumps, bleed air and electrical generators). The propulsive power is provided by that energy remaining after the high-pressure turbine.

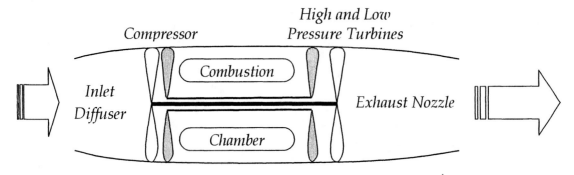

Fig. 17.10 Split-Spool Turbojet

17.1.4.2 **Gas Turbine Configurations.** In a pure jet engine, the hot gas flow is accelerated across a nozzle to produce high exhaust velocities and thrust (Fig. 17.10). In a turbo-fan, a low pressure turbine drives the fan, and the propulsive jet is the combination of the core air passing through the combustion process and the air passing through the fan (bypass air). Low-bypass ratio afterburning turbo-fans have bypass ratios on the order of 0.2 to 0.3 and mix the bypass air with the core air to cool the afterburner chamber (Fig 17.11). Large commercial turbo-fans have bypass ratios of 4 to 6, with the bypass air passing through a separate nozzle (Fig. 17.12). Finally, in the case of the turbo-prop (Fig. 17.13), the air exiting the high pressure turbine passes over a low pressure turbine, whose shaft drives